The New Philosophy
of
Universalism

The Infinite and the Law of Order

**_Prolegomena_ to a Vast, Comprehensive Philosophy
of the Universe and a New Discipline**

First published by O Books, 2009
O Books is an imprint of John Hunt Publishing Ltd., The Bothy, Deershot Lodge, Park Lane, Ropley, Hants,
SO24 0BE, UK
office1@o-books.net
www.o-books.net

Distribution in:

UK and Europe
Orca Book Services
orders@orcabookservices.co.uk
Tel: 01202 665432
Fax: 01202 666219 Int. code (44)

USA and Canada
NBN
custserv@nbnbooks.com
Tel: 1 800 462 6420 Fax: 1 800 338 4550

Australia and New Zealand
Brumby Books
sales@brumbybooks.com.au
Tel: 61 3 9761 5535 Fax: 61 3 9761 7095

Far East (offices in Singapore, Thailand,
Hong Kong, Taiwan)
Pansing Distribution Pte Ltd
kemal@pansing.com
Tel: 65 6319 9939 Fax: 65 6462 5761

South Africa
Alternative Books
altbook@peterhyde.co.za
Tel: 021 555 4027 Fax: 021 447 1430

Text copyright Nicholas Hagger 2008

Design: Stuart Davies
Cover design: Nick Welch

ISBN: 978 1 84694 184 9

A CIP catalogue record for this book is available from
the British Library.

Printed by Digital Book Print

O Books operates a distinctive and ethical publishing philosophy in all
areas of its business, from its global network of authors to production and
worldwide distribution.
This book is produced on FSC certified stock, within ISO14001 standards.
The printer plants sufficient trees each year through the Woodland Trust
to absorb the level of emitted carbon in its production.

The New Philosophy
of
Universalism

The Infinite and the Law of Order

Prolegomena **to a Vast, Comprehensive Philosophy
of the Universe and a New Discipline**

Nicholas Hagger

BOOKS

Winchester, UK
Washington, USA

"And new Philosophy calls all in doubt,
The Element of fire is quite put out."

      John Donne 'The First Anniversarie' (1621,
      referring to Descartes)

"The clash of the infinite is a dilemma that is deeply imbedded in our minds. Wherever we look, we find its manifestation in our thinking about the Universe .... We are drawn to the limits of time, space and matter in our search for answers to the ultimate questions about the Universe. There we find infinities .... The quest to understand the nature of matter and the Universe of space and time may come to rely uniquely and completely upon the beckoning ... of the infinite."

      John D. Barrow, *The Infinite Book*, pp 273-274

"I could be bounded in a nutshell and count myself a king of infinite space."

      Shakespeare, *Hamlet*, II.ii.264

To the memory of Alfred North Whitehead, the most prominent metaphysical philosopher of the 20th century who was open to Nature and science in *The Concept of Nature* and *Science and the Modern World* and would have been very interested in this book as I took up where he left off; for Sir Roger Penrose, with whom I once got lost in Cambridge walking under stars and talking of the universe, in recognition of his landmark work in the 1960s in seeing and calculating that the Big Bang began from a singularity; in memory of Mark Barty-King, who first suggested that I should write such a work; and for John Hunt, discerning Universalist.

*

"Tell them I came, and no one answered,
That I kept my word," he said.

Walter de la Mare, 'The Listeners'

"Another damned, thick, square book! Always scribble, scribble, scribble! Eh! Mr Gibbon?"

William Henry, Duke of Gloucester, when presented with the second volume of Gibbon's *The Decline and Fall of the Roman Empire*, quoted in Boswell's *Johnson*, vol.ii, p2, n.

ACKNOWLEDGMENTS

I was fortunate to have the metaphysical philosopher and friend of T.S. Eliot, E.W.F. Tomlin, as my boss in Japan, and I benefited from many discussions with him on the great philosophers of East and West and on "the Absolute". I probably owe my growing awareness of the infinite to my encounters with Zen Buddhism and Eastern religion at that time.

I have benefited from discussions with Sir Roger Penrose, John Barrow, Sir Fred Hoyle and David Bohm. The Norwegian mathematician Henning Bråten helped with the mathematics of my Form from Movement theory. Mary Morrison, Assistant Professor of Biology at Lycoming College Williamsport, Pennsylvania, discussed the origin of life and inherited instinct with me in the Galapagos Islands. In Antarctica I had lengthy discussions with Charles Wheatley, marine biologist and teacher of oceanography and environmental sciences in San Diego, and Larry Hobbs, for many years research biologist at the US National Marine Mammal Laboratory. I also discussed plate tectonics and aspects of Ice Ages with Patrick L. Abbott, Professor Emeritus in Geological Sciences at San Diego University. I am grateful to two researchers into biology and anatomy, Robin Woodleigh and Jason Cook, for contributing to my coverage of symbiosis and feedback respectively; and to Tricia Moxey, who helped run the Epping Forest Conservation Centre, for discussions on Nature's ecosystem which affected a couple of paragraphs. I am grateful to the Continental Phenomenologist philosopher Christopher Macann for reading the text and generously making suggestions. I am also grateful to David Lorimer, Programme Director of the Scientific and Medical Network, for reading the text.

It must have been in 1995 when Mark Barty-King, then Managing Director of Transworld and publisher of Stephen Hawking's best-selling *A Brief History of Time*, having read some of my work, met me and asked if I would consider writing a book that would counterbalance Hawking's Materialistic emphasis with a more metaphysical approach. But he died before I could make a start. It has taken me fourteen years to get clear of other projects and write the book. I am grateful to John Hunt, who with typical perspicacity recognised its importance at a very early stage, made penetrating points in our early discussions and saw it through.

I am grateful to my wife Ann who travelled with me to the Galapagos Islands and to Antarctica, and to my PA Ingrid who kept up with me though I set a blistering pace.

CONTENTS

PART THREE
THE METAPHYSICAL VIEW OF THE UNIVERSE
IN THE NEW PHILOSOPHY

PART FOUR
ORDER IN HUMAN AFFAIRS:
APPLICATIONS OF UNIVERSALIST THINKING

NOTE ON DATES OF SOURCES

Cosmological sources have to be treated with care as new discoveries are constantly putting sources' data out of date. Since WMAP in 2003, the universe has been 13.7 billion years old, not 10 billion or 15 billion (as it was at different times in the recent past). And the beginning of Cosmic Microwave Background Radiation has been put at 380,000 years after the Big Bang rather than 300,000 years. In 1998 it was found that the universe's acceleration has speeded up. Some of Hubble's calculations are now unsafe, and books published before 1998 do not mention dark energy. From 1998 to 2002 it was estimated that the universe was composed of 4 per cent matter, 30 per cent dark matter and 66 per cent dark energy. Since WMAP in 2003, the percentages for dark matter and dark energy have been revised to 23 per cent and 73 per cent respectively. I have used books published before 1998 and 2003 but have endeavoured to avoid using any material that is now out of date. See Notes and References to Sources on p407.

This book was finished before the results were known regarding CERN's search from September 2008 for the invisible field of bosons predicted by Peter Higgs in 1964. I have given an overview of the anticipated findings of CERN's experiment in the Postscript on p368.

NOTE ON CAPITAL LETTERS

Words such as "reality", "whole", "one", "all", "void", "fire" and "light" can have a metaphysical meaning as well as an everyday, social meaning. When referring to the unitive infinite/finite order I have given these words capital letters.

NOTATION
NUMBERS IN POWERS OF TEN

Very large and very small numbers contain many zeros. I have sometimes referred to them in words, but I have also used the powers-of-ten notation in which, for example, 10 with 6 zeros after it is written as 10^6 and 1 divided by 10 with 6 zeros is written as 10^{-6}. Thus:

one	or	1	is	10^0
ten	or	10	is	10^1
hundred	or	100	is	10^2
thousand	or	1,000	is	10^3
ten thousand	or	10,000	is	10^4
hundred thousand	or	100,000	is	10^5
million	or	1,000,000	is	10^6
ten million	or	10,000,000	is	10^7
hundred million	or	100,000,000	is	10^8
billion	or	1,000,000,000	is	10^9
ten billion	or	10,000,000,000	is	10^{10}
hundred billion	or	100,000,000,000	is	10^{11}
trillion	or	1,000,000,000,000	is	10^{12}
ten trillion	or	10,000,000,000,000	is	10^{13}
hundred trillion	or	100,000,000,000,000	is	10^{14}

And:

one-tenth	or	1/10	is	10^{-1}
one-hundredth	or	1/100	is	10^{-2}
one-thousandth	or	1/1,000	is	10^{-3}
one-ten-thousandth	or	1/10,000	is	10^{-4}
one-hundred-thousandth	or	1/100,000	is	10^{-5}
one-millionth	or	1/1,000,000	is	10^{-6}
one-ten-millionth	or	1/10,000,000	is	10^{-7}
one-hundred-millionth	or	1/100,000,000	is	10^{-8}
one-billionth	or	1/1,000,000,000	is	10^{-9}
one-ten-billionth	or	1/10,000,000,000	is	10^{-10}
one-hundred-billionth	or	1/100,000,000,000	is	10^{-11}
one trillionth	or	1/1,000,000,000,000	is	10^{-12}
one-ten-trillionth	or	1/10,000,000,000,000	is	10^{-13}
one-hundred-trillionth	or	1/100,000,000,000,000	is	10^{-14}

1/10 can be written as 0.1, 1/100 as 0.01 and so on.

LARGER NUMBERS IN POWERS OF TEN

quadrillion or	1,000,000,000,000,000	is	10^{15}
quintillion or	1,000,000,000,000,000,000	is	10^{18}
sextillion or	1,000,000,000,000,000,000,000	is	10^{21}
septillion or	1,000,000,000,000,000,000,000,000	is	10^{24}
octillion or	1,000,000,000,000,000,000,000,000,000	is	10^{27}
nonillion or	1,000,000,000,000,000,000,000,000,000,000	is	10^{30}

decillion or

| | 1,000,000,000,000,000,000,000,000,000,000,000 | is | 10^{33} |

undecillion or

| | 1,000,000,000,000,000,000,000,000,000,000,000,000 | is | 10^{36} |

duodecillion or

| | 1,000,000,000,000,000,000,000,000,000,000,000,000,000 | is | 10^{39} |

And:

one-quadrillionth or	1/1,000,000,000,000,000	is	10^{-15}
one-quintillionth or	1/1,000,000,000,000,000,000	is	10^{-18}
one-sextillionth or	1/1,000,000,000,000,000,000,000	is	10^{-21}
one-septillionth or	1/1,000,000,000,000,000,000,000,000	is	10^{-24}
one-octillionth or	1/1,000,000,000,000,000,000,000,000,000	is	10^{-27}

one-nonillionth or

| | 1/1,000,000,000,000,000,000,000,000,000,000 | is | 10^{-30} |

one-decillionth or

| | 1/1,000,000,000,000,000,000,000,000,000,000,000 | is | 10^{-33} |

one-undecillionth or

| | 1/1,000,000,000,000,000,000,000,000,000,000,000,000 | is | 10^{-36} |

one-duodecillionth or

| | 1/1,000,000,000,000,000,000,000,000,000,000,000,000,000 | is | 10^{-39} |

The above figures assume short scale (e.g. 1 billion = 1,000 million) as opposed to long scale (e.g. 1 billion = 1 million million).

Image of the astronaut-surfer who breasts the infinite on the top edge of our expanding universe (see facing page). For the surfer, see pp69-70.

Image showing how the universe began and the shuttlecock-like shape of our expanding universe.[1]

PROLOGUE

This book calls for a revolution in Western philosophy. The Presocratic Greeks, Plato and Aristotle focused on the Reality behind the universe. I hold that Western philosophy should return to its origin in them. This book seeks to overthrow the Establishment orthodoxy of Western philosophy which has been entrenched in logic and language for nearly a hundred years, ever since Einstein bewildered philosophers with his two relativity theories of 1905 and 1915. It calls for the universe to be let back into philosophy. Explaining the universe and Nature should once again be the business of philosophy.

It is sometimes said of our knowledge of the universe, "It's all theories. We don't know how the universe began. The Big Bang is a theory. All that physicists and mathematicians say about the universe is just theories. Nothing is known." This is too pessimistic. I have tried to sift what can be safely known, and in this book I have built up a picture of what we can know about the universe. I have been careful to distinguish fact from theory. What is missing today is a university course that does this. And yet in the Middle Ages such a course was central to university life.

One of my favourite quadrangles, which I always look in on when I visit Oxford, is the Schools Quadrangle at the Bodleian Library. It was built between 1613 and 1619 as an extension to Thomas Bodley's 1602 Library, and the ground and first floors were used as lecture room for the university students of the seven liberal arts. These "scolae" or "schools" are named in Latin lettering over the stone Quadrangle doorways: Logic, Grammar and Rhetoric (the *trivium*); and Arithmetic, Geometry, Music and Astronomy (*the quadrivium*). As the Renaissance had flourished by 1613, "modern subjects" are over other doorways: Hebrew, Greek, History, Medicine and Jurisprudence. Beside them is the oldest and most important subject: Divinity. As the Quadrangle looked back to the medieval time of Duke Humfrey's 1443 Library, which Bodley restored from 1602, also present in Latin lettering are the three branches of philosophy found in all medieval universities to which the seven liberal arts were introductory: the very important subjects of Moral and Natural Philosophy and Metaphysics, which were the centre of the curriculum of all medieval universities.

Two thousand years previously, philosophy was at the centre of the Greek curriculum in the Academy Plato founded, where Aristotle was a pupil. For nearly two hundred years before then, Greek philosophers had looked at the universe – sky, sea and land – and had attempted an explanation. I was fortunate enough to be introduced to these first Western philosophers at school, where I also spent long hours reading Socrates' arguments in the original Greek of some of Plato's dialogues. Plato continued this metaphysical tradition of attempting to explain the universe, and since then, despite changes soon after the Renaissance, some philosophers have offered quite theoretical explanations involving an invisible Reality. Other philosophers had seen the world of Nature as the real world and had offered more scientific explanations. Aristotle continued this scientific tradition of natural philosophy, and since then many philosophers have been scrupulously evidential and wary of excessive theorising. Some philosophers have been more interested in human conduct, or moral philosophy.

The definition of philosophy has therefore varied in accordance with the interests of particular philosophers, who have defined philosophy to reflect their interests. *The Shorter Oxford English Dictionary* defines philosophy as "the love, study, or pursuit of wisdom, or of knowledge of things and their causes, whether theoretical or practical". It adds, "That more advanced study to which, in the mediaeval universities, the seven liberal arts were introductory; it included the three branches of natural, moral and metaphysical philosophy, commonly called the three philosophies." These are all defined separately in *The Shorter Oxford English Dictionary*. "Moral philosophy" was "the knowledge or study of the principles of human action or conduct; ethics", "the part of philosophy that treats of the virtues and vices, the criteria of right and wrong, the formation of virtuous character." "Natural philosophy" was "the knowledge or study of natural objects and phenomena; now usually called 'science'". "Metaphysical philosophy" was "that department of knowledge or study which deals with ultimate Reality, or with the most general causes and principles of things". "Metaphysics" is more specifically defined as "that branch of speculation which deals with the first principles of things, including such concepts as being, substance, essence, time, space, cause, identity etc.; theoretical philosophy as the ultimate science of Being and Knowing".[1]

In the later Middle Ages the philosophical disciplines of logic (including disputations), ethics (the Aristotelian version of moral philosophy), natural philosophy and metaphysics were central to the curriculum as they were studied as a preparation for the higher calling of theology (or divinity).[2] Little attention was paid to epistemology (how we know) as scepticism was not an issue due to the dominance of theology (or divinity) and the importance of direct intellectual vision with the aid of divine illumination. Likewise, little attention was given to psychology as soul and body were thought of as being dualistically distinct.

The two main traditions of philosophy – the metaphysical and the scientific – developed until in the late 17th century the rise of science began to crowd philosophy out. Newton, whose physics revolutionised science in the 1660s, saw himself as a "natural philosopher", but in the 18th century "natural philosophy" had been renamed "science". Scientists ("natural philosophers") occupied the philosophers' traditional ground including the ground once occupied by metaphysics, and in the 20th century theoretical physicists theorised about the universe as metaphysicians had done – often no more evidentially: there is only circumstantial evidence to support their black holes, dark matter and dark energy, which have not been observed. (However, if they do not in fact exist there will have to be new hypotheses to explain the fact that only 4 per cent of the universe's matter is visible, according to mathematical calculations; which means that 96 per cent of matter is missing without any explanation.)

During the last 350 years of increasing Materialism and reductionism, and of scientists fragmenting the universe into smaller and smaller bits, philosophy has vacated its metaphysical and scientific ground to science. Philosophy has been left with Phenomenology (and its focus on consciousness) on the Continent – Existentialism is now dead – and linguistic philosophy in the Anglo-Saxon world; and moral philosophy. These are all separate and in some degree of conflict with each other. They are all removed from the universe and Nature which first inspired philosophy. There is no prospect within the existing movements and structures that, if left to their own devices, philosophy can be reunited, and a return of the universe into philosophy offers the only prospect of such a reunification. The existing groupings only explore parts of Reality, which for 2,600 years has included both the finite universe that emerged from the Big Bang and the infinite from which it came.

It is time for philosophy to return to its original purpose of looking at the whole of Reality (the finite *and* the infinite) and of seeing man in relation to an orderly universe. The name I give to this way forward and renewed focus on the universe is "Universalism". The term has been used variously in the past to describe the salvation of the whole of humankind (universal salvation), or the recognition that all religions have equal validity (religious universalism or universal religion) or that the whole of humankind should have equal political rights, an increasingly aired view as thinking becomes more global (political universalism or universal suffrage). Thus *The Shorter Oxford English Dictionary* defines a Universalist as "one who believes or maintains the doctrine that redemption or election is extended to the whole of mankind". In this book I use the term "Universalism" to describe the new philosophical worldview and movement that seeks to take philosophy back to its origins and which like Romanticism or Classicism may one day have a life outside philosophy as well as within it. My Universalism can eventually put philosophy back at the centre of the university curriculum which it occupied until the rise of science. For the study of the philosophy of the universe incorporates the findings of all the sciences.

My term "Universalism" incorporates "universe", "universal" and "universality". It is a philosophy focusing on (1) the universe, (2) universal science, (3) a universal principle of order ("universal" in the sense that its effects are found in every aspect of Nature and its organisms) and (4) humankind as a whole and its place in the universe. It also refers to (5) the universal being, the deeper self below the rational, social ego, which is open to (6) universal cosmic energies which stimulate the growth of plants and organisms and convey the principle of order. The universe is the earth, galaxies and intergalactic space regarded as a whole: all that is, all existing things. It has become the fiefdom of Materialist scientists who focus on its matter and energy, using mathematical methods. "Nature" is the system within the universe in which we are integrated.

"Order", in the first of its 20 meanings in the *Concise Oxford Dictionary* (one present in Middle English during the Middle Ages), is "the condition in which every part or unit is in its right place, tidiness", which covers the behaviour of insects, fish, birds, reptiles, mammals and humans within Nature's system. It is also "a state of peaceful harmony under a constitutional authority" (as in "order has been restored"), suggesting organisation, and "the nature of the world society" (as in "the order of things"). Thus "order" suggests an organised, basically tidy world scheme or system in which everything knows its defined or right place (until randomness intrudes). Such a system is described in systematic philosophy, which proposes a Grand Unified Theory of Everything.

The theme of this book, then, is that philosophy should now return to its traditional task of understanding and explaining the universe and the world of Nature, and that it should reunify the existing conflicting groupings and movements. In a sense, this book is an introduction or *Prolegomena* to a vast philosophy of the universe – to a new discipline, for it redefines the scope of philosophy and subsumes many existing scientific disciplines from a new angle. I hope that, seeing the new direction that I offer, young philosophers will come forward and use their skills to articulate Universalist philosophy and turn it into as detailed and wide-ranging a movement as Phenomenology and Linguistic Analysis are.

This book explains why Universalism is needed in philosophy; what it derives from the evidence of science; what the new metaphysical philosophy is; and its applications. In Part One we look at the history of Western philosophy since the Greek Thales, c.585BC. We shall see that for 2,600 years there have been two main emphases in philosophy, stemming from the Presocratic Greeks, Plato and Aristotle. These passed into the Christian era, and since the Renaissance fragmented. These two tendencies have changed their names since the Renaissance, but in essence they have remained the same though they have declined in importance: one metaphysical and focusing on Reality, the other scientific and focusing on Nature. Part Two deals with the scientific view of the universe. I give an up-to-date view of all the main scientific disciplines, and have tried to be scrupulously objective and factual, as if I were trying to explain it to the satisfaction of Aristotle, the founder of the scientific tradition, and of Sir Francis Bacon, the first Empiricist. To synthesise all scientific disciplines and distill a whole view was possible in the time of the Renaissance but is very difficult today as knowledge has become specialised and fragmented. Only the philosopher, contemplating the philosophy of science, is in a position to draw the specialised subjects together and attempt a synthesis, and if he makes mistakes of detail, that is surely a price worth paying for achieving a whole view.

This scientific view of the universe is one wing of Universalism. Part Three deals with the metaphysical view of the universe. It assimilates the hidden Reality of the metaphysical view over 2,600 years. This metaphysical view of the universe is the other wing of Universalism. Universalism now reunifies the two emphases in philosophy by incorporating Plato's tradition of hidden Reality emerging from the infinite and Aristotle's tradition of a scientific approach to the universe. We shall see that the first focuses on the unchanging and infinite, the second on finite, changing Nature. Part Four gives some modern applications of Universalism and integrates moral philosophy's focus on human conduct.

I use a dialectical method which was anticipated by Eastern thinking. This was first explained to me in 1965 by the Japanese poet Junzaburo Nishiwaki, who has been called Japan's T.S. Eliot. (He was a contemporary of Eliot's.) As we drank *saké* (rice-wine) in a bar with sawdust on the floor round the corner from Keio University, he wrote out on a reply card which is now framed on my study wall: "+A + −A = O, Great Nothing." The O is zero; the Great Nothing is the One. In this thinking, two opposites are reconciled into harmony within the One. The opposites can be Being and non Being, Being and Becoming, life and death, the world of time and the world of eternity, the finite and infinite, the metaphysical and the scientific. Universalism as a reconciler as well as a revolution employs this dialectic to unite the two traditions, and within the context of this unity to unite Continental Phenomenology's ontology and Anglo-Saxon Analytical philosophy's focus on language.

In attempting a reunification of the two traditions after 2,600 years, I am by definition attempting to reunite largely untestable metaphysics and testable science – which were united at the origin of Western philosophy.

Categories are very important. The metaphysical emphasis is often theoretical and speculative (as can be seen from Idealism), and has Rationalistic and Intuitional forms. Its reasoning is often untestable. As I have already observed, theoretical physics is in a similar position, with little proof outside mathematics for its various theories, such as superstring theory. Mathematics is not evidential. The scientific emphasis is empirical and testable. Theoretical, speculative reasoning and empirical handling of data must be kept separate. The philosophical category can handle opposites of different categories (+A + −A) so long as it is clear what the categories are. There are other categories – for example, categories of interpretation, faith, myth and culture – and philosophy must be wary of these.

Just as the philosopher uses a philosophical category to consider theoretical reasoning and scientific evidence, so the universal principle of order is a philosophical category. It may, however, also be a scientific hypothesis based on evidence of its cells' operating effects within Nature. There is no anomaly in its being a philosophical category *and* a scientific proposal at the same time. Indeed, it may one day be proved to be a scientific law – perhaps sooner than we realise in view of CERN's search from a hitherto undetected field of bosons from September 2008 – just as for 2,500 years the atom was both a philosophical category and an unproved scientific proposal until it was proved to have operational existence. Similarly, infinity and timelessness are philosophical categories which may similarly be scientifically evident.

Works of philosophy tend not to have endnotes so that their general ideas may appear more enduring. Today ideas by themselves are insufficient to describe the universe. Gone are the days when philosophers can make sweeping generalisations about the universe without a discerning reader rightly wishing to know, "Where's your evidence?" The process of thought, though an indispensable tool, is by itself incapable of approaching the truth about the universe, which demands the allegiance of other disciplines besides thought. The philosopher of the universe must be up-to-date on all the recent advances in thinking and scientific discoveries such as those that might be made at CERN. Today the philosopher of the universe must be evidential and transparent about his sources. No book of modern philosophy about the universe can be taken seriously if it does not offer sources.

The main theme of this book, then, is that philosophy needs to be reconnected to Nature and the universe, and that the evidence of a principle of order in cells will enable philosophy to be reunited and reassume a more central position in Western life. The philosopher can come to be regarded as a man who has explored all the cross-disciplinary scientific knowledge of the Whole, the One, and has an up-to-date view of the structure, substance, cause, first principles and reality of the scientific universe. The philosopher can cease to be thought of as an abstruse, irrelevant figure lost in minutiae of language, and can be regarded once again, as he traditionally was: as a sage. The dictionary definition of a sage is "a profoundly wise man", a fitting description for a "lover of wisdom".

PART ONE

THE STORY OF PHILOSOPHY

"It is incumbent on the person who specialises in physics to discuss the infinite, and to inquire whether there is such a thing or not, and, if there is, what it is."

Aristotle, *Physics*[1]

"The eternal silence of these infinite spaces frightens me."

Blaise Pascal, *Pensées*, 206

"There are more things in heaven and earth, Horatio,
Than are dreamt of in your philosophy."

Shakespeare, *Hamlet*. I.v.166

1

THE ORIGINS OF WESTERN PHILOSOPHY

The tradition of Western philosophy appears bewildering from our perspective today, full of different approaches, "isms" and schools, and it is hard to detect any shared theme or consensus among the many philosophers. However, if we track the tradition back to its source we do find a shared theme and a startling consensus, which the Universalist philosopher must note, that philosophy should reflect and explain the true nature, underlying order and fundamental unity of the known universe.

The Presocratic Greek Philosophers

Western philosophy began with the Presocratic Greek philosophers of the 6th and 5th centuries BC. Deriving from the poetry of the 8th-century-BC Homer and Hesiod whose cosmos was a flat earth surrounded by the Ocean, these Ionian and South-Italian philosophers who preceded Socrates are credited with being the first in the Western tradition to move beyond traditional mythology to a rational account of Nature.

These early Greeks (some of whom were actually contemporaries rather than predecessors of Socrates, who lived from 470 until 399BC) sought to explain the physical, natural world. They focused on an ordered "cosmos". The Greek word *kosmos* means "the universe as an ordered whole", "the ordered world". They were called "cosmologists", students of the ordered universe. Ionians, Greeks occupying the Western coast of Anatolia (modern Turkey), had lived under Persian rule from c.600BC until the Ionian revolt which was crushed with the Persian capture of Miletus in 494BC. The Ionian philosophers were imbued with the monism of Ahura Mazda, and brought their monism into their rational philosophy. They suggested that the phenomena of Nature are manifestations of one Reality which is boundless and infinite.

The first of the Presocratics was widely recognized as Thales of Miletus, who flourished c.580BC and held that everything came out of water. According to tradition, he founded Greek philosophy on 28 May 585BC[1] by predicting an eclipse of the sun, a compelling instance of the early Greek philosophers' involvement with Nature. His disciple Anaximander of Miletus, who flourished c.570BC, saw the world as emerging from the eternal, infinite, "boundless" (*to apeiron*), an eternal and eternally moving boundless Reality from which the universe began as a finite germ (*gonimon*).[2] The "boundless", he said, has no distinguishable qualities. Anaximander was amazingly close to the view of physicists in our time who have asserted that the universe began from a singularity or point which contains infinity. Anaximander's disciple Anaximenes of Miletus, who flourished c.550BC, saw air rather than water as the origin of things. He held that matter condensed from air.

The philosophers of the school of Miletus – the Milesians – were followed by the Southern-Italians. The non-Milesian Eleatics were poets. They wrote their philosophy in verse (a feature which can only appeal to the philosopher-poet today.) The link between Southern Italy and Ionia was Xenophanes of Colophon in Ionia. He was born some time between c.580 and 560BC and emigrated to Elea in Southern Italy c.546/5BC, when the Medes conquered Colophon. (See map in Appendix 6.) He flourished c.530BC and sources maintain that he lived to be a hundred. A poet, he travelled around Greece reciting his poems, and his philosophy was expressed in his poetry and has survived in fragments. He believed that "from earth all things are…,from earth and water come all of us".[3] He supported Anaximenes' reservations regarding theology and was especially critical of the anthropomorphic polytheism of Homer and Hesiod. He asserted that there could only be one God, the ruler of the universe, who must be eternal: "One god….Always he remains in the same place, moving not at all."[4] Xenophanes saw "that which is" as motionless unity.

Aristotle, the 4th-century Greek philosopher, unable to understand Xenophanes' motionless god, complained that he "made nothing clear" and that Xenophanes "gazing up at heaven (*ouranos*) as a whole (or with his eye on the whole world or material universe) declared that the One (or Unity) is god (or God)".[5] Scholars are divided as to whether Xenophanes was an Ionian monist who asserted the unity of Being or a pantheist who asserted the unity of material world of Nature. However he is regarded, Xenophanes was concerned with the One and focused on God and the universe. Theophrastus summed up Xenophanes' teaching: "Xenophanes…supposed that the all is one, or there is one principle, or that what exists is one and all….This 'one and all' is god."[6]

Apart from these Presocratics, but one of them, had been Pythagoras of Samos, who, after travelling in the East and Egypt, left the Greek island and emigrated to Croton in Southern Italy. He flourished c.530BC, and evolved a metaphysic of number in which "all things are numbers",[7] a mathematics of ratios that surely began with his study of the stars. His theorem came from making triangles out of different stars. It is interesting that in the 6th century BC a mathematician such as Pythagoras devised his science of numbers by studying the universe's night sky. He also focused on the transmigration of souls, a philosophy he acquired in the East.

Back in Ionia, Heracleitus (or Heraclitus) of Ephesus, who flourished c.500BC, held that Reality was fire, which had always existed. In this he disagreed with Thales' focusing on water and Anaximenes' focusing on air. He saw the world order as an "ever-living fire kindling in measure and being extinguished in measure"[8] that "ever was, and is, and shall be", and he included the ether ("*aither*") in the upper atmosphere in this fire. The "ever-living fire" was the metaphysical Fire as he felt the soul consists of eternal Fire. To Heracleitus the world was in perpetual flux: "*panta rhei*".[9] The river is never the same as it flows on: "Upon those who step into the same rivers different and ever different waters flow down."[10] Heracleitus detected unity behind all opposites. All opposites are reconciled in an underlying unity, so that "the way up and the way down are one and the same".[11]

Before Xenophanes died c.478BC, Parmenides of Elea, Xenophanes' disciple, developed Xenophanes' thought and perhaps derived his view of the unity of Being from him. Also a poet like Xenophanes, he was younger than Heracleitus and differed from him. He was probably born c.515BC. Plato's *Parmenides* is about the "very young" Socrates' meeting with Parmenides and Zeno of Elea (who was then "nearing forty"). Socrates says to Parmenides in this dialogue, "You assert, in your poem, that the all is one" (or "all things are really one"). And: "You assert unity."[12] Socrates was over 70 when he was put to death in 399BC. If "very young" means "twenty" the meeting between Socrates and Parmenides took place in c.450BC when Parmenides would have been about 65 and Zeno nearly 40.[13]

Parmenides' surviving work is fragments totalling 150 lines of a philosophical poem about the Way of Appearance (or Seeming) and the Way of Truth. In this poem Parmenides asserted that "what is" has always existed. It could not have come into being and cannot pass away as it would have come out of nothing or become nothing. Furthermore, it has no motion for it would then be a motion into something that is, which is impossible, or a motion into something which is not: "There still remains just one account of a way, that it is.... Being uncreated and imperishable it is, whole and of a single kind and unshaken and perfect. It never was nor will be, since it is now, all together, one, continuous.... If it came into being, it is not: nor is it if it is ever going to be in the future.... Nor is it divided, since it all exists alike: nor is it more here and less there, which would prevent it from holding together, but it is all full of being. So it is all continuous: for what is draws near to what is."[14] Commentators ask, "Is Parmenides claiming that all reality is one?" The answer seems to be: yes.[15] Parmenides' poem continues: "But changeless within the limits of great bonds it exists without beginning or ceasing.... Remaining the same and in the same place it lies on its own and thus fixed it will remain."[16]

To Parmenides the phenomenal world is an illusion and behind it is an immobile being. Parmenides is very modern for his One anticipates the modern singularity and he saw space as a plenum:[17] "Nor is it divided, since it all exists alike....It is all full of being." This appears to contradict Heracleitus' "flux". Parmenides asserted the unity of Being and was a monist.[18] Interested in biology and astronomy, he held that the earth was spherical – he was the first Greek to assert this – and that the moon shines with reflected light. Greeks had looked at the stars above the North Pole, including *Arktos*, the Bear, and, seeing that they revolved round a pole, proposed that the world must be round, not flat, and that a southern continent opposite to the one known in the north (the Arctic) must exist, one that would come to be known as Antarctica. Antarctica was not sighted until 1820.[19] For Parmenides reality was different from the world of appearances, and in the second part of his poem the Way of Truth is in conflict with the Way of Appearance.

In one view, Parmenides saw the One as infinite and a single "Whole".[20] Melissus of Samos, who interpreted Parmenides in prose c.440BC took this view: "It has no [spatial] beginning or end, but is infinite (or always was and always will be)."[21] Melissus saw the Eleatic One as coming from infinity (*to apeiron*).[22] In another view, Parmenides saw the phenomenal world as finite, yet immobile. Thus, "Parmenidean entities, unlike Melissan entities, are finite in extent."[23] If this second reading is correct, Parmenides is pointing to the limits of the world (an aspect of the Way of Appearance) but the One is still present behind it. The individual in a finite world is still face-to-face with the infinite.

Parmenides is regarded as holding a philosophy that was the opposite of Heracleitus's. To Parmenides the One was immobile. To Heracleitus, the world was in a perpetual flux. To Parmenides, this world of flux was an appearance, an illusion. Both saw unity behind opposites and so although they differed regarding the nature of the world, they were in fundamental agreement about the unity that reconciles all contradictions. Parmenides, then, held that in essence the universe consisted of one thing, which was a single, timeless and changeless Reality. At the level of Reality, nothing moved.[24] To Heracleitus "god" was immanent in things or the sum total of things.[25]

Anaxagoras of Clazomenae (which was in Ionia, near Izmir), who flourished c.480-470BC, followed Parmenides, but was a pluralist. He held that everything is contained in everything but in infinitely small parts. *Nous* or intelligence set them into whirling motion.

Zeno of Elea, the younger friend of Parmenides who flourished c.460BC and met Socrates with Parmenides perhaps c.450BC, defended Parmenides' immobility with more than 40 paradoxes, such as the one about the race between Achilles and a tortoise, which starts ahead of Achilles. When Achilles reaches where the tortoise began, the tortoise has moved forward, and so on *ad infinitem*, and so Achilles never overtakes the tortoise. Zeno's paradoxes are logical nonsense which have of course been endlessly refuted, but they made a contribution to epistemology, to understanding how the real world differs from the world of appearance which we perceive.

Empedocles of Akragas (who featured in Matthew Arnold's 'Empedocles on Etna') focused on all four elements,[26] drawing together the thinking of Thales, Anaximander, Xenophanes (who championed earth) and Heracleitus. In reconciling four opposing views that preceded him, Empedocles was something of a Universalist cosmologist.

The Atomists introduced a Materialist dimension to the Presocratics. Leucippus, who flourished c.440BC held that there was empty space and filled space, which is composed of indivisible atoms. The Greek "atom" means "uncuttable". Democritus, who flourished c.420BC, carried forward Leucippus's thinking by writing of atomic structure. These two Materialist philosophers were modern in anticipating matter.

It is worth pointing out in passing that the Atomists Epicurus (4th century BC) and Lucretius (1st century BC) anticipated the multiverse. They believed in an infinite number of worlds, because they believed there were an infinite number of atoms in an infinite universe. Aristotle, however, believed the universe was composed of a finite amount of matter with a centre and symmetry essential to the nature of things, and so the universe was finite and there were not an infinite number of other worlds. He advocated one universe of purposeful change, and

Christian theology went along with it.[27]

Greek philosophy passed to the Sophists, who "made a living out of being inventive and clever" (*sophizesthai*), and who were the precursors for sophists such as Jacques Derrida today; and eventually to Socrates, the giant figure of Greek philosophy who humanised Presocratic scientism. He opposed wrongdoing and stood for what is right, "the Good". He used Sophist techniques to question the opinions and assumptions of others. He wrote nothing, but his dialogues were recorded by his pupil Plato, who wrote that "the Eleatic school, beginning with Xenophanes and even earlier, starts from the principle of the unity of all things (or explains in its myths that what we call all things are actually one)".[28] Plato himself taught at his academy in Athens and reflected Parmenides' view that reality differs from appearance in his image in *Republic* of a fire differing from the flickering shadows on a cave wall.[29] Both Socrates and Plato were concerned to find out how one should live the good life in accordance with truth rather than illusion and appearance.

According to tradition Plato believed that Greek philosophy came from "the fountains (or springs) of the West", meaning Italy and Sicily which lay to the west of Athens and which he visited c.388BC. However, as we have seen, these "Western" philosophers who were his immediate predecessors had derived from Xenophanes' migration from Ionia, which lay to the east of Athens, and Greek philosophy should be seen as being inspired from the east: "*ex oriente lux*".

The achievements of some of the early Greek thinkers have a stunningly contemporary ring. In the 6th century BC Anaximander of Miletus discovered the evolution of the species, holding that life began in water with animals representing sea urchins. In the 5th century BC Democritus of Abdera proposed atomic theory. In the 5th century BC Hippocrates of Cos devised the ethical code which still governs the medical profession. In the 3rd century BC Aristarchus of Samos, a Postsocratic, became the first to demonstrate and maintain that the earth rotates and revolves round the sun.

The Presocratics were astonishingly modern. Their subject was the universe, and they looked out and tried to understand the nature of reality and the scientific perceptions they came up with anticipate the findings of physicists today, seeing the universe beginning from "a boundless germ" (Anaximander), seeing Reality as Fire or one substance and anticipating matter through atoms. But the fundamental perception of Parmenides' One should be seen as a statement about a Reality that is invisible. It is the Oneness behind this world that echoes down to us today, combined with Heracleitus's vision of moving Fire.

The origins of Western philosophy are in these early philosopher-poets who reflected their questioning and probing of the universe. These thinkers from ex-Persian Ionia glimpsed a Reality *before* creation (the Big Bang) which is continuous, infinite and timeless: an eternal and eternally moving boundless infinity (*to apeiron*) which threw up a finite germ (*gonimon*).[30] They also saw the phenomenal world *after* creation in terms of "God and the universe",[31] and as cosmologists searched for an order in the cosmos which would last until the end of the universe.

The Presocratic focus on the One substance of the universe, and on the infinite, call us to return to the origins of Western philosophy in Nature and the universe. In fact, their perspective had already been anticipated by the philosophies of the early religions.

The Philosophies of Early Religions

The origins of religion can be tracked back to shamanism,[32] the religion of Upper Palaeolithic times which surfaced in Central Asia c.50,000BC. Shamanism was found among the peoples of Siberia, the Soviet Arctic, Mongolia, Manchuria, Nepal, Bhutan, Sikkim, and Tibet where shamanistic Bon-Po was the indigenous religion c.25,000BC. Much later shamanism can be found among the Kurgan people, traditionally the first Indo-Europeans. The Kurgans, who took their name from *kurgan*, the Russian and Turkic for "burial mound", as they buried their dead in barrow-style mounds, originated in South Russia and came into Central Europe in three waves after c.4500BC.[33]

In all shamanistic societies the fundamental unity of the universe was emphasized: the three worlds of Underworld, Earth and Sky were united by a World Tree, which had its roots in the Underworld and its crown in the sky, and sometimes by a Cosmic Mountain, a high place where the shaman (which means "one who sees" or "one who knows" in Tunguso-Manchurian), besides being a visionary and medicine-man, acted as a kind of priest. There was a widespread belief that all animals had souls or spirits, and when early hunters hunted them for food or clothing they had to placate their souls in a ritual involving blood and hair. Hence the rock-carvings of animals in Palaeolithic caves in France. The philosophies of these early pre-literate societies can be deduced from their religious rituals and beliefs, and from their cave art. The shamanistic universe was an ordered unity.

The early post-literate ancient civilisations were all influenced by this early shamanism and expressed their awareness of ordered unity through sacred books and creation myths linked to their religions. They represented the ordered, unified universe through their main gods. The Sumerians, the first Mesopotamians whose religion came from Central Asia and Siberia, expressed the unity of the universe in terms of their sun-god Utu c.2600BC. Utu became Shamash when the Akkadians dominated Sumer c.2400BC.[34] The Egyptians expressed their unified universe in terms of their sun-god Ra, who was well established by the fourth dynasty, c.2600BC. The Egyptians expressed the unity of the universe through Ra in the Eyptian *Book of the Dead*, which assembled 200 religious texts, spells and prayers written between 1600 and 900BC, though many began as Pyramid texts and go back to 2400BC or before.[35]

The Indo-Europeans, including the Kurgans, expressed their unified universe by worshipping a Sky Father, Dyaeus Pitar ("shining father"), who ruled over a sun-god. The Indo-European Dyaeus became Zeus after a second wave of Indo-European Kurgans entered Greece c.2200BC, perhaps from the Baltic via Anatolia, and became the ancestors of the Mycenaeans c.2000BC.[36] The Indo-Europeans who settled in Europe also settled in Iran c.2250BC and their god Mitra passed to the Indian civilisation.[37]

The Indo-European Aryan speakers who migrated from Iran and arrived in the Ganges Valley c.1500BC wrote the *Rig Veda*. In that epic the concept of order in the universe is expressed as *rta:* the harmony in Nature which, the early Indians believed, is a reflection of the divine harmony. *Rta* is cosmic order in Jeanine Miller's *The Vision of Cosmic Order in the Vedas*, and in passages written between c.1500 1200BC such as: "The wise seers watch over their inspired intuition refulgent as heavenly light in the seat of *rta*" (*Rig Veda* X. 177.2); and "by the song borne of *rta* the sun shone forth" (X.138.2d).[38]

By c.600BC Vedism had developed into Brahmanism, in which the Atman (the divine within man) united with the eternal Brahman. The vision of a unified universe is found in the *Upanisads* of the 9th century BC, for example: "Beyond the senses, beyond the understanding, beyond all expression,…is pure unitary consciousness, wherein awareness of the world and of multiplicity is completely obliterated…. It is the supreme good. It is One without a second." (*Mandukya.*) The concept of the ordered, unified universe spread into Hinduism and Buddhism.[39] The vision of the Buddha, who achieved enlightenment c.525BC, is of the harmonious Whole, the Oneness of the universe. In the Indian Avatamsaka school which spread to China as Hu-Yen and Japan as Kegon, all objects and energies are under one law, the One behind each of the many, sameness behind difference; a metaphysical philosophy that later passed into Zen.

The Iranian concept of an ordered, unified universe was expressed through the god of the Achaemenian Kings, Ahura Mazda, whose messenger was Zoroaster (between 660 and 500BC). Though there were twin spirits of light and darkness (Spenta Mainyu and Angra Mainyu) the *Gathas* ("songs" or "odes"), thought to be of the 7th century BC, make it clear that when we see correctly, the universe has unity: "Grant me the power to control this mind, this Lower Mind of mine,…and put an end to all Duality, and gain the reign of One." (32.16). Ahura Mazda was the sole god who unified the universe. An inscription by Darius I (522-486BC) in Old Persian, Akkadian/Babylonian and Elamite in Persepolis says: "The Great Ahura Mazda, the greatest of the gods – he created Darius the King; he bestowed on him the kingdom. By the favour of Ahura Mazda Darius (is) the King." Ahura Mazda may be shown in Persepolis in carvings of the *Faravahar,* the Achaemenian crest of a human figure with eagle's wings, legs and talons, that were carved during the reign of Artaxerxes I (after 465BC), but this may be a misunderstanding as there is a strong body of scholarly opinion that no image was ever made of Ahura Mazda. Nevertheless the concept of Ahura Mazda in the 6th century BC influenced the Ionian philosophers of Miletus and the Greek philosophers who moved to southern Italy.[40]

The Anatolian, Syrian and Israelite civilisations adopted different forms of the Indo-European Sky Father. The Indo-European Hittites arrived in Anatolia c.2000BC, and expressed the ordered, unified universe through a Storm-and-Weather god during the Hittite Old Kingdom, which lasted from c.1700 to c.1500BC. The Hittites ruled Syria as a Hittite colony on and off, and the Syrians expressed unity through their cloud-ruling Baal from c.1500BC. The Semitic Hebrews were influenced by Indo-European fire-cults and the *Old Testament* expressed the ordered, unified universe through their storm-god Yahweh from the 20th to the 4th centuries BC. Yahweh unified and ordered the universe.[41] The Indo-European Dyaeus became the Greeks' Zeus after 2200BC, and the shamanistic journey to the Underworld was reflected in the Eleusinian grain mysteries from c.1800BC, the climax of which was the display of a reaped ear of corn.[42]

The early Chinese were influenced by shamanistic Central Asia, which was active in nearby Mongolia. The early Chinese Supreme Ruler from c.1766BC united man and Heaven (*T'ien*). In due course the Chinese civilisation expressed the ordered unity of the universe in the *Tao*. The founder of Taoism Lao-Tze or Lao-Tzu, is traditionally thought to have been born c.570BC. The *Tao* (Way) is a Void out of which all creation came. It is the One, the unity under the plurality and multiplicity of the universe. "We look at it and do not see it; its name is The Invisible. We listen to it and do not hear it; its name is The Inaudible. We touch it and do not find it; its name is The Subtle (formless)....Infinite and boundless, it cannot be given any name." (*Tao Te Ching, Book of the Way and its Power*, poem 14.) It anticipates the Greek "infinite and boundless". And: "*Tao* produced the One. The One produced the two. The two produced the three. And the three produced the ten thousand things." (Poem 42.)

We are enjoined to live in harmony with the *Tao*, the One, and with eternal energy, *ch'i*. The *Tao* "flows everywhere like an ocean" (poem 34). It is Non-Being or *wu*: "All things in the world come from being. And being comes from non-being." (Poem 40.)[43]

The Indo-European Dyaeus was also reflected by the Celts, who were first distinguishable from the mass of Indo-Europeans c.1200BC and who expressed their unified universe through their Indo-European Sky Father, Du-w ("the one without any darkness").[44]

All these religious traditions, which were all to the east of Athens ("*ex oriente lux*"), were behind, and may have influenced, the philosophical view of the Greeks, and they continued their focus on the ordered unity of the universe in their respective countries alongside the development of Western philosophy. Thus, while Western philosophers strove to improve on the philosophies of the Presocratics the One could be found in Hellenistic pagan Hermetic texts such as the Greek *Poimandres*, the most Gnostic of 1800 discourses in the *Corpus Hermeticum*; and in the Alexandrian Neoplatonism of Plotinus in the third century AD: "We may believe that we have really seen when a sudden light illumines the Soul: for this light comes from the One and is the One....If then a man sees himself become one with the One he has in himself a likeness of the One." (*Enneads* VI, 9, 9.)[45]

Many traditions of different civilisations arose after the Greek Presocratics and expressed the ordered unity of the universe within the religious texts of the Roman, Byzantine-Russian and Germanic-Scandinavian civilisations while the Western philosophers were developing Western philosophy. Thus, in the Tibetan civilisation the *Six Teachings or Six Doctrines* brought to Tibet in the 8th century AD by Padmasambhava presents the Clear Light of the Void. The Tibetan *Book of the Dead* is a commentary on the part of the *Six Teachings* that deals with the Yoga of the After-Death State and the Clear Light, written down in the time of Padmasambhava. The art of dying is to see the Clear Light, which unites earthly existence and the after-death state, expressing a unity of the universe that transcends both life and death.[46] In the Arab Islamic tradition Sufis spoke of *Tawhid*, a feeling of the timeless unity within multiplicity. The Spanish mystic Ibn al-Arabi taught the Unity of Being and the Unity of all existence before he died in 1240: "Phenomenal existence…is the concealment of His (God's) existence in his oneness."[47] In the 19th century Wordsworth saw the universe as "One Spirit" (in *The Prelude*) and Shelley wrote: "The One remains, the many change and pass;/Heaven's light forever shines. Earth's shadows fly;/Life, like a dome of many-coloured glass,/Stains the white radiance of Eternity…." ('Adonais'.)[48]

Many of these ancient traditions could have influenced the early Greek philosophers and their later Western counterparts. In the ancient traditions, philosophy and religion were the same thing. The philosopher looked at the phenomena of Nature and its testable data, and was open to concepts that lay behind or beyond them, the underlying order which religious texts suggested. The Mesopotamian creation myths, the Egyptian *Book of the Dead*, the *Rig Veda*, the *Upanisads*, the *Old Testament*, the *Gathas* and the *Tao Te Ching* all proclaimed the truth that the Reality that existed before creation is a continuous, infinite and timeless Void or sea of moving energy without end, from which the finite universe of Nature was thrown up. Eastern philosophy again and again emphasises that the Void was more of a Fullness rather than an Emptiness as it threw up the universe. The Gnostic Plenitude echoed this perception. Like the Presocratic Greeks, the philosophers of the ancient religions focused on the One, the infinite, the boundless, Nature and the universe, and the Presocratics were influenced by the Indo-European Iranians as well as their own Indo-European Greek tradition of Zeus. Their vision could also have been shaped from Mesopotamia, Egypt (which is reputed to have shaped Plato's thought during his visit there soon after 399BC and given him, or confirmed, his Ideas), Indo-European Anatolia and China, from whom they may have derived fragments of their awareness of the One, the infinite, the boundless, Nature and the universe.

Philosophical Summary

What can the Universalist philosopher take from the philosophies of the Presocratic Greeks and early religions? Most importantly, the eternal, moving "boundless", the infinite of Anaximander and the singularity of his "germ" within the boundless which became the universe. Also the idea of the One in Xenophanes and Parmenides and their view that space is a plenum, a fullness rather than an emptiness. Also Heracleitus' moving "ever-living Fire" and his *aither* (ether) and process philosophy of flux. And he should note the Atomism of Leucippus and Democritus, which anticipated modern scientific Materialism. From the early religions he should take the concept of an ordered, unified universe, the idea that the *Tao* is like an ocean, a Void or sea of moving energy, "infinite and boundless". Lao-tze and Anaximander both flourished c.570BC, when the idea of the "infinite and boundless" was about in both West and East. The early philosophies either anticipate or echo and reinforce the ideas of the Presocratic Greeks.

To sum up, the Universalist philosopher can distill from the early Greeks and religions:

the eternal, moving "boundless", the "infinite";
the Void as a sea of moving energy;
moving "ever-living Fire";
the "germ" within the "boundless" that became the universe;
the One;
the ordered, unified universe or *kosmos* ("ordered whole"), concept of order;
philosophers focused on the universe and Nature;
aither and process flux;
Atomism.

The Universalist philosopher will find these themes in the Western philosophical tradition of the next 2,500 years.

2

THE DECLINE OF WESTERN PHILOSOPHY AND THE WAY FORWARD

The Presocratic Greeks speculated about the infinite, the "boundless", and attempted to explain the universe scientifically. They were half-mystic and half-scientific. In their first philosophical fumblings can be found the two differing emphases of later philosophy: two emphases that gave rise to two traditions which underlie the historical pursuit of philosophy. For many centuries they were unified, but later they came into conflict and became disunited.

Plato and Aristotle: Metaphysical and Scientific Emphases

Philosophers can be classified by their intentions and interests, and each tradition has its own motive and focus. The first tradition, which is found in the works of Plato, stems from religion and seeks a pattern in the universe, general truths about Reality and permanent values which influence one's way of life. The second tradition, which is found in the works of Plato's pupil Aristotle, is close to the spirit of science and desires clarity about Nature and the mind of man. These two traditions became prominent in the 4th century BC.

Plato, the establisher of the first tradition, was an admirer of the itinerant Socrates and was about 24 when Athens was defeated in the Peloponnesian War. He supported the oligarchy of the Thirty Tyrants, which was led by his relative Critias, and when it was overthrown in 399 he supported the restored democracy which to his dismay executed Socrates. In due course he "was forced to admit that things would not become better in politics unless the philosophers became rulers or the rulers philosophers".[1] He travelled to Italy and in Syracuse met Dion, the brother-in-law of the ruling tyrant Dionysius I, who promised to implement his political ideas. He returned to Athens and wrote *Politeia (Republic)*, his view of an ideal state ruled by philosophers.

In 387BC he founded his Academy, whose ruins can be visited today in a garden square in Kimonos Street, where Platonism flourished through the teaching of his *Dialogues* (mostly written after 380BC). The early dialogues reflect the teaching of Socrates, the later ones probably contain more of his own thinking. Influenced by the mathematics of Pythagoras, the Platonists taught a universal science centred on mathematics at the Academy. Such was the primacy of the immutable laws of mathematics to Plato that the following words were reputedly engraved over, or at, the door of his Academy: "Let no one ignorant of geometry enter here."[2] The Academy remained open for over 900 years (a longer life span than has been enjoyed by any educational institution in the West), from 387BC to AD529 when it was suppressed by the Byzantine emperor Justinian. Plato went back to Syracuse in 367 to tutor Dionysius II but his ideas were not adopted. In the end he was disillusioned with both democracy and tyranny, and also the world of the senses.

Plato, focusing on the ultimate nature of Reality and the suprasensible world, rejected the changeable, deceptive reality of the sensible world in favour of an unchanging and therefore more truthful hidden Reality behind sensible phenomena, a world of Ideas or Forms. All the objects humans perceive with their senses are but imperfect copies of eternal Ideas, Plato thought, which pre-existed their physical copies. So the Idea of Perfect Beauty is behind all beautiful things and the most important Idea is the Good, which is behind everything we call good. The world of the Ideas did not find favour with Aristotle and though the Ideas are called "universals" (in the sense that they are principles of grouping or classifying that can be applied to more than one particular thing, which Plato regarded as having their own entity) the world of Ideas is not the hidden Reality Universalists identify. Nevertheless it is "beyond the physical", and established the first tradition, which can be described as "metaphysical".[3]

The term "metaphysics" goes back to Aristotle's early students, who spoke of "*ta meta ta phusika*", what comes "after physics" – the subject dealt with chronologically after the subject of physics in the sequence of Aristotle's works. Aristotle himself called metaphysics "first philosophy" or "theology". His *Metaphysica* was in three parts: ontology (the study of Being and existence, the classification and properties of entities, and change); natural theology (the study of God, the existence of the divine, creation and religion); and universal science (the study of first principles and laws). Aristotle differentiated three theoretical sciences – metaphysics, mathematics and physics (or natural science); two practical sciences (ethics and politics); and two productive sciences (art and rhetoric).

Over the centuries metaphysics came to be "that branch of speculation which deals with the first principles of things including such concepts as being, substance, essence, time, space, cause, identify etc." or "theoretical philosophy as the ultimate science of Being and Knowing" (*The Shorter Oxford English Dictionary*). From this concern with Being and ontology metaphysics came to be regarded as what is "beyond" physics: "the science of things transcending what is physical or natural" (*The Shorter Oxford English Dictionary*), meaning what is beyond physicalism, what is supersensible. It came to mean what is "behind" or "hidden within" physics: the study of Being or Reality, or "the study of the first principles of Nature and Thought" (*Chambers' Twentieth Century Dictionary*). It became the branch of philosophy which studies Being or ultimate Reality as a cause that controls the first principles of the universe and Nature.

The second tradition, which can be described as "scientific", was established by Aristotle, a follower of Socrates and Plato who had been a member of Plato's Academy for 20 years. Though deeply concerned with problems of metaphysics (substance, form, matter, potentiality and actuality, and cause), Aristotle was more tolerant and more down-to-earth than Plato, more practical and concerned with common sense and experience in the natural world. His 47 extant works such as the *Nichomachean Ethics* and *Analytica Posteriora* are in the form of lecture notes, and he was a logician, analyst and classifier of knowledge. Aristotle was less positive about Plato's Ideas, arguing that Ideas or forms, or universals, exist but only "in" the particulars in which they are observed. Thus ideal Beauty was in a beautiful face and did not pre-exist it. Like Plato, Aristotle thought that universals are real, but whereas Plato advocated "*universalia ante rem*" ("universals before the thing") Aristotle advocated "*universalia in re*" ("universals in the thing") or "*universalia in rebus*" ("universals in things"). The Latin phrases were later summaries, but Aristotle's emphasis *was* more scientific.

In *Metaphysics* (IV, 1-2/1003a) Aristotle held: "There is a science which studies Being as Being and the attributes which belong to this in virtue of its own nature." This science or first philosophy (i.e. metaphysics) studies that which is common to everything that exists, the principles or causes of all things, that which makes everything what it is. The study of Being as Being is in fact a study of substance for to be is to be a substance. "Substance" describes what might be called the underlying Reality.

If Plato founded Western philosophy, Aristotle determined the orientation of Western civilization. The son of the court physician to the King of Macedonia, Aristotle received a grounding in Greek medicine and biology when very young. He arrived at Plato's Academy in Athens in 367 and left it in 347 to travel, and tutored the future Alexander the Great at the Macedonian court for three years. In 335 he opened the Lyceum, his equivalent of Plato's Academy, in a gymnasium attached to the Temple of Apollo Lyceus, which was in a grove outside Athens and had been visited by Socrates. It was known for its teaching of biology and history, and lasted nearly 300 years. The Lyceum was also known as Peripatos as teaching took place in the *peripatos* or covered walkway of the gymnasium and grove, instruction taking place while students walked about. Hence the school of Aristotle was known as Peripatetics (from the Greek *peri*, "around", and *patein*, "to walk").[4]

Plato's Academy and Aristotle's Lyceum dominated the philosophy of the ancient world. They were the equivalents of Oxford and Cambridge, or Harvard and Yale. In due course Alexandria became the new intellectual centre, but there is truth in Whitehead's remark that "the European philosophical tradition... consists of a series of footnotes to Plato" (*Process and Reality*). Platonists at the Academy and Aristotelians at the Lyceum shaped the schools of Hellenistic and Roman philosophy that succeeded Plato and Aristotle in the Hellenistic and Roman world.

These schools continued the metaphysical emphases of Plato and Aristotle. As the Greek city-states decayed after the death of Aristotle in 322BC and fell under the sway of the Hellenistic kings who succeeded Alexander, life became troubled. The founder of the Stoics (so-named because he taught in the Stoa Poikile in Athens), Zeno of Citium, drew on Heracleitus, Socrates and Plato in advocating living for virtue and putting up with fortune's vicissitudes. The founder of the Epicureans, Epicurus, drew on Plato (via his schoolmaster, the Platonist Pamphilus of Samos, and Xenocrates, head of the Academy after Plato) and also on Aristotle in advocating simple pleasures as the essence of a happy life. Hence Epicurus retired to his garden, cultivated friendships and warned against participation in public life. The founder of the skeptics, Pyrrho of Elis, drew on Plato, Aristotle and the Stoics by challenging the claims of Platonism, Aristotelianism and Stoicism. He held that we cannot be certain that what we perceive is real and not illusory. The founder of the Neoplatonists, Plotinus, drew on Plato,[5] Aristotle the Stoics and Middle-Eastern religion. He held that there were several levels of Being, the highest of which is that of the One, or the Good, which are identical. Plotinus's focus on the One is very important for Universalism's focus on the One.

After Plotinus, Greek philosophy ceased to be creative. Not long afterwards, the Roman Empire fell, and a new Christian culture slowly took shape in western Europe. The Middle Ages began with the fall of Rome at the end of the

4th and beginning of the 5th centuries AD and continued until the 15th-century Renaissance. This long period of about a thousand years was dominated by Christianity. Quite simply, Greek philosophy flowed into Catholic thought and theology, and the best-known philosophers of the medieval time were churchmen. Any independent thinking earned a rebuke from superiors.

Catholic philosophers such as Ambrose, Victorinus and Augustine absorbed Neoplatonism into Christian doctrine to put forward a rational interpretation of Christian faith.[6] However, medieval philosophy retained the methods of Plotinus. In the Middle Ages, philosophy and religion were at one, philosophy serving theology to make possible a rational understanding of faith. The consensual unity of the Presocratics was continued in the consensual unity of the Christian philosophers who were inspired to develop new philosophical ideas by the Christian faith.

Plato's metaphysical emphasis continued in Augustine's Neoplatonism.[7] The scientific emphasis of Aristotle only had a very slight impact on Augustine. Augustine had led a debauched life in his youth in Carthage before going to Rome, and in his *Confessions* (c.400) he tells us he prayed during these excesses, "Lord, give me chastity, but not yet."[8] The reformed Augustine drew on Plato and Plotinus to show that beyond the world of the senses there is a spiritual, eternal realm of truth which the human mind can know and to which humans can strive. This truth he saw as Christianity's God. To Augustine, Plato's Ideas or Forms, or universals, were archetypes in the mind of God. Truth and Beauty are encountered with the mind, not the senses, he stressed. The mind receives the intelligible Light in which humans see truth. Proof of a necessary, unchangeable truth is proof of the existence of God.

Augustine followed Plato in seeing the truths of mathematics and ethics as unchangeable and eternal, and beyond the mind. They are received during illumination when the metaphysical Light enters human minds, bringing immutable truth or God. Humans see Truth with their immortal soul, which is superior to the perishable body. However, as the body is to be eventually resurrected, it is a very important part of a human being. Nevertheless, like Aristotle, who in what is known of his lost dialogue *Eudemus* (*On the Soul*) saw the soul as being bound to the body in an unnatural union, like a prisoner bound to a corpse by Tyrrhenian pirates, Augustine saw soul and body as separate substances, a view that was taken up later on by Descartes who formalised the split between body and mind.

27

The metaphysical emphasis continued with Boethius, who translated the Neo-platonist Porphyry's *Isagoge* and in an accompanying commentary raised the fundamental question about universals: "whether *genera* and species are substances or are set in the mind alone; whether they are corporeal or incorporeal substances, and whether they are separate from the things perceived by the senses or set in them." In his *De Consolatione Philosophiae (Concerning the Consolation of Philosophy)*, c525, he follows Plato in seeing universals as innate Ideas. The Platonic-Augustinian position was followed by the Greek Fathers of the Church such as Origen, Gregory of Nyssa, Nemesius, pseudo-Dionysius the Areopagite and Maximus the Confessor, and later in the 9th century by John Scotus Erigena, from whom the One descends into manifold creation.

Anselm, an Italian who became abbot of the French monastery of Bec and later Archbishop of Canterbury, in his *Monologium*, 1077, offered "proofs" for the existence of God which derived from Neoplatonic thought: the multiplicity of good things share in the same Good, which (being a universal with its own entity) is Good in itself, God; and the multiplicity of beings that have a measure of perfection share in the One, the supreme and perfect Being. In *Proslogium*, 1077/8, he offered another "proof": God is the greatest being that can be thought, but God exists in reality, and a God that exists in both reality and understanding is greater than one that can only be thought.

The Cistercian Bernard, who established a monastery at Clairvaux, was influenced by his Platonist friend William of Champeaux who taught that universals are realities apart from the mind. In his letters and sermons he offered a doctrine of mystical love which derived from Plato: "The Father is never fully known if he is not loved perfectly."[9] In other words, God is Perfect Love, and we imperfect copies have to perfect ourselves in order to know God. This Platonism inspired him to send the first Knight Templars to the Holy Land and promote the Second Crusade for the French King Louis VII and Pope Eugenius III, and to challenge the rationalistic dialectician Peter Abelard, who expounded Aristotelian logic in a new school of Peripatetics (named after Aristotle's Peripatetic school). Abelard opposed the Platonic Realism of Bernard's Platonist friend William of Champeaux, asserting that particular individual things have reality, not universals (which are merely mental concepts).

After the Muslim conquest of Syria and Egypt, texts that had been studied in the Greek philosophical schools were translated into Arabic, including Plato's dialogues, Neoplatonic treatises and the works of Aristotle and Alexandrian commentaries on them. Through these translations Platonism influenced Islamic philosophy, particular Avicenna, and medieval Jewish philosophy, which especially in Spain had a strong Neoplatonic link as in the works of Ibn Gabirol. At the same time Aristotle's writings spread through the Middle East and influenced Arab thought through Avicenna, al-Ghazali and Averroës, and Jewish thought through Ibn Gabirol and Maimonides.

In the 12th century a cultural revolution swept through Europe, replacing the old education that emphasised the liberal arts, grammar and reading the classics with logic, dialectic and scientific disciplines. Platonism declined in Europe and Aristotelianism was taken up.[10] Translations were made of Aristotle's logical treatises. The first chancellor of Oxford University, Robert Grosseteste, translated Aristotle's *Ethica Nichomachea* from Greek to Latin, and introduced Empiricism in the form of an experimentation principle which either verifies or falsifies a theory. Extremely interested in science, he used mathematics and optics in his treatise *De Luce* (*On Light*), 1215-20, which states a metaphysics of light: God is the primal Uncreated Light, and physical light is the basic form of all things. Grosseteste's pupil Roger Bacon carried on his mathematical and experimental methods, and his view that experience gives reasoning certainty through the senses and inner illumination.

Platonism was defended against Aristotelianism and Avicenna's Aristotelian commentaries. William of Auvergne, a follower of Plato and Augustine, opposed the God of Avicenna who creates the universe eternally through 10 Intelligences (which came from the 10 *sefirot* of the Kabbalah). The Franciscan friar Bonaventure preferred the metaphysical tradition of Plato, Plotinus and Augustine to the natural science of Aristotle, who, he argued, denied the existence of divine ideas. Bonaventure was scrupulous in not confusing philosophy (knowledge of Nature and the innate soul) with theology (knowledge of heavenly things based on faith and divine revelation). His *Itinerarium mentis in Deum* (*The Soul's Journey into God*), 1259, follows Augustine's path to God from the external world of Nature to the interior world of the mind, and then from the temporal to the eternal above the mind.

Aristotelianism triumphed in the 13th century. The Dominican Albertus Magnus advanced botany, zoology and mineralogy in encyclopaedic writings that made use of recently translated Greek and Arabic scientific and philosophical texts. He synthesised Aristotelianism and Neoplatonism and blended the philosophies of Aristotle, Avicenna, Ibn Gabirol, Augustine and pseudo-

Dionysius. He was a truly Universalist reconciler of conflicting philosophies into one whole, uniting Greek, Arab, Jewish and Christian philosophies.

His Dominican pupil Thomas Aquinas championed Aristotle and Arab and Jewish thinkers. He realised that the theologians of his day were Christian Neo-platonists because Arab commentators had distorted Aristotle's philosophy. In his writing he tried to show the value of Aristotle's philosophy to his Christian contemporaries. In *Summa Against the Gentiles*, 1258-64, and *Summa Theologiae*, 1265/66-1273, he says that philosophy seeks the first cause of things whereas theology inquires into God. He drew on both Plato and Aristotle but differed from them, showing God to be primary mover and (unlike Aristotle) affirming the soul's immortality. He saw God as pure being, and creation as participating in being according to creatures' essence.

The 13th century saw growing Rationalism as "Averroists" such as Siger de Brabant used rationalistic Aristotelian doctrines to contradict faith. In this they followed conceptualist arguments of Roscelin, Abelard and William of Ockham. Christian theologians condemned such developments. In the later Middle Ages philosophy became formalised inside schools: Dominicans, Averroists, Franciscans and Thomists (followers of Aquinas). John Duns Scotus criticised the Rationalists and Aristotle, who, he said, was ignorant of the Fall. He saw God as infinite being rather than as primary mover or first cause (the Aristotelian Greek-Arab view). He held that God's will was supreme and that universals are abstract concepts but are based on "common natures" which exist in individuals as "thisness" (*haeccitas*).

In the late 14th century there was a trend away from Aristotelianism. William of Ockham rejected the "old way" (*via antiqua*) of Thomism and Scotism and followed the "modern way" (*via moderna*). Like Scotus he opposed the Aristotelian Greek-Arab emphasis on God as a first cause, but stressed God's freedom. There are no "common natures" he claimed, but only individual things. Everything could have been different – stars, fire, morality – if God had so willed, but creation was done with economy of thought and the bare minimum number of beings necessary, the principle of Ockham's razor. Nicholas of Autrecourt rejected Aristotelianism and sought to return to the Atomism of the Presocratics. Meister Eckehart rejected Aristotelianism for Christian Neoplatonist mysticism: the soul ascends to God in Neoplatonic terms, purifying itself from the body, transcending being and knowledge until it is absorbed in the One, or God as the being of all things. Nicholas of Cusa was also a Neoplatonist rather than

Aristotelian. He saw Aristotle as an obstacle to the ascent to God and showed through mathematical symbols how in infinity contradictions are reconciled. God is absolutely infinite, the universe is relatively infinite (he said) and man has limited, approximate knowledge of the infinite God.

At the end of the Middle Ages Aristotelianism was being abandoned and Christian philosophers were again turning to Neoplatonism. The Platonism of the 15th-century Renaissance continued the Christian Platonism of the Middle Ages. The Renaissance was largely a consequence of the rediscovery of classical texts following the fall of Constantinople in 1453. The Byzantine libraries were looted and classical texts that were hitherto unknown began to arrive in the west. These reversed the trend of medieval learning and restored admiration for Plato. Gemistos Plethon, the Byzantine philosopher and humanist scholar, had visited Florence in May 1439 as part of the Byzantine Emperor's delegation to bring the Orthodox and Catholic faiths closer together. Cosimo de' Medici, who had begun the Medici Principate in 1434, had received him. Gemistos spoke "On the Difference between Aristotle and Plato" and gave the Italian humanists a new interest in Plato. Speaking as a Platonist rather than as an Orthodox cleric, Gemistos had proposed that all Plato's works should be translated. He inspired Cosimo to revive Plato's Academy, which had run from 387BC to AD529.

The Academy opened at a villa in Careggi, Florence, in 1462. Cosimo put Marsilio Ficino, his doctor's son, in charge of it and of translating Plato's entire works into Latin. Ficino's Neoplatonism influenced Botticelli in 1478 when he was painting the *Birth of Venus* and *Primavera*, in which the creative principle of spring is shown as the earth goddess Venus, an embodiment of Plato's Idea of Perfect Beauty. Leonardo da Vinci reputedly succeeded Botticelli as Grand Master of the Priory of Sion in 1510 and would have known Botticelli well. It is likely that his *Mona Lisa*, begun in 1503 and endlessly worked on until he died in 1519, was his equivalent of a representation of the Platonist Idea of Perfect Beauty which Botticelli had shown as Venus. Leonardo's version was an earth goddess who was also spiritual Venus – hence her enigmatic smile.

A member of the Academy, Pico della Mirandola, gave his *Oratio de Hominis Dignitate (Oration on the Dignity of Man)*, which emphasised the humanist centrality of man in the universe, at the age of 24 in 1486. Ficino, Pico and Giordano Bruno, the late 16th-century Italian philosopher, introduced Platonism to Renaissance metaphysics. Ficino was also a Kabbalist and Hermeticist, having absorbed Egyptian Hermeticism which was entangled with Egyptian

Neoplatonism. Through their influence Plato's account of moral virtues became the Renaissance Ideal of courtly virtue and the gentleman's code. Plato's mathematics was linked to Pythagoras's attempt to discover the secrets of the sky through numbers, and impacted on Renaissance science in the work of Nicolaus Copernicus, Johannes Keplor and Galileo

Now Platonism influenced the form of Renaissance philosophy. The treatise – the *summa* used by Thomas Aquinas – gave way to Platonic dialogues in the works of Bruno, Galileo and Niccolo Machiavelli (in *Dell'arte della guerra, The Art of War*, 1521). Humanism was a literary and moral, as well as a philosophical, movement (like Universalism), and the humanist philosophers were well-read men with experience of public life, as were the Dutch Erasmus (scholar, translator and traveller to European courts), the English Sir Thomas More (Henry VIII's chancellor) and the French Michel de Montaigne (mayor of Bordeaux). New translations from the Greek Atomistic Materialists and Skeptics, and Roman Stoics, had an impact and revived the ancient Greek schools. Translations of Democritus and Lucretius influenced Galileo and Bruno. Translations of the Skeptic Sextus Empiricus influenced Montaigne. Translations of the Stoics Seneca and Epictetus also influenced Montaigne, and later Decartes and Spinoza.

Plato's *Republic* was behind a rise in political theory following the rise of nationalism as countries reacted against the sway of Church Scholasticism and Aristotle. In the 16th century, the Italian Niccolo Machiavelli described how to seize and retain power in *Il Principe (The Prince),* 1512-13. The French Jean Bodin later held that the State must have absolute power. Later still, the English Thomas Hobbes advocated a social contract in which all surrendered their rights to the sovereign, who then protected them under the rule of law.

But this surge of Platonism was its last gasp. The Plato-Aristotle monolith was now crumbling. It had dominated philosophy throughout pagan times and then throughout the Christian era, and had lasted 1,900 years (c.400BC-c.1500BC), and it would linger on at least another 200 years in the universities to c.1700. The two metaphysical and scientific emphases which began in Socrates's dialogues and became established in the Academy in Athens were to last a staggering 2,100 years or more in a unified descent that formed the backbone of Western philosophy until modern times.

The Fragmentation of Modern Philosophy

Around 1500 a great change began to take place. Over the coming decades

Christian metaphysics was no longer based on Platonism interspersed with spells of dominance by scientific Aristotelianism. Where the Middle Ages had found the universe hierarchical, organic, ruled by God and subject to divine intentions, the Renaissance found it pluralistic, mechanistic, ruled by mathematical laws and subject to physical causes.

Now humanism was focusing on the centrality of man. Now a separation took place between the metaphysical and scientific emphases, the Christian context weakened, and the metaphysical tradition passed to Rationalists, while the scientific tradition passed to Empiricists. Rationalist philosophers freed themselves from the Christian legacy. They were still metaphysical, and concerned with what lies behind or beyond the universe, the ultimate Reality as a cause that controls the first principles of the universe and Nature; but they began to devise their own metaphysical systems, using their own rational experience. The same applied to Empirical philosophers. Both Rationalist and Empirical philosophers used instrumental or mechanical science for testing, and mathematical explanation. (It should be pointed out that the Rationalists, such as Descartes and Leibniz, were in general far better scientists than the Empiricists, who were for the most part men of letters.)

Leonardo, as rounded a Renaissance man as any, being a painter, sculptor, architect, engineer, anatomist and mechanical inventor, caught the first indication of this change in three theses he wrote in the early pages of his *Notebooks*:

"Since experience has been the mistress of whoever has written well, I take her as my mistress, and to her on all points make my appeal....Instrumental or mechanical science is the noblest and above all others the most useful, seeing that by means of it all animated bodies which have movement perform all their actions....There is no certainty where one can neither apply any of the mathematical sciences, nor any of those which are based upon the mathematical sciences."[11]

These three propositions caught the new Renaissance worldview. Leonardo resolved to follow (1) Empiricism; (2) mechanistic science; and (3) the mathematical sciences. (Leonardo warns in his *Notebooks*: "Let no man who is not a mathematician read the elements of my work.") For a while philosophers included all three in their work, as did Leonardo himself, but over the coming decades, during the next century or two, these three separated and philosophers advocated one but not all of them.

An early indication of this change can be found in the empirical and anatomical dissections and drawings of the Belgian Andreas Vesalius, who revealed the vascular, neural and muscular systems; and in the mathematical work of the Italian astronomer Galileo, who used new experimental instruments (the lens and telescope) to apply mathematical theory to astronomy and physics.

From now the Renaissance philosophers were not saints or university men but gentlemen with roles in society, members of an *élite* writing philosophy in their spare time. They were separated from formal centres of learning. Hobbes was contemptuous of the Aristotelians at Oxford and Decartes of the medievalists at the Sorbonne, while Spinoza refused a professorship at Heidelberg.

The Renaissance philosophers continued to define themselves broadly so that philosophy overlapped with religion and science, as it did for the Presocratic Greeks. The early Presocratics were philosophers *and* theorists of the physical world of Nature. Pythagoras and Plato were philosophers *and* mathematicians. Aristotle was a philosopher *and* a natural scientist. In the same way Galileo and Decartes were philosophers *and* mathematicians *and* physicists (or "natural philosophers" as physicists were called in their day). Newton was a "natural philosopher" or physicist as well as a theoretical thinker.

The first of these new Renaissance philosophers, Sir Francis Bacon, a lawyer, judge, courtier and chancellor as well as an amateur philosopher, took up the Empirical cause of Leonardo's first proposition and became the founder of modern Empiricism.[12] In his *Advancement of Learning*, 1605, Bacon defined knowledge in terms of history, which depends on memory; poetry, which depends on the imagination; and philosophy, which depends on the reason. He saw the reason as serving "the fresh examination of particulars", and he devised a method of ordering facts in a way that could inductively establish the true causes of phenomena in Nature and the true "forms" of things in metaphysics. He held experience to be the only source of valid knowledge, and he wanted to create a new system of philosophy based on a true interpretation of Nature that would replace Aristotelianism.

Leonardo's second proposition, mechanism, flourished in Hobbes, who saw the corporeal universe as the real world and restated Greek Atomistic Materialism. He saw philosophy as comprising physics, moral philosophy (or psychology) and civil philosophy, and excluded the metaphysical emphasis. In Hobbes the scientific had already separated from the metaphysical.

Leonardo's third proposition, mathematical sciences, flourished in René Descartes, the French mathematician, who invented analytic geometry, conducted anatomical experiments, and drew on the work of Galileo. He established metaphysical Rationalism and was the founder of modern Rationalism.[13] He looked back to the Greek Skeptics Pyrrho and Sextus Empiricus; the Roman Stoics; Augustine; Anselm; and Aquinas. He began by systematically doubting all assumed knowledge. The only basis for certainty was "*Cogito ergo sum*", "I think, therefore I exist". Seeking a principle of absolute certainty and basing certainty on the reason, he argued that there were two independent worlds of mind and body, a dualistic position not acceptable to Universalists. All three of Leonardo's maxims can be found in Descartes' work: Empiricism (*Discours de la méthode, Discourse on Method*, 1637); mechanism (*Principia Philosophiae, Principles of Philosophy*, 1644, and *Les Passions de L'âme, The Passions of the Soul,* 1649); and mathematics (*Meditationes de prima philosophia, Meditations on the First Philosophy,* 2nd edition 1642 – the "first philosophy" being metaphysics – and *Regulae ad Directionem Ingenii, Rules for the Direction of the Mind,* 1628, published 1701). But Descartes was primarily a mathematician.

In his *Principia* Descartes referred to philosophy as a tree whose roots are metaphysics, whose trunk is natural philosophy (physics) and whose branches are medicine, mechanics and morals. Descartes was most concerned with the trunk (physics) and he was only concerned with metaphysics to provide a firm foundation for the trunk. The metaphysical was already being weakened in philosophy. His *Discourse on Method* was thus a physics founded on metaphysics rather than (as in Aristotle and Whitehead) a metaphysics founded on physics. He sought certainty in the self, regarding it as the only innate idea unshaken by doubt, but such subjectivism raises its own questions. From the self he deduced the existence of a perfect being or God with clear ideas. This approach was anti-empirical, and Bacon remarked: "Reasoners resemble spiders who make cobwebs out of their own substance." The substance he was referring to was Descartes' self.

Cartesianism dominated the intellectual life of Europe until c.1700 but Scholasticism was still taught at the universities. Oxford University banned the teaching of Descartes and in 1663 the Catholic Church put his books on the Index of Forbidden Books. At this time philosophy still included mathematics and science, and by and large philosophy had not split from religion: the philosophers all professed to be Christians. Cartesianism rooted Renaissance science in the medieval view of God and the human mind, but it created dualism between God the creator and a mechanistic creation, and between spiritual mind and matter. The unified world of Plato and Aristotle was now divided.

This split into dualism had ramifications in literature, and T.S. Eliot wrote of the "dissociation of sensibility" that took place in the 17th century, in which thought and feeling separated after being united in the poetry of Donne.[14]

Continental Rationalism was continued by the Dutch Jewish Benedict de Spinoza and the German Gottfried Wilhelm Leibniz.

Spinoza saw philosophy as a personal quest for wisdom and the achieving of human perfection. Drawing on Descartes' terminology and mathematics, and using a deductive system based on Euclid, in *Ethica* (*Ethics*), 1677, Spinoza saw the universe as being formed of a single infinite substance, a monism which he called God. He believed that wisdom involved seeing the universe in its wholeness through the "intellectual love of God", which merges the finite human into eternal unity. Achieving the search fills one with joy, he held.

Leibniz's work is in expositions and fragments. A mathematician who invented the infinitesimal calculus, a codifier of laws, diplomat, royal historian, court librarian and inventor, he saw logic as a mathematical calculus. He distinguished "truths of reason" from "truths of fact", necessary propositions of logic and mathematics from "contingent" or empirical propositions of science. He escaped Descartes' dualism and Spinoza's monism by proposing a pluralism of spiritual substances ("monads"). In *Principes de la nature et de la Grâce fondés en raison (Principles of Nature and of Grace Founded in Reason)*, 1714, he captured metaphysical Rationalism: "True reasoning depends upon necessary or eternal truths, such as those of logic, numbers, geometry, which establish an indubitable connection of ideas."

The prestige of reason reached new heights with the rational Enlightenment. Sir Isaac Newton, its founding father in view of his discoveries of the properties of light, gravity, calculus and mechanics in 1664-6, produced *Philosophiae Naturalis Principia Mathematica* (*Mathematical Principles of Natural Philosophy*), 1687. It applied reasoning mathematics to Nature and crowned the astronomical works of Copernicus, Kepler and Galileo, and the work of empirical Bacon and rationalistic Descartes. It demonstrated that the reason was able to unravel the secrets of the universe and explain them in terms of physical forces.

The British Empiricists John Locke, George Berkeley and David Hume turned from the metaphysical, broke away from medieval religious writing and accepted the growing influence of science on philosophy. They adopted a common-sense approach to philosophical problems, concentrating on sense-impressions, experience of physical objects and existence. Locke, another founding father of the Enlightenment, in his *Essay Concerning Human Understanding*, 1690, turned from the Reality behind Nature to the structure of the mind and the origin of reason. In his *Treatise of Human Nature*, 1739-40, Hume completed Locke's focus on man by applying experimental reasoning to moral subjects. These two works, supported by Berkeley's work on perception, shifted philosophy from the late Renaissance's preoccupation with the natural world and the mathematical to sensory knowledge, ideas and experience. They created a new epistemology that dominated Western philosophy until Kant. Now philosophy shifted from being metaphysical and Rationalistic to being epistemological and Empiricist.

Locke offered a new criterion of truth: the origin of ideas, their certainty and where this certainty becomes uncertainty. Descartes had regarded ideas as self-evidently certain ("*Cogito ergo sum*"). Locke focused on mind as a passive receiver of impressions and sensations at the expense of mind as an original thinker. He was reductionist in reducing conceptual experience to sensory building-blocks, suggesting that our ideas are fundamentally sensations. He thought that all our knowledge is ultimately derived from experience (Leonardo's first proposition). He argued (with Galileo and Newton) that there are "primary" qualities of matter and "secondary" qualities of mind – preserving Descartes' dualism. Locke's sensationalist Materialism was taken up by followers of Descartes in France and produced Voltaire's skeptical Empiricism, and his political theory that all citizens have freedom and equality – which denied Hobbes' emphasis on the divine right of kings and the power of the sovereign – influenced Rousseau.

Berkeley, a Church of England bishop, opposed this dualism. In his *Treatise Concerning the Principles of Human Knowledge, part 1,* 1710, and in other works he argued that primary qualities of matter are reducible to secondary qualities of mind. Nature is what humans perceive with their senses, and sense data are to be considered "objects of the mind" rather than "qualities adhering in a substance", an epistemological position known as phenomenalism. He asserted that a thing's existence depends on its being perceived ("*Esse est percipi*", "*to be is to be perceived*"), and that all reality is mental. As objects always exist there must be a God who perceives them. Berkeley was an Empiricist in focusing deeply on perception, but he was also a subjective Idealist because he reduced reality to "spirits" (his name for subjects) and to the ideas they entertain. In 1764 Dr

Johnson said of Berkeley's philosophy, Boswell tells us, "striking his foot with a mighty force against a large stone, till he rebounded from it, 'I refute it *thus*.'" Johnson demonstrated the solidness of matter with a hefty kick.[15]

Hume also found the origin of knowledge in sense-impressions, but found "an associating quality of the mind". He reckoned that the mind's connections of things is accidental, that the mind is "a bundle of perceptions" without unity. All our ideas are copied or developed from our sense-impressions, he thought. No "ought" or value-statement can logically be derived from an "is" or statement of fact, and so there is no possible proof for ethics. There is no order in Nature or in the mind, and no substantial or unified self, and there is therefore contingency in both. Hume reduced the metaphysical concept of the self to experience, and causation to habit. He replaced the metaphysical with skepticism and left philosophy in crisis, unsure as to what is certain. Hume therefore added to philosophy's fragmentation.

Reason was not finished. The Enlightenment's fundamental ideas – dedication to reason, belief in intellectual progress, seeing Nature as a source of inspiration, creating political and social institutions that expressed tolerance and freedom – found their way into 18th-century philosophy and helped forward advances in chemistry and biology, and developments in psychology and social sciences, ethics and aesthetics, history, economics, sociology and law. Now philosophy passed from amateur gentlemen back to university "specialists" such as Christian Wolff, who defined philosophy and wrote textbooks. Journals appeared and there were histories of philosophy in Germany.

Metaphysical Rationalism incorporated the Empiricists' focus on the mind in the unifying works of Immanuel Kant, a professor at Konigsberg in East Prussia, who crowned the Enlightenment. In true Universalist mode, he sought to synthesise Leibniz's Rationalism and Hume's Empiricism by relating the *a priori,* or deductive, and sensory aspects of knowledge. He redefined philosophy. Philosophy's sole task, he said, is to determine what reason can and cannot do. And so philosophy "is the science of the relation of all knowledge to the essential ends of human reason". Philosophy should therefore "outline the system of knowledge arising from pure reason" (which would satisfy followers of Leibniz) and "expose the illusions of a reason that forgets its limits" (which would satisfy the followers of Hume). Philosophers are thus "lawgivers of reason".

Kant felt that a new philosophical method was required to determine the source, extent and validity of human knowledge and the limits of reason. This new method should reject the Rationalists' dogmatic assumptions and return to Descartes' initial skepticism. This method was the critique: his critical examination of reason in thinking (i.e. science in his *Kritik der reinen Vernunft, Critique of Pure Reason,* 1781); in willing (i.e. ethics in *Kritik der praktischen Vernunft, Critique of Practical Reason,* 1788); and in reason in feeling (i.e. aesthetics, in *Kritik der Urteilskraft, Critique of Judgement,* 1790). In each critique there are three parts: analytic (which analysed reason's right functioning); dialectic (which showed where reason falls into error); and methodic (which gives rules for practice).

Kant felt that philosophy should see what reason could achieve when all experience is removed. He found that mind imposes its own pattern on the outside world. In his *Critique of Pure Reason* he presented a transcendental Idealism, so-called because he thought of the self as a "transcendental ego" (an answer to Hume's fragmented self or "bundle of perceptions"), a metaphysical entity which constructs knowledge out of sense-impressions, upon which the self imposes universal concepts he called categories. In *The Metaphysic of Morals,* 1799, he established an objective ethic, his "Categorical Imperative" that one should always act in such a way that the principle of one's action could become a general law. He asserted that certain statements about the world are "necessarily true" (an answer to Hume's skepticism). Throughout his work there is a distinction between the "phenomenal" world of appearance and the "noumenal" world of Reality, the Reality behind Nature on which his own metaphysic is based. Ultimately Kant was an Idealist, and he stated a philosophy of Idealism that gives a central role in interpreting experience to the ideal or spiritual rather than the sensory and empirical.[16]

The Enlightenment ended with Kant's death in 1804, and his synthesis between Rationalism and Empiricism and their views of the structure of the mind became the base for a new century in which there was further fragmentation. Romanticism, initially a poetic revolt in favour of irrational feeling, challenged reason and influenced German Idealism. The social deprivations caused by the Industrial Revolution led to social reform and to Utilitarianism. The 1848 Revolution in France, Germany, Italy and the Austrian Empire led to class divisions (the proletariat opposing the bourgeoisie) and to Marxism. Darwin challenged the Christian tradition with biological evolution and led to Pragmatism. The metaphysical and the scientific continued to be in conflict, and metaphysical Idealism was challenged by Realism; irrational Intuitionism by Positivism; and liberalism by Marxism.

Kant's establishing of the limits of reason had also established reason's inability to penetrate ultimate Reality without coming into conflict with scientific method. In the early 19th century the metaphysical Rationalism of Spinoza and Leibniz returned through the German Idealism of university professors whose fathers had been Protestant pastors in Jena and Leipzig (the father of Johann Fichte) and Tubingen Seminary (the fathers of Friedrich Schelling and G.W.F. Hegel). Attention shifted from Kant's *Critique of Pure Reason* (in which he focused on natural science and denied certainty in metaphysics) to his *Critique of Practical Reason* (which was on the nature of the moral self) and his *Critique of Judgement* (which was on the purpose in the universe). The Idealists held that the main metaphysical fact and certainty is the human self and self-consciousness, that the universe is spiritual and has a cosmic self, and that the will and moral faculty count for more than intellectual reason.

Fichte taught at the University of Berlin and saw the philosopher as journeying from self-consciousness and its moral will to the cosmic totality or Absolute, from his mind to ultimate Reality. Hegel succeeded Fichte at the University of Berlin after holding posts in Jena, Nuremberg and Heidelberg. Like Kant he examined the place of reason in Nature, experience and reality. Hegel found that reason is in the world and not imposed by the mind (as Kant had argued). Hegel projected reason as Fichte had projected consciousness, and declared that "the rational is real" and that "the truth is the whole". This was in keeping with Idealism's belief that Reality is spiritual.

Hegel asserted that the Absolute or Whole is a universal entity and is not static but develops in time through a dialectical process. Thus he held that reason is historical rather than eternal, and philosophy must include cultural history. The philosopher should approach the Absolute through consciousness and recognise it as Spirit, the "World-Spirit". (Wordsworth's invocation in *The Prelude*, book II, to the "Wisdom and Spirit of the universe", which was probably written during his visit to Germany with Coleridge in 1798/9, appears to echo Hegel.) To Hegel, man's subjective spirit (habit, appetite and judgment) are in dialectical opposition to his objective spirit (his laws and social and political institutions), and both are synthesised in Absolute Spirit, in which matter is assimilated to mind (in art, religion and philosophy). Hegel's metaphysics of the Absolute was thus a philosophy of human culture.

The metaphysical German Idealists had ignored the scientific emphasis. Now the empirical, scientific view returned through Realist philosophers who carried the practical common-sense approach of the Empiricists to an extreme. Their aims were realistic and scientific, and their philosophies included psychological or sociological truths.[17]

In France Realism took the form of Positivism through Auguste Comte's *Cours de Philosophie Positive* (*The Positive Philosophy*), 1830-42, which drew on Bacon and British Empiricism and asserted that knowledge that is not derived from the methods of science is invalid. Comte held that the sciences had emerged in a strict order: mathematics, astronomy, physics, chemistry and biology – and that sociology (Comte's new philosophy) was next. He saw thought as developing from religious and metaphysical phases to a scientific phase, and introduced a militantly anti-metaphysical note into philosophy that continued in the 20th century.

In England the empirical, scientific view was continued through John Stuart Mill, whose Utilitarianism (presented in *Utilitarianism*, 1861, and in *Examination of Sir William Hamilton's Philosophy*, 1865) drew on Bacon, Hume and Jeremy Bentham. Utilitarianism equates value with utility, and in ethics advocates the seeking of pleasure and happiness. The aim of society is to produce the greatest amount of happiness for its members. Moral choices are therefore ethical calculations to bring pleasure or happiness. Mill had advocated scientific methodology in *The System of Logic*, 1843, but he principally followed Locke as a political liberal social reformer.

Karl Marx, a more radical social theorist and social revolutionary, a German Jew, took over Hegel's dialectic but in reverse: mind is assimilated to matter. Hence his philosophy was called "Dialectical Materialism". In *The Communist Manifesto*, 1848, he called for the violent overthrow of the established order. Realistic in appreciating historical causes, he applied Hegel's dialectic to the class struggle between the proletariat and the bourgeoisie, who are synthesised in the social change of a proletarian Communist State. Marx saw ideas as not being purely rational but as being dependent on the social order in which they arose.

Metaphysical Idealism underwent a Hegelian renaissance in England in the works of T.H. Green, F.H. Bradley and Bernard Bosanquet. Bradley followed Hegel (without accepting his dialectic) in *Appearance and Reality*, 1893 (which T.S. Eliot read in 1913 and was influenced by). Bradley saw the Absolute, the totality of Being, as the only reality. It can only be approached by clear intellectual judgment as opposed to the psychology of thought-processes, he held, and in seeing the spiritual as having a central role in interpreting experience he was a metaphysical Idealist.

In America there was a reaction against Idealism by the Pragmatism of C.S. Peirce and William James, a professor at Harvard and the brother of the novelist Henry James. Pragmatism defines truth as what has useful consequences. Peirce held that the meaning of a concept is in the practical consequences of the meaning. James applied Peirce's view to truth and held that humans had a right to believe and that belief must be measured by the practical consequences of their belief. They should therefore have "will to believe". James supported religion (for example, in *The Varieties of Religious Experience*) on the grounds that it leads to happiness and he claimed that religion was true on these grounds.

Towards the end of the 19th century appeared the Intuitionists, philosophers who had lost faith in the powers of reason and rejected the systems of the Rationalists. They felt that intuition was a better guide than reason, and that glimpses of Light, moments of illumination, give the feel of the universe rather than its structure. Their irrational philosophies contradicted the reason of Rationalism, the Enlightenment and Hegel (who had continued Enlightenment rationality despite his Idealism). Nevertheless they continued the metaphysical emphasis intuitionally. The main Intuitionists were the Danish Christian thinker Soren Kierkegaard, the German Idealist Arthur Schopenhauer and the German philosopher Friedrich Nietzsche.

Kierkegaard held that truth depends on an individual's mind or emotions ("Subjectivity is truth") and is in an individual's feeling of his own existence. He challenged the logical claims of Hegel's system, championing concrete human existence and states of consciousness such as anxiety and despair against Hegel's abstract theorising.

Schopenhauer asserted that the irrational is real. He held like Bradley that there is a real world behind appearances that the reason cannot penetrate. In *Die Welt als Wille und Vorstellung* (*The World as Will and Idea*), 1819, he saw the world as dominated by a universal cosmic Will that expresses itself in strife and conflict, through which humans make contact with ultimate Reality.

Nietzsche saw the task of the philosopher as destroying old values and creating new ideals and a new civilisation. He emphasised strength of will in *Also sprach Zarathustra* (*Thus Spoke Zarathustra*), 1883-92, and *Jenseits von Gut und Böse* (*Beyond Good and Evil*), 1886, and these works together with *Der Antichrist* (*The Antichrist*), 1895, an attack on Christianity, *Der Wille zur Macht* (*The Will to Power*), 1901, were seized on by the Nazis, who claimed him as an inspiration for

Fascism whereas he was in fact a master of aphorism and epigram in bold psychological generalisations.

(Kierkegaard and Nietzsche are more commonly regarded as Existentialists, and Schopenhauer as a post-Kantian who was a precursor of Nietzsche.)

The Intuitionists, writing as accomplished men of letters outside the universities, were all operating in fragmented isolation. They saw the mind as hidden and deep, a view that would be taken up in the 20th century in the psychology of Freud and Jung, who focused on the unconscious mind.

The Intuitionist metaphysical emphasis continued in the new century with the French philosopher Henri Bergson, the American John Dewey and the English philosopher Alfred North Whitehead. Bergson and Dewey were influenced by Darwin and both saw the mind as an instrument of man's adaptation to his environment during biological evolution. Bergson and Whitehead both advocated a process philosophy, echoing Heracleitus's river image and sense of flux (*"panta rhei"*, "everything is in a flux"). They saw ultimate Reality as flowing like a river, dynamic, a time process.

Bergson turned away from the static intellect, which he thought was inadequate. He saw continuous movement, change and creation. In the *Introduction à la metaphysique* (*An Introduction to Metaphysics*), 1903, and *L'Évolution créatrice* (*Creative Evolution*), 1907, Bergson distinguished knowing by analysis (the method of science, using the static intellect which makes artificial divisions) and knowing by intuition and intellectual sympathy (the method of his dynamic approach that grasps the Whole, which is also dynamic). Intuition was a kind of self-conscious instinct. Bergson held that all metaphysical truths are grasped by intuition, which is how we know our deepest selves, the time process, the essence of things and the *"élan vital"*, the vital spirit or force which creates Nature. His view of time influenced Proust.

Whitehead, a mathematician, was more speculative. He rejected Idealism and also Realism and thought of himself as proposing metaphysics (which he called "speculative philosophy"). He defined metaphysics as the effort "to frame a coherent, logical, necessary system of general ideas in terms of which every element of our experience can be interpreted".[18] He surveyed the natural world with generalised understanding in *Science and the Modern World*, 1925; *Process and Reality*, 1929; and *Adventures of Ideas*, 1933. He was a metaphysician and philosopher of culture, like Bergson, and had an unwavering belief that he could come to understand existence.

Dewey, on the other hand, was pragmatic. He dominated American metaphysics, ethics and all forms of philosophical knowledge but focused on practicality and moral purpose: applying intelligence to social affairs. His amelioration and Lenin's more radical, revolutionary Marxist thinking continued the outlooks of Mill and Marx. Lenin was a metaphysical materialist who saw material things as existing independently of mind, and ideas and sensations as copies of independently existing things. He had a strong practical side that turned political theory into Communist practice, and philosophy into ideology. His Marxism was a further example of the fragmentation of philosophy.

The metaphysical emphasis of the Intuitionists was opposed by Logical Analysis, a scientific emphasis that dominated the universities of the English-speaking world. Once again the philosopher was a specialist at a university rather than a gentleman or man of letters. Logical analysis, or analytic philosophy, had two branches: Logical Positivism, or Logical Empiricism, which derived from an interest in mathematics and symbolic logic in the works of Gottlob Frege and Russell; and Linguistic Analysis which focused on the logic of common language in the works of G.E. Moore and Ludwig Wittgenstein.[19]

The Logical Positivists were inspired by Hume and by the new logic of Frege, who, when teaching mathematics at Jena, showed that mathematics could be deduced from a few principles of logic and that arithmetic was simply deductive logic. His work attracted the attention of Bertrand Russell, who, in collaboration with his tutor Whitehead, formulated it in *Principia Mathematica*, 1910-13. A school of Logical Positivism was declared at the University of Vienna in a seminar in 1923, and meetings continued there as the Vienna Circle until 1938. Rudolf Carnap arrived at the University in 1926 and helped shape the Circle's 1929 manifesto, *Wissenschaftliche Weltauffassung: Der Wiener Kreis* (*Scientific Conception of the World: The Vienna Circle*). The title of the manifesto announced that philosophy should now be scientific. It should produce clear thinking, not complicated pictures of the world. (Einstein's revolutionary concepts in physics had made the world very complicated, and it can be argued that the Vienna Circle were running away from getting to grips with the new cosmology.)

The Viennese Wittgenstein, Russell's student, stated in *Logische-philosophische Abhandlung* (*Tractatus Logico-Philosophicus*), 1921: "The object of philosophy is the logical clarification of thoughts. Philosophy is not a theory but an activity. A philosophical work consists essentially of elucidations. The result of philosophy is not a number 'philosophical propositions', but to

44

make propositions clear." Logical Positivists focused on logic, language (semantics) and perception. Rebelling against the woolly language of previous philosophers, they held that meaningful discourse consisted of formal sentences of logic and mathematics or the factual propositions of the sciences; that the factual only has meaning if it is possible to say how it might be verified; that metaphysical assertions are meaningless; and that all statements about moral and aesthetic or religious values are meaningless. Whitehead, seeing where the Vienna Circle was heading, had broken away. He emigrated to the US in 1924 and instead of completing a fourth volume of *Principia Mathematica* worked on developing a system of metaphysics.

Whereas the Presocratics, Plato and Aristotle had studied the universe and Nature, in Vienna the universe was in effect declared a no-go area for philosophy. Metaphysics and assertions of cultural value were held to be meaningless, and culture was demolished and nothing replaced its ruins. Philosophy was further fragmented, much of if in broken pieces.

The Vienna Circle had founded a journal in 1930 and had reached an international audience. When the Nazis occupied Austria, Carnap, Hans Reichenbach (a Berlin Positivist) and others moved to the US and from there dominated Anglo-American philosophy. Carnap brought out books on perception, language and then logic. A.J. Ayer, a disciple of the Vienna Circle, wielded influence in Britain. Universalists now regard Logical Positivists as having been too extreme in writing off metaphysics – one of the two perennial strands of philosophy since the Greeks – and in refusing to discuss the post-Einsteinian universe.

The Linguistic Analysts, or Oxford philosophers, were Realist and common-sensual. G.E. Moore, professor of philosophy at Cambridge from 1925 to 1939, criticised Bradley in 'The Refutation of Idealism', 1903, arguing that objects of perception have independent existence. He focused on ordinary language and experience, commenting on the linguistic mistakes of his contemporaries to elucidate ethics in *Principia Ethica* (*Principles of Ethics*), 1912. He emphasised sense-impressions. Wittgenstein, who succeeded Moore to the chair of philosophy at Cambridge, attended the Vienna Circle meetings, believing with Russell, his tutor, that all propositions could be reduced to basic or "atomic" propositions. Later he became skeptical of the foundations of mathematics and logic, and turned to ordinary natural language in his *Philosophical Investigations* (published posthumously in 1953). This was the bible of Linguistic Analysis, and it showed how linguistic experience shapes one's world. He claimed that "all philosophy is critique of language", and that the crucial philosophical question was not to understand the world or the universe, but to understand the mechanisms of linguistic use which would be therapeutic and solve long-standing philosophical problems. He had many followers in the US, and as many in England, where Gilbert Ryle, J.L. Austin and P.F. Strawson examined category mistakes and perception words, and focused on the language of ethics, aesthetics and religion. (The usually cited dynasty runs from Wittgenstein to Austin to J.R. Searle, and Moore's work was continued by R.M. Hare, who focused on the language of ethics.)

In 1948 Bertrand Russell took part in a BBC Third Programme debate with Father F.C. Copleston, and the following exchange took place which sums up the cosmological bankruptcy of the Vienna Circle:

COPLESTON: ...But your general point then, Lord Russell, is that it's illegitimate even to ask the question of the cause of the world?

RUSSELL: Yes, that's my position.

COPLESTON: If it's a question that for you has no meaning, it's of course very difficult to discuss it, isn't it?

RUSSELL: Yes, it is very difficult. What do you say – shall we pass on to some other issue?

It is of course culpably dishonest for a philosopher to ignore the creation of the universe because it cannot be explained in purely physical terms, and doubly dishonest to blame the ignoring on language. To Universalists, this "position" is completely unacceptable.[20]

In 1959, while I was at Oxford, Ernest Gellner published *Words and Things* which attacked linguistic philosophy. The book contained an epigraph from Russell who had either forgotten about his 1948 broadcast or else was indirectly recanting: "The later Wittgenstein...seems to have grown tired of serious thinking and to have invented a doctrine which would make such an activity unnecessary. I do not for one moment believe that the doctrine which has these lazy consequences is true....The desire to understand the world is, they think, an outdated folly." Universalists heartily agree with Russell's point about the later Wittgenstein's not wanting to understand the world, and regard later Wittgenstein's turning away from the universe to language as "losing the plot" and forgetting what philosophy's main task is. Wittgenstein himself may have come to see this: "There is indeed the inexpressible. This *shows* itself; it is the mystical....Whereof one cannot speak, thereof one must be silent." But whether he came to see the deficiencies of Linguistic Analysis or not, in ducking the detection of natural and supernatural realities and all moral advice, analytic philosophy had led to a dead end. Language and logic were completely divorced from earlier metaphysical content to the impoverishment and further fragmentation of philosophy. The much-vaunted linguistic revolution in philosophy had changed nothing and had come to nothing.

In passing we should note that Linguistic Analysis had an impact on literary criticism, and was reflected in the New Criticism and the linguistic and verbal approaches of the new criticism of I.A. Richards (who taught at Cambridge), his pupil William Empson and Empson's *protégé* Christopher Ricks (who taught at Oxford), and later in some literary theories (for example, structuralism, which was generated by Ferdinand de Saussure's linguistics) and semiotics. As an Oxford undergraduate I was not immune from this influence, but made sure that the universe remained in the foreground of my perception, with linguistic ambiguities and wordplay in a subordinate role.

Analytic philosophy has had little impact on the Continent of Europe, where the metaphysical emphasis has remained strong through the Phenomenology[21] of Edmund Husserl and Maurice Merleau-Ponty and the Existentialism[22] of Martin Heidegger and Jean-Paul Sartre, who were also Phenomenologists.

Phenomenology's founding father was Husserl, who devised a method for restricting the philosopher's attention to the data of consciousness, ego and personal world ("*Lebenswelt*","life-world"), without any influence from metaphysical theories or scientific assumptions. In his *Ideen zu einer reinen Phänomenologie und phänomenologische Philosophie* (*Ideas, General Introduction to Pure Phenomenology*), 1913, Husserl "brackets out the object" to focus on the data in the human consciousness. Husserl edited a *Jahrbuch* (annual) from 1913 to 1930, which published the works of his two main followers: Max Scheler on ethics and emotion in moral values, and Martin Heidegger's *Sein und Zeit* (*Being and Time*), 1927, which explored human existence in terms of being-in-the-world, dread, care and being toward death. Both Scheler and Heidegger denied that philosophy is empirical but claimed that it sees into the structure of experience. Merleau-Ponty took over Husserl's perspective in *Phénoménologie de la perception* (*The Phenomenology of*

Perception), 1945. In our time, a British philosopher living in France and teaching at the University of Bordeaux, Christopher Macann, has produced a work in the metaphysical Phenomenological tradition of Husserl and Heidegger, *Being and Becoming*, 1998-2007, in which, in keeping with Universalism, he reconciles transcendental (Husserlian) and ontological (Heideggerian) phenomenology.

Existentialism continued the Intuitionists' metaphysical emphasis through the German philosopher Karl Jaspers and the French man of letters Jean-Paul Sartre, who drew on Kierkegaard and Nietzsche in analysing human Being and choice. Jaspers' ontology deepened Existentialism as a practical, personal philosophy of human existence. To Jaspers, philosophising reveals Being, what man is and can become. Through thought and "inwardness" man becomes aware of the deepest levels of Being in extreme situations such as conflict, guilt, suffering and death. Sartre, the author of novels and plays on freedom, focused on decision: the choices of the free will. In *L'Être et le Néant* (*Being and Nothingness*), 1943, he points to human dread before Nothingness, and awareness of freedom when experiencing Being. To Sartre, man is responsible, and denying his responsibility is bad faith ("*Mauvaise foi*").

Now Continental philosophy's unifying methodology of Phenomenology has fallen apart into structuralism, post-modernism, pluralism, relativism and feminism. Saussure-generated linguistic structuralism has become the new Continental orthodoxy and has in effect succeeded Phenomenology.

So today philosophy is split between metaphysical Continental philosophy and scientific Anglo-Saxon philosophy, and each is fragmented within and has cut adrift from the first two thousand years of the Western philosophical tradition. Each has lost the vision of the Presocratic Greeks, the sense of the universe emerging from the boundless and infinite. Each has lost the Renaissance's preoccupation with Nature and the external physical world. Each has lost the Enlightenment's focus on the rational mind that knows the external world. Philosophy has become concerned with expression: language, symbolism and social communication. This preoccupation with expression rubbed off on the metaphysical Intuitionists: Heidegger turned to the language of poetry and etymology for the revelation of Being. Jaspers pondered the meanings in human speech. Whitehead wrote *Symbolism: Its Meaning and Effect*, 1927. Ernst Cassirer, a German follower of Kant, wrote *Philosophy of Symbolic Forms*. Dewey was involved in social communication.

In turning away from the Greeks' universe, the Renaissance's Nature and the

Enlightenment's rational mind to focus on symbolic manipulation and the characteristics of words, quite simply philosophy has failed humankind and become irrelevant to our attempts to understand the universe. The philosophers have allowed themselves to become irrelevant by defining philosophy's task in terms of logic and language. Philosophy has taken a side-road and is confronted with dead ends. In which direction should it now be travelling?

The Way Forward: Reunification

We have seen that throughout the history of Western philosophy there have been two traditions or emphases: the metaphysical and the scientific; that for two thousand years philosophers moved backwards and forwards between Platonism and Aristotelianism; and that soon after the Renaissance the metaphysical and scientific appeared in pairs under a new guise: Rationalism and Empiricism (17th to 18th centuries); Idealism and Realism (18th to 19th centuries) and Intuitionism and Logical Analysis (later 19th and 20th centuries). Essentially the entire history of philosophy is a perpetual flow of energy between these two poles, like the flow of an electric current between two opposite charges.

The way forward from the current fragmentation is reunification. There is no prospect of a reunification within the existing conflicting structures of philosophy of the Continental and Anglo-Saxon schools of philosophy if left to themselves, and a new philosophy that can integrate the existing philosophies is the best hope for the future of philosophy. The way forward is to use the dialectical method to assimilate the opposite metaphysical and scientific emphases into a new synthesis. This synthesis should not merely be a reconciliation of 20th-century Continental and Anglo-Saxon philosophy, two polarised and mutually hostile camps. It should be more fundamental and reconcile the two poles of the last 2,600 years.

The task of philosophy is therefore, (1) to state a new metaphysic for our time, which continues the concerns of the metaphysical emphasis regarding Reality during the last 2,600 years; and (2) to state a new scientific interpretation of the universe and Nature and a universal principle of order that will continue the concerns of the scientific emphasis during the last 2,600 years and act as a bridge between the metaphysical and the scientific. Within the context of the reconciliation between the metaphysical and scientific there can be a reconciliation between the ontology of Phenomenology and Existentialism and the logic and language of Logical Positivism and Linguistic Analysis, which can

49

both serve the new philosophy's view of the universe and Nature. The purpose of modern philosophy is now to investigate the Reality within the universe and man's place in Nature.

What, then, is the best way to fulfil the two aspects of philosophy's task? We need to consider the two aspects separately:

(1) From the point of view of the metaphysical emphasis, philosophy frames and rationally states a system of general ideas which can interpret *all* our experience of the universe, including *all* known concepts. There must be a new framing of a system of general ideas which includes *all* known concepts, including the metaphysical. This must be done *after* a new scientific interpretation of the universe and Nature, so the metaphysical can rise naturally out of the scientific. It should therefore be done in Part Three. I have already quoted Whitehead's words at the beginning of *Process and Reality*, and they are so appropriate that I must here quote them again, only more fully: "Speculative philosophy" (i.e. metaphysics) "is the endeavour to frame a coherent, logical, necessary system of general ideas in terms of which every element of our experience can be interpreted. By this notion of 'interpretation' I mean that everything of which we are conscious, as enjoyed, perceived, willed, or thought, shall have the character of a particular instance of the general scheme."[23] "Every element of our experience" means *every* experience, as Whitehead himself wrote in *Adventures of Ideas*:[24]

"In order to discover some of the major categories under which we can classify the infinitely various components of experience, we must appeal to evidence relating to every variety of occasion. Nothing can be omitted, experience drunk and experience sober, experience sleeping and experience waking, experience drowsy and experience wide-awake, experience self-conscious and experience self-forgetful, experience intellectual and experience physical, experience religious and experience sceptical, experience anxious and experience care-free, experience anticipatory and experience retrospective, experience happy and experience grieving, experience dominated by emotion and experience under self-restraint, experience in the light and experience in the dark, experience normal and experience abnormal."

Whitehead could have included: experience of the Greek boundless and infinite, and experience of hidden Reality. It must be stressed that only some of these experiences are available within science. Science cannot help us with the Romantic poets' enjoyment of Nature or regarding Nature's aim. Science does not deal with *all* the evidence of human experience. In including every known concept, philosophy does not just include concepts that science can prove and approve of, but *every* concept that has ever been thought, including the possibility that there is a hidden Reality and an after-life. Whether or not they are testable, if they are concepts they must be included.

(2) From the point of view of the scientific emphasis, philosophy looks back to Bacon, the first Empiricist, and to his ambition to create a new philosophy that would interpret Nature correctly. This requires a new up-to-date interpretation of the universe and Nature. There must be a new scientific interpretation of the universe and Nature and a new statement of a universal principle of order which will act as a bridge between the metaphysical and the scientific. As we have just seen this should precede the system of general ideas so the metaphysical can rise naturally out of the scientific, and should be done in Part Two. The new synthesising philosophy will then be able to reconcile the metaphysical and the scientific.

A synthesiser must follow one very important principle. A human being has many faculties, and to single out one – the reason, for example – is as erroneous as the erroneousness of the Presocratics who singled out one element – water, or air – to explain the universe. The reason, observation, spirit, social practicability, intuition, mystic love, logic and use of language – all have their place, as does the world of phenomenal appearance *and* the world of hidden Reality. Philosophy should not focus on one aspect of a human being such as logic and language, which form only a tiny part of the systematic workings of the universe and Nature, but on a human being's total role in the whole universe and Nature, using all his or her faculties as a creature of finite possibilities and probabilities.

A synthesising Universalist philosopher is thus open to what is hidden in the universe as well as to an overview of testable data; to being, substance, essence and cause in the Greek boundless infinite, and to time, space and identity in the observable world of Nature. A human being must be seen to connect with the universe through *all* his or her faculties.

So where should Universalism's synthesis start?

There is a historical case for holding that I should return to around 1910, to the time before the First World War when the metaphysical emphasis was in the hands of Bergson, T.E. Hulme, Husserl, Whitehead and William James, all of whom had different perspectives in philosophy's fragmented state. They were all concerned to reunify the finite and the infinite, nature and mind, before the too-extreme debunking of metaphysics by the Vienna Circle. Bergson's *élan vital,* Hulme's essays on the Renaissance in *Speculations* and Whitehead's "concept of Nature" are all starting-points, clearings round which the finite jungle later closed, foundations on which Universalism can be built.

However, as we have just seen the rational case demands that a new scientific interpretation of the universe and Nature should precede a new metaphysic and system of general ideas. It is more coherent to begin in the present and to address the universe that Universalism is letting back into philosophy, to reflect current scientific thinking about the universe and evidence of order within Nature. From this consideration will emerge the best cosmological model of the universe that Universalism should adopt – a cosmological model sanctioned by science that can carry the scientific emphasis of the latter-day Aristotelians and Baconians with it. I can then combine a scientific view of the finite universe with its infinite context that, though essentially metaphysical, is less theoretical than the Rationalists' and more scientifically-based, like the Empiricists', thus unifying the two traditions.

We must bear in mind during the next two Parts that Universalism seeks to reunify the two traditions by embracing the main thrust of each: the metaphysical tradition's proposals that there is an ultimate Reality or Being that is hidden behind or within the phenomena of Nature and cosmology, determining man's place in the universe; and the scientific tradition's view that philosophical proposals should be tested by observation and empirical experience.

We are starting with the scientific tradition, but first a word on the metaphysical tradition as it stands today. It is inevitably the result of historical growth and development. I have described (on pp24-25) what metaphysics was in Aristotle's time, and in the *Prologue* I have described (on p1) the philosophical disciplines that were central to the curriculum of the later Middle Ages. By the 18th century (the time of Christian Wolff) metaphysics had become ontology. Psychology and epistemology had become more important to philosophy and included moral philosophy. For our moral judgments are psychological, and if the Good is objective as in Plato, also ontological or metaphysical. Natural philosophy had given way to science, which in turn had become cosmology.[25] Theology remained outside metaphysics and although for centuries it had been a high calling for which the study of metaphysics prepared, it is now seen as being within a different category from ontology, one more to do with faith than rational study.

Today metaphysics can be said to have four subdivisions, which in descending hierarchical order are:

> ontology (the study of Being), which includes traditional metaphysics;
>
> psychology (what can be experienced of Being) which includes some traditional moral philosophy;
>
> epistemology (what can be known about Being), which includes some ethics; and
>
> cosmology (the structure of the universe), which includes traditional natural philosophy.

In modern metaphysics, cosmology means philosophical cosmology – the philosopher writing down his view of the universe. No longer can the philosopher sit behind his desk and tell us about the universe without reporting on science. No longer can he write down a rational system that ignores science. Universalism must propose a philosophical cosmology that reflects the scientific view of cosmology. Similarly it must propose a philosophical epistemology, psychology and ontology that reflect the scientific views on these subjects, working with the material of science without imposing its own Rationalist scheme on science.

The most satisfactory way to proceed, therefore, is to consider what all the sciences have to say about the structure of the universe and to arrive at a consensual view of "what is" according to science, and of what all the sciences assume can be known and experienced about Being, the metaphysical hidden Reality, the infinite and "boundless" of the Greek Presocratic philosophers. In particular, we need to be alert to all the sciences' view on the degree of order in the universe, and the extent to which "what is" is self-organising. This means we must go to cosmology, astrophysics, physics, biocosmology, biology, geology, palaeontology, ecology and physiology. Having scrutinised the scientific position I can then construct a metaphysical cosmology, epistemology, psychology and ontology that are fully grounded in science, that blend metaphysics and science. The newly constructed philosophy will reunite the philosophical traditions that have been divided and fragmented since the Renaissance.

Philosophical Summary

What can the Universalist philosopher take from the Western philosophical tradition? Plato's metaphysical tradition of hidden Reality. Aristotle's scientific tradition, "science which studies Being". Plotinus's One. Augustine's "Intelligible Light", Grosseteste's metaphysical "Uncreated Light". The limitations of Empiricism, Rationalism and Idealism. Leibniz's monads. Bergson's life principle. Whitehead's philosophy of organism and concept of nature, his focus on experiences. Phenomenology's focus on consciousness, Existentialism's on being. The four subdivisions of metaphysics today.

To sum up, the Universalist philosopher can distill from the Western philosophical tradition:

truth about hidden Reality;
"science which studies Being";
the One;
Intelligible or Uncreated Light;
limitations of Empiricism, Rationalism and Idealism;
focus on consciousness and being;
four subdivisions of metaphysics today.

To return to my argument, the starting-point is what science has discovered about the universe. What can Universalist philosophy take from science regarding the origin of the universe? In due course I shall be asking whether science can detect a self-organising system and a principle of order, and ordered unity, in the origin of the universe, and we must keep this thought at the back of our minds.

PART TWO

THE SCIENTIFIC VIEW OF THE UNIVERSE AND THE ORDER PRINCIPLE

"It was (Sir Francis) Bacon's ambition to create a new system of philosophy, based on a right interpretation of nature, to replace that of Aristotle."

> Sir Paul Harvey, *The Oxford Companion to English Literature*

"We find in the objective world a high degree of order."

> Albert Einstein

"A certain order prevails in our universe. This order can be formulated in terms of purposeful activity."

> Max Planck

"This is a dangerous idea that I am simply unwilling to contemplate."

> Albert Einstein, on the multiverse

"The overthrow of the 'random universe' by contemporary science is the great unnoticed revolution of late-twentieth-century thought."

> Patrick Glynn, political scientist

3

COSMOLOGY AND ASTROPHYSICS: THE BIG BANG AND THE ORIGIN OF THE UNIVERSE

The Big-Bang theory which explains the origin of the universe and of more than a hundred billion stars and which has been supported by two satellite probes and telescopes that can see to a distance of over 13 billion light years was arguably the 20th century's greatest intellectual achievement. There is now general agreement that the universe began with the Big Bang, a fireball, about 13.7 billion years ago (to a precision of 1 per cent). Calculations based on the Wilkinson probe in 2003 refined earlier estimates that it began 15 billion years ago.

The hot beginning formed clouds of hydrogen and helium, which gravitational forces collapsed into stars. The stars then converted hydrogen and helium into carbon, nitrogen and oxygen, the building blocks of life on earth. Exploding stars (supernovae) returned these elements to interstellar space. They formed clouds of water, carbon monoxide and hydrocarbons which collapsed into new stars and solar systems. Our solar system began 4.6 billion years ago after one or more supernovae explosions. It is thought that 500 planetesimals (asteroids and comets) occupied the region where Venus, the earth and Mars are now. Over some 100-200 million years 4.55 billion years ago planetesimals accumulated and formed the earth. The hot earth cooled and formed water 4 billion years ago, and by 3.5 billion years ago life had emerged.

How did the universe begin? Where did it come from? Where did it emerge? What is its structure? Will it end? These are questions asked by cosmologists and answered by astrophysicists. Cosmologists provide a theoretical understanding of the beginning, life and end of the universe. Astrophysicists use the techniques and concepts of physicists to study the stars – to measure distances, brightness and the size of redshifts (wavelengths shifting towards long wave) which indicate that galaxies are receding from earth. Whereas cosmology is "the study and description of the physical structure, science or theory of the universe" astrophysics is "a branch of astronomy concerned with the physics and chemistry of celestial bodies" (i.e. of the stars of the galaxies). Both disciplines focus on the universe.

Despite the confidence of theoretical mathematicians whose findings are often not supported by observations and evidence, knowledge is still very sketchy.

The Big Bang and Infinite Singularity

The notion of the Big Bang began as a cosmological theory and has received some confirmation from astrophysicists' evidence. The theory of the Big Bang goes back to Einstein. One class of potential solutions to Einstein's equations allowed for the possibility that the universe was expanding or contracting. Einstein himself dismissed this possibility as there was a general belief that the universe was not in motion but static.

It is amazing that Einstein laid the foundations of the Big Bang theory despite believing in a static rather than an expanding universe and without knowing about the strong and weak forces, which had not been discovered. In 1905 he transformed physics with his special theory of relativity, his theory on the equivalence of mass and energy, his theory of Brownian motion and his photon theory of light.

In his special theory of relativity Einstein held that the speed of light is the same for all observers, and that all physical laws are the same for two frames of reference moving with uniform speed relative to each other. This theory was verified in 1919. Meanwhile his general theory of relativity of 1915 showed that space, time and gravitation are based on frames of reference – for example the earth's surface or an interplanetary spaceship – whose motions are accelerated in relation to each other.

Einstein confirmed Newton's laws for bodies travelling at low speeds but gave different results for bodies travelling at high speeds, such as stars, galaxies and elementary particles. In the general theory Einstein found that when an observer or frame of reference accelerates in relation to another when gravitational forces are present, gravity is seen not as a force, but as a curvature in space-time. (The geometrical concepts for this theory had all been developed in the 19th century and Einstein was able to adopt them ready-made.) Thus the earth orbits the sun in a curved path because the sun distorts space-time in the sun's vicinity. The general law explains why Mercury's orbit differs from how Newton saw it in his gravitational theory. Newton's theory did not take account of the bending of stellar light and the gravitational redshift (in which an atom near a star has a longer wavelength and looks redder than an atom on earth). Einstein's theory that large objects can distort space and time around them was confirmed to within 95.95 per cent certainty in 2006 when Jodrell Bank researchers showed that a beam of radio waves from one pulsar (the remains of an exploded star) was affected by a distortion of time and space caused by another pulsar.

Einstein's theory of gravity introduced a new conception of space-time.

According to Einstein gravity is an effect of the curving of space-time.[1] If matter, which is formed by the mass and energy it contains, is of too great a density (more than 6 hydrogen atoms per cubic metre of space, emptier than the most "empty" vacuum), the universe will be curved up into a finite volume while emptier spaces can extend forever. A universe not exceeding a density of 6 hydrogen atoms per cubic metre of space can expand forever.[2]

The Big-Bang hypothesis was first put forward in 1927 by Georges Lemaître.[3] It was opposed on the grounds that a universe created by a Big Bang would be less than the age of the earth suggested by geological evidence, and the theory fell out of favour.

In 1929 the Russian cosmologist Alexander Friedmann worked out that the universe was not static. Seven years earlier he had formulated a model of the universe deriving from Einstein's general theory of relativity, in which the average mass density is constant and all parameters are known except the expansion factor. In late 1922 and 1924 he postulated a "Big-Bang" model for the evolution of the universe. Now, realising that the universe was identical in whichever direction we look, he came up with three models. Either the universe expands and then collapses and comes to an end in a Big Crunch, in which case space is finite. Or the universe expands forever, in which case space is finite until the expansion reaches a point beyond which it will last forever and so becomes infinite. Or, if the universe's density is just less than the critical 6 hydrogen atoms per cubic metre of space, the critical rate of expansion, space is flat, balanced between expansion and collapse, in which case space is finite but will be infinite if the equilibrium lasts forever.[4]

The first evidence of the Big Bang came in 1929 when the American astrophysicist Edwin Hubble found that the universe was expanding, and that the recession speed of a galaxy is proportional to its distance from the earth. Until then astronomical distance had been measured by triangles, the apex being a distant star and the base a known length with known angles, and by "brightness candles" ("Cepheids"), which allowed enormous distances to be estimated. Hubble discovered that redshifts of nebulae (which the US astronomer Vesto Slipher had been measuring) were receding from our galaxy, and that brightness candles were correlated to the recession velocities. (The relationship became known as Hubble's law.) The Hubble constant states that there is a constant of proportionality between distance and velocity. Hubble's data showed that the universe was expanding at an accelerating rate to within a 10-per-cent uncertainty margin. The picture was still far from clear. Calculations based on Hubble's work bewilderingly suggested that the age of the oldest stars (up to 20 billion years old assuming an even accelerating rate, a wrong premiss) was older than the universe (13.7 billion years old).[5]

In the 1940s the Russian-born, US-based nuclear physicist and cosmologist George Gamow proposed a series of reactions from neutrons which produced the light elements after a hot beginning. He and two colleagues predicted the background temperature of the universe. John Archibald Wheeler, the American physicist, coined the phrase "Big Bang" as it corresponded to Einstein's general theory of relativity. (He also coined the phrase "black hole".) Fred Hoyle, the British cosmologist, is also credited with the first use of the term in the 1950s after becoming an advocate of the Steady-State theory formulated by the Austrian-British mathematician Hermann Bondi and the Austrian-British Thomas Gold in 1948. The Steady-State theory holds that the universe is expanding and that new matter is being created at a rate that keeps its mean density constant and balances the expansion. To Hoyle, the Big Bang was a term of derision for he was deeply critical of the Big-Bang theory.[6]

In 1964 the US radioastronomers Arno Penzias and Robert Wilson discovered a constant background noise level that originated in the cosmos while they were trying to improve microwave antennae at their laboratory. The cosmic microwave background was interpreted as the residual radiation from the hot beginning. It had a temperature of 2.725K (about $-270°C$).[7] The COBE satellite measurements confirmed the uniformity of this radiation to a millionth of a degree. In other words, the radiation is the same everywhere throughout the universe. This suggests that the Big Bang had no centre in the universe, for otherwise the sky around its centre would glow with primordial radiation and the sky a long way away would be cold.[8]

By 1970 the Big Bang had won over most of the supporters of the Steady-State model proposed by Bondi, Gold and Hoyle. This was a variation of Friedmann's third model, the flat universe. It had infinite space and no beginning or end.

Einstein's theory of gravity requires that the universe exploded from a state of infinite density. At the beginning of the Big Bang, according to Friedmann's original 1922/1924 model, the density of the universe and the curvature of space-time were infinite, and the distance between galaxies was zero. Mathematics comes to a halt in the face of the infinite, and the general theory of relativity predicts that at its beginning the universe was contained in a point, in which the theory of relativity breaks down. This point is called a singularity (an atom-size point in space-time at which the space-time curvature becomes infinite). Interestingly, Dante writes of the infinitesimal "point" in *Paradiso*, canto 28.

The Idea of a singularity stemmed from the US theoretical physicists Robert Oppenheimer and Hartland Snyder, who had shown in 1939 that a spherical collapse of a massive star could lead to a central space-time singularity (or black hole) at which the classical theory of relativity is stretched beyond its limits.[9] It was taken up in the British mathematician and physicist Roger Penrose's 1965 theorem, which stated that any body undergoing gravitational collapse must eventually form a singularity.[10] For a singularity or black hole the size of an atom to form, 10^{36} atoms must be squeezed into the space occupied by one atom.[11]

Reading this in 1965, the British theoretical physicist Stephen Hawking realised that if time were reversed Penrose's theorem would still hold, and prove that an expanding universe must have began with a singularity.[12] In 1970 Penrose and Hawking proved in a theorem that the general theory of relativity implied that the universe must have a beginning and possibly an end and that at the beginning there must have been a Big-Bang singularity of infinite density and space-time curvature. The initial space-time singularity (the Big Bang) represents the creation of space-time and matter in "a white hole" (the hypothetical time-reverse of a black hole) rather than its destruction in a black hole. Thus the Big Bang began in one atom-size point and happened everywhere within that point. Everything was connected in that one point, and was One.[13]

It is important to be clear about our definition of "infinite". The German mathematician Georg Cantor (1845-1918) distinguished three kinds of infinity:

"The actual infinite arises in three contexts: first when it is realized in the most complete form, in a fully independent, other-worldly being, *in Deo*, where I call it the Absolute Infinite or simply Absolute; second when it occurs in the contingent, created world; third when the mind grasps it *in abstracto* as a mathematical magnitude, number or order type. I wish to make a sharp contrast between the Absolute and what I call the Transfinite, that is the actual infinities of the last two sorts, which are clearly limited, subject to further increase, and thus related to the finite."[14]

Absolute Infinity is the totality of everything, which is beyond mathematics.

Infinity in the physical universe (such as a singularity) is transfinite, Cantor claims, but not necessarily Absolute. Mathematical infinity is transfinite numbers constructed which may be in a series of differing "sizes" all of which are infinite. It is possible to believe in one, two or three of Cantor's infinities at the same time, in eight permutations as John Barrow's table of historical figures points out:[15]

	Mathematical infinity	Physical infinity	Absolute infinity
Abraham Robinson	No	No	No
Plato	No	Yes	No
Thomas Aquinas	No	No	Yes
Luitzen Brouwer	No	Yes	Yes
David Hilbert	Yes	No	No
Bertrand Russell	Yes	Yes	No
Kurt Gödel	Yes	No	Yes
Georg Cantor	Yes	Yes	Yes

These three infinites – the Absolute, physical and mathematical infinites – must be distinguished, and no semantic imprecision should cloud the concept of the Absolute Infinite. Nevertheless there *is* an infinite timelessness or Void (the Absolute in Cantor's terms) from which the point or singularity emerged and on which space-time is overlaid, and we have seen that in the physical singularity the laws of space-time and gravity break down. Beneath space-time is the timeless infinite, the Absolute, and so it is reasonable to associate the absence of space-time in the physical infinity of the singularity with the timeless infinite Absolute into which it reverts. Cantor's distinction between the physical singularity and the Absolute infinite is academic as the Void of the Absolute is present around and within the singularity and has caused space, life, time and gravity to break down.

The infinity within the singularity is not a part of finite space-time but is part of the totality of the infinite Absolute. This may be an uncomfortable fact for some cosmologists and physicists who want the Big Bang to have begun within the finite, beginning finite space-time, but to the philosopher that is a fact, a truth that cannot be ignored as being inconvenient. Mathematician engineers doing calculations involving air speed find that a speed faster than the speed of sound

(750 mph), which causes a sonic boom, appears in equations as an infinity, and particle physicists' equations are frequently part-finite and part-infinite. They subtract the infinite part in a process called "renormalisation". The mathematical infinities may be due to their being constructed in the human mind, perhaps in a clumsy way of seeing things.

To put it bluntly, the distinction between metaphysical infinity and mathematical infinity allows mathematicians to study geometry and arithmetical series without being enjoined to reserve the term "infinite" for the Absolute.[16] But mathematical infinity leads to metaphysical infinity because the infinity in the singularity is essentially a metaphysical infinity as it pre-exists space-time and the physical world, and is therefore beyond physics ("metaphysical").

The mathematics of cosmologists, then, suggest that the Big Bang began as a singularity in one point of infinite density in which all matter and energy were concentrated. The trouble is, a singularity is an unphysical state[17] and there does not seem to be a physics of singularity. Hawking therefore changed his mind and came to believe that there was no singularity at the beginning of the universe once quantum effects are taken into account. He argued that just as there is nothing north of the North Pole, so there is no singularity before the Big Bang.[18] He maintained that the gravitational field was so strong that quantum gravitational effects became important, and cited the US physicist Richard Feynman's proposal to formulate quantum theory in terms of a sum over histories. His mathematical construction, in conjunction with the American James Hartle, of a quantum state, for which there was no observation or evidence, sees time as finite in the past, not stretching back into timelessness, yet having no beginning. It sees the universe as a natural event in terms of a framework that is supranatural (or supernatural).[19]

In classical theory of gravity, the universe either existed for an infinite time or it began as a singularity at a finite time in the past. In the quantum theory of gravity, space-time can be finite in extent yet have no singularities that form a boundary or edge, on which the laws of science break down, and no edge of space-time. Hawking argued that the universe would be self-contained and have no beginning or end, but just be.[20] Penrose did not accept Hawking's view and wrote: "Though ingenious, his suggestion involves severe theoretical difficulties, and I do not myself believe that it can be made to work."[21] Universalists, anyway sceptical of aspects of "random" quantum theory, do not accept Hawking's view, which ignores the beginning of the universe.

Einstein, Penrose and Hawking have regarded singularities as arbitrary as the laws of physics break down when a singularity appears. Space and time are destroyed at the places where infinite densities appear, and the laws of gravity cease to hold at physical infinities. A physical infinity is more serious than mathematical infinity as encountered by particle physicists and engineers. Cosmologists are reluctant to admit physical infinities into the universe and prefer to see them as mathematical problems, but the fact remains that the universe began in an infinity when there was no space-time or gravity, merely an infinite Void.[22]

Then, in the 1990s there was a development. The cosmic microwave background radiation seemed too smooth and uniform, and suggested a smooth universe 380,000 years after the Big Bang which would have resulted in a universe with no stars or planets, or people. In 1992 the COBE (Cosmic Background Explorer) satellite launched by NASA found ripples in the cosmic microwave background radiation, showing that 380,000 years after the Big-Bang singularity there were already wispy clouds of matter stretching in excess of 500 million light years in length. These could be ripples or clouds that collapsed in upon themselves, broke up and formed clusters of galaxies. The size of the clouds seemed to confirm that 96 per cent of the universe is missing and may be in the form of invisible matter or energy, which is different from bright stars and galaxies,[23] and that there must be a hundred times more invisible matter in the universe than the matter we can see.[24]

In short, there was not enough visible matter to support the experientially-measured expansion rate. In fact, the hypothetical concepts of dark matter and dark energy accounted for the missing mass of the universe: 23 per cent dark matter and 73 per cent dark energy. (These figures were estimated by the WMAP – Wilkinson Microwave Anisotropy Probe – mapping team of 2003 which revised the previously accepted 30 per cent dark matter and 66 per cent dark energy.) It is thought that dark matter was needed to help galaxies to form. In September 2008 it was announced that the European space probe Pamela had discovered a surge of high-energy particles (positrons, a form of antimatter) from the heart of the Milky Way which closely matches the radiation signature predicted for dark matter.[25]

The Shaping of the Universe

There is a consensus on the known facts and timings of the early universe.[26] Many of the timings are minute fractions of the first second. In one second the infinitely small, hot and dense singularity expanded like a balloon inflating to the size of our universe, and outside the balloon there was no space, matter or light. The inflation was not centred within the universe, as we have seen. Rather, the universe *is* the inflation, which is why the cosmic microwave background radiation is uniform throughout the universe. Within that same second the expanded singularity cooled from trillions of degrees K to a few billion. (K is an abbreviation for Kelvin, a scale of temperature named after the British physicist Lord Kelvin with absolute zero as zero.)

Particle physicists have focused on the first three minutes after the Big Bang. The four fundamental forces (the electromagnetic, strong and weak forces and gravity) were unified at this stage. The gravitational force separated at 10^{-43} seconds after the Big Bang (Planck time, the shortest time interval that can ever be measured, an infinitesimal amount of a split second), when the temperature was 10^{32}K (the beginning of the Grand Unified Theory era). Mass-energy became the observed universe at 10^{-43} seconds. Space-time is regarded as having begun when the universe began.

Within 10^{-36} seconds (a trillionth of a trillionth of a trillionth of a second) of the Big Bang the microscopic compact, high-density point that resembled an atom expanded extremely rapidly and inflated large enough to encompass the solar system's size, the horizon we see, while keeping gravitational and kinetic energy in balance. It doubled in size every 10^{-34} seconds, so there were a hundred doublings within the space of 10^{-32} seconds. It has continued to expand ever since.

The expansion of the universe was like the surface of a balloon being blown up, and if the galaxies were spots on the balloon's surface, they all receded from each other as the balloon inflated.[27] Another analogy is raisins in a loaf. As the dough rose the raisins were carried away from each other.[28] This rapid expansion was attributed to "inflation" in a book by Alan Guth, *The Inflationary Universe* (1981). Spacecraft are being prepared to test this theory of inflation empirically by looking for primordial microscopic vibrations. During this period of early rapid expansion within the first minute, the density and temperature of the universe decreased and radiation and matter evolved. For the first 10^{-35} seconds everything was hotter than 10^{28} degrees.

The temperature had dropped quickly, and around 10^{-35} seconds after the Big Bang, at a temperature of 10^{27}K, it seems there was very rapid, exponential expansion and inflation by a factor of 10^{54}, faster than the speed of light. This was the quark era. Leptons and quarks were formed from radiation energy as energy passed into matter in accordance with Einstein's $E=mc^2$ (the beginning of the quark era). Further cooling to $10^{15}K$ around 10^{-12} seconds (a trillionth of a second) separated strong and electroweak forces and quarks combined to form protons, neutrons and their antiparticles. Radiation in photons lacked the energy to break up the new particles. Matter and antimatter collided and annihilated each other, leaving a residue of matter which makes up the universe. At 10^{-5} seconds there were temperature fluctuations that seeded galaxies. Further expansion and cooling around 10^{-4} seconds after the Big Bang resulted in the electroweak force separating into electromagnetic and weak forces (the beginning of the lepton era). Lighter particles (such as electrons, muons, taus and neutrinos) were prevalent.

Ten seconds after the Big Bang the temperature was $10^{10}K$ (10 billion degrees). There was rapid cooling. From 10 seconds after the Big Bang, radiation dominated the universe (the beginning of the radiation era). After a brief balance between radiation and matter when there was an equal number of particles, antiparticles and photons, antimatter was eliminated and the universe consisted of photons and neutrinos.

Three minutes after the Big Bang the temperature was 10^9K (1 billion degrees).

From 1 to 20 minutes after the Big Bang nucleosynthesis took place: colliding protons and neutrons fused and formed light nuclei (deuterium, lithium, helium and hydrogen, the building blocks from which all elements would eventually form). Nucleosynthesis ceased after a few minutes, after which the ratio of hydrogen to helium was 3:1, as it is today. The Big Bang, or expanding singularity, was over within 30 minutes.

Some 380,000 years after the Big Bang the universe cooled to 3,000K, a temperature at which helium nuclei could absorb electrons. Electrons remained attached to nuclei and atoms formed (the beginning of the matter era), giving out light which is the cosmic microwave background radiation we can monitor today. It had uniformity of direction, suggesting that all matter in the universe was uniformly distributed. Before atoms formed, matter existed in a plasma (or gas of positive ions and free electrons with roughly equal positive and negative charges). The plasma of free nuclei and electrons absorbed photons, which gave radiation and matter the same temperature. Now atoms could travel without colliding with photons, clumps of atoms formed, and matter decoupled from energy. Atoms collected together through gravitation into clouds of gas.

Just over 13 billion years ago the energy in cosmic microwave background radiation cooled from 3,000K and became less than the energy in matter in the matter-dominated universe. The background heat now pervading the universe is 2.725K (about −270C, absolute zero being −273.16C).

380,000 years ago photons (15 per cent) outnumbered atoms (12 per cent) and neutrinos (10 per cent), with dark matter forming 63 per cent of the matter in the universe – according to NASA's WMAP Mission Results. There were now three kinds of matter: radiation, including photons and neutrinos, massless or nearly massless particles that move at the speed of light and exert a large positive pressure; baryonic matter or "ordinary matter" composed of protons, neutrons and electrons that exert virtually no pressure; and dark matter, non-baryonic matter that interacts weakly with ordinary matter and which, though never directly observed in a laboratory, is suspected to exist. (Dark energy, a fourth kind of matter or perhaps a property of the vacuum, which has a large negative pressure, only took over after 8 billion years, see p68.)

A billion years after the formation of atoms, atoms continued to be brought together by gravity. Gravitational attraction between hydrogen atoms collected them into clouds of gas that condensed as stars grouped in clusters and galaxies. Gravity exerted pressure on their cores, creating mass and high temperature that produced helium as colliding hydrogen atoms fused. Over a period of 30 million years stars contracted and achieved stability as their cores were heated when protons and also carbon were converted to helium to become hydrogen for 10^{10} (10 billion) years. As the hydrogen fuel reserves were used up in thermonuclear fusion, gravity caused the cores to contract and as gravitational potential energy became kinetic energy, the cores became hotter and stars became red giants that burnt helium. The cores of smaller stars burnt to carbon and shrank to white dwarfs (stars with closely-packed matter and a high temperature but low luminosity) and then could be black dwarfs (hypothetical, unevidenced stars thought to have cooled and to have no luminosity and to be about to "go out" near the end of their stellar evolution). Our sun will "go out" as a black dwarf in 5 billion years' time.

In larger stars, thermonuclear fusion continued and the cores' temperature increased, combining electrons with protons to form neutrons. The stars then collapsed, blowing off their outer layers in giant supernovae explosions (like the Crab nebula, which was seen by the Chinese in 1054 and is still seen today). The radio waves from the rotation of neutron stars are detected as pulses; hence they are known as "pulsars". Sometimes two stars moved in orbit round a common mass, and these are known as "binary stars".

When massive stars contracted, their density was somehow so high and gravity so great that they became black holes (hypothetical, unevidenced cosmic bodies with such intense gravitational fields that no matter, radiation or light can escape them, thought to be often formed by the death of a massive stars). A black hole with a radius of just 1 cm could contain the mass of the earth. Black holes are thought to have been at the centre of some galaxies and to have powered luminous objects such as quasars and active galactic nuclei. Stars have their own life cycle in which their beginnings and endings are relatively short periods.

About 4.5 billion years ago our sun condensed from a cloud of gas and dust. We know it is a second-generation star as elements heavier than iron are present in its black body radiation spectrum. The sun gravitationally attracted a rotating gas cloud into a flat disc. The cloud rotated faster, larger particles captured smaller ones and most of the material was swept into the sun. Smaller entities became planets, comets and asteroids (rocks that fell to earth as meteors or shooting stars if they crossed the earth's path). Comets are dust embedded in ice, formed from water and methane. (Halley's Comet, seen in 1986, returns every 76 years.) The solar system's planets rotated round the sun in the same direction and roughly in the same plane. The inner planets, the earth, Venus and Mars are formed of small rocky high density materials. The outer planets, Jupiter and Saturn, are large, gaseous and low-density.

The sun grew larger and nuclear fusion took place. There were nuclear explosions, and material was blown outwards against gravitational forces. The sun radiated energy and a "solar wind" also blew smaller particles (charged particles such as hydrogen and helium, which are light elements) to the outer regions of the solar system. Sudden surges in the sun's energy have continued as solar flares and sunspots.

The planets were too small for nuclear fusion to take place and so they have gradually cooled and are only visible through reflected light from the sun. Their original atmospheres were lost when thermonuclear fusion took place in the sun. Their secondary atmospheres resulted from the release of gases from the planets' interiors. Earth's atmosphere has been modified by biological processes.

Galaxies are generally elliptical and spiral in shape. Our own galaxy, the Milky Way, is spiral with a central bulge and a disc of stars, dust and gas, which rotate round the galactic centre. In spiral galaxies new stars appear blue. In elliptical galaxies there is no disc and stars look red. Some galaxies have a luminous quasar or active galactic nucleus at their centre. Radio waves are used in mapping the stars at the centre of our galaxy as interstellar dust is transparent to radio wavelengths. The rotating stars' speed remains constant at all distances from the galactic centre, which does not have enough mass to exert a pull to draw them in. Stars are travelling too fast to remain gravitationally bound to the Milky Way.

The Swiss astrophysicist Fritz Zwicky proposed that there must be an extra gravitating material that contributes to the pull. This hypothetical, unevidenced material known as dark matter does not emit electromagnetic radiation like stars or galaxies. Dark matter could be present in MACHOs, massive compact halo objects such as brown dwarfs (hypothetical crosses between a planet and a star, failed stars with too little mass to ignite hydrogen), black dwarfs and black holes; and also WIMPs, weakly interacting massive particles such as neutrinos with few islands of atoms that pass through everything. We have seen that dark matter is hypothetically thought to make up 23 per cent of the universe according to the WMAP estimate.

The universe now contained 10^{18} (a billion billion) stars, one per cent of which (10 million billion) are probably similar to our sun. In the observable universe there were then 10^{10} (10 billion) galaxies. (There are now 10^{11}, a hundred billion galaxies.) The number of stars in all the galaxies in our observable universe is about 10^{23}.[29] The total number of grains of sand is about 10^{23}.[30] There is one star in our observable universe for every grain of sand.

Acceleration, the Surfer and the Infinite

In 1998 it was discovered that the universe's expansion has been accelerating during the last five billion years.[31] Having accelerated greatly immediately after the Big Bang, the expansion had become less rapid. After some 8 billion years of deceleration from inflation, renewed acceleration began and no more galaxies could form – the expansion is moving too fast for matter to resist and gather into galaxies. The mysterious speeding-up contradicted the second law of thermodynamics (which states that a system begins ordered and that its passage through time brings increasing disorder).[32]

What shape does this speeded-up universe have? In geometry there are three possible shapes the universe could have: spherical with curved space (called a closed universe); a flat space extending in all directions with zero curvature (known as a flat universe); and a hyperbolic ("plane curve") universe with no curved space where space expands forever (known as an open universe).[33]

Einstein's equations tell us that the universe may have no edge as space is gently curved around upon itself, making the equivalent of the surface of a sphere. The universe may appear a closed sphere even though it is open. Yet as it is accelerating, the curvature does not form a perfect sphere but rather a curved shuttlecock – curved as the feathers are curved on a shuttlecock, the base of which is the point or singularity and the top of which is open. The universe is not a closed sphere that can be walked around forever, like a fly walking around a snooker ball, and therefore that has no boundary.[34]

The universe is finite and shuttlecock-shaped (or trumpet-shaped, shaped like a trumpet swelling to its bowl, or shaped like a loudspeaker).[35] If the finite universe is accelerating it must be accelerating into something, and therefore must have an edge or boundary. What would happen if one could peer over the edge?[36]

Let us ask ourselves, what would it be like to stand on the edge or tip of the shuttlecock-shaped universe? The rim of the shuttlecock-shape is forever surging forward, so one would be like a surfer in a spacesuit crouching, arms out, knees bent on the crest of a surging wave. What is the surfer advancing into on the edge of our expanding, accelerating universe? It cannot be material space for space-time was created with the Big Bang and *is* the expanding, accelerating universe. There is no space or time outside the universe, yet the universe is surging forward into something, which is by definition not material space or time.

The something can only be the infinite which pre-existed the universe of space-time and from which the point or singularity emerged. It can only be the metaphysical Reality beyond space-time, and the spacesuited surfer, crouching on the crest of the wave that is forever thrusting forward like an endless *tsunami*, ever surges – hurtles – into the infinite, the Void. The infinite is metaphysical because it is outside the wave of expanding, accelerating space-time and outside our physical universe.

Hitherto, the infinite has been a theoretical concept in metaphysical philosophy. But, within my rational concept, what the surfer is *experiencing is* the infinite. The infinite is thus evidential, for the surfer has practical experience of it. This is a crucially important idea that advances the reconciliation between the metaphysical and scientific emphases. The infinite is both theoretical and, to the surfer, scientifically evidential and its existence would be confirmed by scrutinising scientists such as Aristotle and Bacon if they were to join the surfer on his endless *tsunami* wave.

So if the surfer throws a stone from surging space-time out into the infinite, across the edge or boundary, what will happen? Does it go into the infinite? Does it cease to exist?[37] The infinite is outside the boundary of space-time. And so the stone, and all space-time, must cease to exist outside space-time when three-dimensional space ceases to exist, where the edge ever surges forward into the infinite?

We are accelerating into Absolute infinity. From the start of radiation (10 seconds after the Big Bang) to around 380,000 years later, the generally accepted end of the radiation era, the expansion was controlled by the gravitational pull of radiation, which had a slowly decelerating effect. In fact, the expansion was still controlled by the gravitational pull of radiation 379,000 years after the Big Bang. For about 380,000 years the expanding gas de-ionized and nuclei and electrons combined into atoms.[38] The beginning of the matter era maintained this reduced expansion. We have seen that about 5 billion years ago it is thought that the hypothetical dark energy took over and that as deceleration changed to acceleration no more galaxies could form as the expansion was moving too fast for matter to resist.

Our universe is still, after 13.7 billion years, expanding. Hubble time ranges from 12 to 20 billion years, older than but corrected by the 13.7 billion years calculated as a result of NASA's Wilkinson Microwave Anisotropy Probe in 2003 (see illustration on p74), which refined the 15 billion figure that was previously the perceived norm for the age of the universe.[39] Hubble time measures the expansion of the universe by a simple sum: the present distance of a galaxy from the earth divided by its present recession speed from the earth = the time since the Big Bang, with adjustments for cosmic deceleration. On this measurement our universe expands closer than 2 per cent to the dividing line that separates indefinite expansion from eventual contraction. Our universe is a finite universe that at present therefore looks as if it will expand forever.[40]

The mysterious force causing the acceleration is thought to be the hypothetical dark energy, which Einstein predicted in 1917. He called it the "cosmological constant". He saw this energy as enmeshed in the fabric of space-time and not having a source. It is thought that dark energy (assuming it exists) is responsible for making space-time flat.[41]

To put it another way, the total number of protons, the building blocks of matter, is 10^{78},[42] a factor of ten billion trillion times too small.[43] A proton's lifetime exceeds 10^{33} years.[44] It is conjectured that our universe must contain two-thirds of its energy density in the form of dark vacuum energy. What the universe's energy density should be is larger than what is observed by a factor of 10^{120}.[45] The hypothetical dark energy in the present vacuum is thought to have an energy scale that is 3 trillion quadrillion (3×10^{27}) times smaller than that of inflation.[46]

Einstein added his "cosmological constant", which for him was an anti-gravity constant, to his static universe to counter gravity, and later regarded the move as his worst mistake. For soon afterwards, in 1929, Hubble showed that the universe was expanding and not static.

In 1998, Einstein's solution was back in vogue as his constant was regarded as supplying the force or energy within the vacuum that has speeded the expansion up. Of Friedmann's three models, the first one that showed the universe ending in disorder with a Big Crunch could now be discounted. On balance it looked as if the universe would expand forever but the hypothetical dark matter may act as a brake and slow it down towards flatness.

Flatness, we have seen (see pp68-69), is a geometric concept. The period of rapid expansion, the inflation phase which began at 10^{-35} seconds, flattened the geometry of the universe. This meant that the rules of Euclidean geometry are observed in the cosmos, that light not being bent by gravity travels in straight lines to infinity, not in curves, and the angles of a triangle still add up to 180 degrees. Space is expanding at such a rate that it is Euclidean rather than curved, giving an expanding but flat universe.[47] We have seen that Einstein held that space appears curved, but an insect crawling on the inside of a fully-inflated balloon would experience the surface as flat even though the expansion makes it appear curved.

We have seen that the critical density of the universe is 6 hydrogen atoms per cubic metre. The number of this critical density is Ω (omega). In a closed universe (see pp68-69) Ω is greater than 1, in a flat universe $\Omega = 1$ and in an open universe Ω is less than 1. In a closed universe the density of matter is greater than the critical density and causes gravitational forces to reverse the expansion and bring about a Big Crunch. Our observable universe is accelerating because of the hypothetical presence of dark energy, and because of the hypothetical braking dark matter Ω works out at 0.2 hydrogen atoms per cubic metre, or 0.3 of the density needed to halt expansion.[48] Our observable universe is flat and looks as if it will expand forever.

As our observable universe has a density of matter that is *less* than critical density (0.3 in relation to 1), space is not curved into the finiteness of a spherical ball but is an essentially flat surface curved by expansion like an inflated balloon. It obeys the laws of flatness but has been curved by inflation. The visible, observable universe is finite. To a multiversalist our universe could be in a bubble and there could be a multiverse outside our bubble; and our local conditions within our bubble may be different from the conditions in the totality of the multiverse outside our bubble. The universe could in theory be infinite. However, as it is in fact shaped like a shuttlecock (or trumpet or loudspeaker) as we have seen, it is finite and advancing into infinity (or Absolute infinity in Cantor's terms).

The universe, then, is now expanding faster than five billion years ago, and this can only be explained by the existence of the mysterious anti-gravity force that opposes gravity hypothetically known as dark energy, which, it is argued, transformed slowing-down into speeding-up. This anti-gravity force balances gravity. Einstein's constant was a universal repulsion or anti-gravity force which balances gravity, and we shall see that Newton also sought an anti-gravity force – in light – which would balance gravity. The 10^{78} atoms in the universe form only 4 per cent of matter, and it must be emphasized that there is no evidence other than imaginative mathematics for the existence of dark matter and dark energy, which WMAP estimated comprise 23 per cent and 73 per cent of the matter in the universe respectively.

Overview of the Big-Bang Model

How firmly established, then, is the Big-Bang theory? We are now in a position to take an overview. The evidence for the Big Bang arising from an atom-size point or singularity is:

(1) Distant galaxies are red-shifted, which means they are moving away and so the universe must be expanding: the Hubble expansion of all galaxies in the universe away from the earth in all directions.

(2) Big-Bang theory predicted that there would be radio waves in space that echo the Big Bang. The cosmic microwave background radiation, first detected in 1964/5, is from 380,000 years after the Big Bang, and it has temperature fluctuations (hot and cold spots) of 1 part in 100,000 that were first detected in 1967. These may suggest seeding sites for galaxies. Ripples in the hot and cold spots suggest evidence of past inflation. There has been support for the theory from the 1992 and 2003 satellite probes.

(3) Electromagnetic signals from distant galaxies report on what the galaxies

were like when the signals left them, allowing us to look back in time, for example back to 380,000 years after the Big Bang when the cosmic microwave background radiation was hissing.

(4) There are light nuclei such as the isotopes of hydrogen, helium and lithium which formed between one and twenty minutes after the Big Bang, when protons and neutrons combined. Production of light elements corresponds to a hot plasma phase of nucleosynthesis existing everywhere in space.

All this evidence corroborates the findings of the 2003 Wilkinson probe that the universe is 13.7 billion years old, to a precision of 1 per cent, and that it will expand forever provided it is not dragged back from expansion through dark energy by gravity in the form of dark matter (assuming that both dark energy and dark matter exist).

The evidence produced *against* the Big Bang is:

(1) Some claim that the age of the universe is in doubt by at least 50 per cent, stretching it back to 20 billion years ago, as some superclusters seem to be older than a universe that began 13.7 billion years ago.[49] However, the discovery in 1998 of the increased acceleration has undermined the dating to 20 billion years ago.

(2) The idea that matter is spread evenly throughout a universe expanding uniformly came from Einstein's cosmological principle which applies the equations of the general theory of relativity to the whole universe.[50] However, the discovery in 1992 of ripples in the cosmic microwave background radiation and the discovery in 1998 that 96 per cent of the matter in the universe is missing have undermined the notion that matter is spread evenly throughout the expanding universe.

Nevertheless, the mapping of astronomers and cataloguing of galaxies on the 1992 COBE satellite map (see top illustration below) showed a spherical, symmetric distribution of matter in all directions from earth. This did not fit with a uniform expansion produced by Einstein's general theory of relativity, and the distribution was not one predicted by the Big Bang.

Furthermore, the 2003 WMAP found that there were also walls and voids, clusters and superclusters that resembled fractal patterns and structural patterns one would not expect to fit in with Big-Bang model. The Sloan Great Wall[51] is

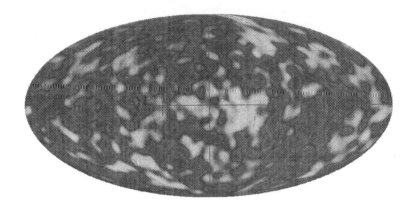

1992 COBE (Cosmic Background Explorer) map of the cosmic microwave
background radiation, produced by NASA.

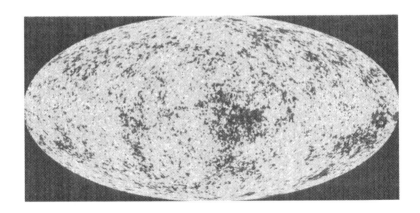

2003 WMAP (Wilkinson Microwave Anisotropy Probe) of the cosmic microwave
background radiation, produced by NASA
(with 35 times better resolution than the COBE map).

1.37 billion light years long (weirdly, exactly one-tenth of the accepted age of the
universe, 13.7 billion years), the CfAZ Great Wall[52] 500 billion light years long,
and the Great Void[53] 1 billion light years wide and formed 6-10 billion years ago.
All were only discovered in 2003, and while they can be seen as the ripples found

in 1992, the universe does not now look smooth. The distribution of galaxies in galactic walls is periodic, regular and symmetric, and the same in all directions away from us, which may not conform to the fall-out from an explosion. It has been questioned whether 13.7 billion years allowed enough time for superclusters and galaxies to form, as they might need double that time, or perhaps even 140 billion years. It is argued that there has not been enough time for gravity to sculpt structures larger than 30 million light years in length or width, let alone 1 billion light years long or wide. Doubts have been expressed as to whether the general theory of relativity explains the distribution of matter in the universe. Penrose, for one, believes that inflation is a "fashion the high energy physicists have visited on the cosmologists"[54] and that our understanding of the Big Bang must await a new breakthrough that will transform quantum theory and gravity, and perhaps consciousness as well.

(3) Quasars are distributed in a spherical annulus (or ring) surrounding the earth, which seemed to be moving radially away from an origin near Virgo about 70 million light years from earth[55] (which would contradict the uniformity of cosmic microwave background radiation), suggesting that the universe is not expanding at all (which would contradict the 1998 finding of increased acceleration).

(4) About 96 per cent of the mass of the universe is missing, and rather than proposing hypothetical, unevidenced dark energy and dark matter so as to make the Big-Bang theory work, it would be more honest to show the missing matter as evidence that it does not work.

Later models have been discredited or lack evidence:

(1) Bondi, Gold and Hoyle's Steady-State model. This was proposed because the proposers could not see how hydrogen could be replenished from matter to escape a central black hole. The Steady-State model was discredited after Hubble's expansion was confirmed by cosmic microwave background radiation and perhaps further discredited by the regular symmetric pattern of walls and voids. The Big-Bang theory had predicted cosmic microwave background radiation, but this turned out to be between ten and a thousand times less powerful than the prediction. The Steady-State theory had predicted a background of radiation but expected it to be in the form of starlight, not radio waves. If the Steady-State theory were to be proved right, the starlight would have had to be changed into

radio waves by something. If the Big-Bang theory is to be proved right, the weakness of the cosmic microwave background radiation has to be accounted for – hence the hypothetical supply of cold dark matter and dark energy.[56]

(2) The multiverse. This has two meanings: several successive universes but always only one at a time (an idea in ancient Hindu thought which sees our universe as one of a succession, each of which is created and destroyed); and several co-existing universes. The idea of this second meaning of the multiverse had been touched on in Presocratic philosophy as we have seen (see p15), and by Gottfried Leibniz, the 17th-century German philosopher who suggested that our world is but one of an ensemble of worlds. It was raised by Robert Oppenheimer and Hartland Snyder in 1939. In our time, the idea of the multiverse grew out of Penrose's collaboration with Hawking at the end of the 1960s when they realised that the bursting-out of the universe at the time of the Big Bang was a mirror-image of a collapse of a massive object into a black hole. Perhaps a previous universe had collapsed into a black hole and "bounced" out into our universe as a Big Bang, and perhaps our universe was just one of a multitude of universes.[57] There is no evidence that there is more than one universe (just as there is only mathematical evidence that black holes exist), which is why Universalism, a philosophy of *this* universe, declines to engage and lays the hypothetical idea to one side.

(3) String Theory. In 1984 the British physicist Michael Green and the US physicist John Schwarz hypothetically proposed superstring theory (or string theory for short) in which the elementary ingredients of the universe are not particles but tiny filaments that look like points and vibrate like rubber bands. The typical length of a string, it is claimed, is about 10^{-35} metres and it stores energy equivalent to 10^{38} proton masses, the numbers relating to the Planck mass and Planck length.[58] They operate in eleven dimensions. The idea alters Einstein's general theory of relativity to make it compatible with the laws of quantum mechanics.[59] There is no evidence that superstrings exist or that there are any more than four dimensions (three of space and one of time), which is again why Universalism lays the idea to one side.

(4) The multiverse Big-Wave model publicised by the American nuclear physicist Roger Rydin who drew on the work of three American physicists, William Mitchell, Alan Dressler and Lawrence Krauss. (The wave is plasma or ionised gas from an unhot beginning.) This theory is a variation and refinement of (2). It holds that the multiverse has a cycle of birth, death and rebirth of

universes and as in (2) that our universe began from the black hole/Big Crunch of a previous universe, spewing out what was previously sucked in, the Big Bang reversing a previous Big Crunch.

The Big-Wave model holds that our universe is spherical and exists alongside other spherical universes in a multiverse. It contradicts the general theory of relativity and also Friedmann's three models in assuming a closed universe heading for a new Big Crunch. It also rejects the formula $\Omega = 1$ as a hypothetical postulation. To that extent it is extreme and controversial. It claims to explain how galaxies formed over a reasonable period of time. It claims that matter at the origin of the cosmic microwave background radiation must be 30 billion years old for it to have cooled to 2.725K above absolute zero (which is −273.16C) within the observable universe. It claims to explain why quasars are in a spherical annulus surrounding the earth and incorporates the idea that the universe is a centreless expanding balloon which all observers see no matter where they are and reconciles it to our ability to see the beginning of the universe in all directions (in the cosmic microwave background radiation). It denies dark matter.[60]

The idea of a Big Wave from a point or singularity is of interest, but as with (2) attributing the Big Bang to a previous universe rather than to infinity within the singularity raises the question as to how the first universe of a multiverse started. It does not eliminate the question, it merely pushes it back. Einstein found the concept distasteful: "This is a dangerous idea that I am simply unwilling to contemplate."[61] Universalists focus on the universe we can see and leave aside theoretical, mathematical proposals for a hypothetical multiverse which cannot be proved evidentially.

So we retain the Big-Bang model, which has consensual support, but note the disquiet over some contradictions. It may be that the walls and voids can be explained in terms of ripples in the cosmic microwave background radiation. It may be that the annulus is a broken-up Great Wall expanding away from the earth in pieces. The uniformity of the cosmic microwave background radiation may be consistent with inflation having whooshed around 10^{-35} seconds after the Big Bang, spreading its hissing echo everywhere simultaneously in equal measure. Clearly cosmology still has many questions to answer.

There is still a problem with how the expanding, flat universe formed. For there is no getting away from it, the point or singularity must have come from a pre-existing infinite something before space-time existed. Bewildered theoreticians such as Hawking who have hypothesised on so much have said that we cannot think of what happened *before* the Big Bang because the Big Bang produced space-time and with it the beginning of time, and we cannot think of what happened *before* time began. Logical philosophers such as Russell (see p46) have made the same point. But this is a convenient, casuistic argument. It is not good enough to say that the universe began from a singularity, but that as singularities lie half-in and half-out of physics, and therefore partly outside physics, we need not think about what pre-existed a singularity.

In stating "a new scientific interpretation of the universe and Nature", the Universalist philosopher has to include the infinite state that existed *before* the Big Bang and *before* time began with the beginning of the universe. As space-time is inseparable it can be said without contradicting Einstein that time is a succession of spatial events. Time is a stream of material change, Heracleitus's flux, a succession of spatial events or qualifications in which the present is added cumulatively to the past so that new layers are endlessly added to previous layers. This view is a refinement of Whitehead's ideas on process. There are 10^{22} spatial events or qualifications per second (10 thousand million million million events per second).[62] Time is derived from a succession of spatial events, and to say we cannot think about what happened before time began is tantamount to saying we cannot think about what happened before spatial, material events began. But what, if anything, is behind the reality we see has always been the concern of philosophy. The Big-Bang model suggests that there was a pre-existent Reality from which the point or singularity, space-time and matter all emerged. This pre-existing Reality is what our surfer breasts on the crest or rim or edge of our expanding universe.

Philosophical Summary

So, having looked at the structure of the universe, what can the Universalist philosopher take from cosmology and astrophysics to structure a philosophy about the universe? Universalism is based on *this* universe, and a very important plank in the new philosophy is the Big Bang, with the various reservations and qualifications I have raised. Equally important is the infinite from which the first singularity emerged. Also, Universalism has no truck with the multiverse. Universalism accepts the account of inflation and the shaping of the universe, and of the 1998 discovery that acceleration has increased. The image of the surfer is very important to Universalist philosophy.

To sum up, philosophy can take from science the following aspects of the origin of the universe:

the infinite;
the first singularity;
no multiverse;
the Big Bang;
inflation and shaping of the universe;
increased acceleration;
the surfer.

I have suspended judgment on the degree of order and self-organising that can be detected until the end of ch. 5.

The "new scientific interpretation of the universe and Nature" must include physics as well as cosmology and astrophysics – the very small as well as the very large. The Universalist philosopher now needs to see if science can detect a sea of infinity in the subatomic and dimensional world of physics. Again, we need to keep at the back of our minds the extent to which science can detect the workings of a principle of order.

4

PHYSICS: THE QUANTUM VACUUM
AND THE EXPANDING FORCE OF LIGHT

Just as cosmology and astrophysics came out of a point, so in physics the subatomic and galactic worlds are one and interconnected in the point. Now I must introduce a complication to our account of the Big Bang: quantum physics and the workings of the quantum vacuum.[1]

In a sense, quantum theory began with the discovery of radiation. In 1896 the French scientist Henri Becquerel found that uranium compounds emitted radiation, and after a search for other radioactive substances Marie Curie extracted polonium and radium. The wave theory of light could not explain the interaction between electromagnetic radiation and matter. In 1900 Max Planck proposed that emission of radiation was intermittent rather than continuous, in multiples of an atom or quantum of energy. Using his constant, Planck proposed that the energy of a quantum is inversely proportional to the wavelength of radiation but directly proportional to the frequency. Quantum mechanics began with Planck's theory of black-body radiation in 1900.

Quantum theory was based on seeing radiation as multiples of quanta or particles as well as rays. Einstein extended Planck's ideas in 1905. He saw light and electromagnetic radiation emitted as quanta, and showed that light behaves as if all its energy is concentrated quantum packets of energy called photons. Light could be both a wave and particles: this electromagnetic radiation could behave like waves in some circumstances and like particles in others. Quantum theory developed with Niels Bohr's proposal of the quantum theory of spectra in 1913.

In 1923 Louis de Broglie (pronounced "de Broy") suggested that matter might also have wave-particle duality and have wave properties. Within a few years his speculation was confirmed by experiments: moving particles of matter also have wave-like properties that can be described in quantum theory.

In the 1920s physicists found that in some cases an electron acts like a particle, occupying only one position in space at a time, whereas in other cases it acts like a wave, occupying several places in space at the same time. The wave-particle duality of matter means that there is uncertainty in Nature (as to whether matter is going to act as particles or waves) and Nature therefore can only be described by probabilities according to quantum mechanics. But the uncertainty impacted on gravity. As we have seen, Einstein's theory of gravity requires that the universe exploded from a state of infinite density, but high densities of matter present problems for traditional classical physics.

True quantum mechanics[2] appeared in 1926 with the matrix theory of Max Born and Werner Heisenberg, the wave mechanics of Louis de Broglie and Erwin Schrödinger and the transformation theory of P.A.M. Dirac and Pascal Jordan, all of which were different aspects of a single body of quantum law. Heisenberg, a supporter of Bohr, held that Schrödinger's model of an atom surrounded by waves was wrong as what an atom looks like, and its speed and position at any one time, are unknowable and can only be understood through mathematics.

In 1931 Dirac's equation describing the motion and spin of electrons incorporated both quantum theory and the special theory of relativity. Dirac laid the foundation for quantum electrodynamics (QED),[3] a quantum theory of the interactions of charged particles with the electromagnetic field. In the 1940s Richard Feynman developed QED together with Julian Schwinger and Shinichiro Tomonaga with his idea that charged particles (electrons and positrons) interact by emitting and absorbing photons, the particles of light that transmit electromagnetic forces. (Photons can produce electrons and positrons in pairs so long as the photons' energy is greater than the total mass-energy of the two particles.)

This theory led to knowledge of virtual photons, and the way they may behave in a quantum vacuum where the emptiness of the vacuum in fact contains virtual photons that can emerge and become real.

Newton's Ordering Light

Einstein was surprisingly influenced by Newton, whose view of the universe he is thought to have obliterated with his own work. Newton saw light as associated with ether. Ether was thought to fill space. Having made his Materialist discoveries in gravity, light and calculus in 1664-1666, from 1669 when he was twenty-six Newton spent nearly thirty years researching into alchemy to find an expanding force of light that would counteract and balance gravity, an idea that would interest Einstein. Gravity drew bodies together. Newton's expanding force would drive objects apart and keep the universe looking static. In 1687, in his *Principia*, Newton set out an absolute space and time and the laws of motion and gravity.

At the same time he was working on light. Newton deduced that "every Ray" of light had immutable properties and "may be considered as having four Sides

81

or Quarters". In Query 30 of *Opticks* he wrote of "Particles of Light", which seems to anticipate wave-particle duality: "Are not gross Bodies and Light convertible into one another, and may not Bodies receive much of their Activity from the Particles of Light which enter their Composition?"[4] Newton held that this expanding force operates in the radiation of light, in chemical composition and biological growth, and also governs the mind – consciousness – and behaviour of human beings.

In Query 31 Newton asked: "Have not the small Particles of Bodies certain Powers, Virtues or Forces by which they act at a distance, not only upon the Rays of Light for reflecting, refracting, and inflecting them, but also upon one another for producing a great Part of the Phaenomena of Nature?"[5] In an alchemical manuscript now in the Burndy Library he states: "Ether is but a vehicle to some more active spirit & the bodies may be concreted of both (i.e. ether and spirit) together, they may imbibe ether well (sic) as air in generation & in the ether the spirit is entangled. This spirit is the body of light because both have a prodigious active principle."[6] In other words light, whose body is spirit, combines with ether, and there is a single unified system which includes both internal (spiritual) and external (physiological) systems.

Ether had been regarded as a tenuous substance that filled the vacuum and was a means of carrying light since the Presocratic Greeks' *aither* (see p13). By the 19th century physicists had proposed that the entire universe was permeated by "luminiferous ether" (i.e. light-bearing ether) which acted as a medium for carrying light. Lord Kelvin, the Victorian scientist who gave his initial to the temperature measurement K, suggested its properties:

"Now what is the luminiferous ether? It is matter prodigiously less dense than air – millions and millions and millions of times less dense than air. We can form some sort of idea of its limitations. We believe it is a real thing, with great rigidity in comparison with its density: it may be made to vibrate 400 million million times per second; and yet be of such density as not to produce the slightest resistance to any body going through it."

In other words ether was very strong yet insubstantial. Theosophy took it up, referring to "the etheric level" on which telepathic thoughts could be transmitted.

However, in 1880 the idea came to grief when Albert Michelson, the US's first Nobel Laureate in physics, set up an experiment to prove its existence involving two perpendicular beams of light, one of which was to be slowed by ether. The two beams of light arrived at the same time, and the experiment failed to detect ether. In 1887, seven years later, Michelson announced his results, which ironically "proved" that ether did not exist, a result Einstein had predicted. The science writer Banesh Hoffmann wittily wrote:

"First we had the luminiferous ether,
Then we had the electromagnetic ether,
And now we haven't e(i)ther."

But what if ether, the Greek *aither*, is a network of dark energy which is composed of tinier particles than neutrinos, perhaps of photinos? Could it be that Newton was right in associating light with ether, and that Michelson's experiment was inadequate and disproved nothing? We shall return to the possibility of an undetected tiny-particle form of light in due course.

Einstein took up Newton's idea of a balancing force in 1917 when he proposed his "cosmological constant": an expanding force that repels and balances contracting gravity. He saw this as a repulsive force of unknown origin which exactly balances the attraction of gravitation in all matters and keeps the universe static (as, like Newton, he thought the universe to be). Einstein believed the constant gave all particles stable interconnectedness for all time. We have seen that he abandoned this constant in 1929 when Hubble established that the universe is expanding and not static, and that he later called his abandonment "the biggest blunder of my life". We have seen that in our time Einstein's constant has returned as hypothetical, unevidenced dark energy, the force which repulses gravity within our accelerating universe. Dark energy may have been the expanding force Newton was seeking. No one knows if this dark energy is conveyed by light (perhaps mixed with "ether" and "spirit", Newton thought), which, Newton believed, governs human and plant growth.

Perhaps Newton was right and the expanding force that counterbalances gravity is light. Perhaps it contains a universal principle of order which stimulates chemical composition through DNA and growth through photosynthesis, and affects consciousness and human behaviour. Perhaps there is a wider spectrum of light than we realise. The electromagnetic spectrum (see p400) is currently known to have gamma rays at the short-wave end, radio waves at the long-wave end and physical light in the middle. Perhaps beyond gamma and virtually undetectable is metaphysical Light (Newton's "ether" and "spirit").

The mystical traditions claim that metaphysical Light is infinite Being from before the Big Bang, from within the timeless nothing that preceded the quantum vacuum, long before the cosmic microwave background radiation. According to many mystical traditions it can be received in consciousness below the rational, social ego. Does metaphysical Light manifest into the universe from timeless infinity, travel with physical light of space-time and, arriving among hosts of gamma rays, ultraviolet rays and radio waves, pervade Nature with order that is intermixed with physical light? If so, there can be a synthesis of physics, mysticism and metaphysics.

Hidden Variables and Order

Einstein's work on his cosmological constant must be seen alongside his unfinished search for "hidden variables". After a quarter of a century of improvised quantum theory, in 1925 Werner Heisenberg proposed a new quantum mechanics. In the new quantum theory, randomness or indeterminacy is fundamental, and identical electrons in identical experiments behave with apparent randomness. Many physicists claim that all nuclei were originally in an identical state, and that nuclei decay or do not decay on a random basis.

Einstein did not accept the randomness – "God does not play dice with the universe"[7] – and he proposed a hypothetical, unevidenced, unknown property, referred to as a "hidden variable" which causes some nuclei to decay but not others. He held that this hidden principle underlies the indeterminacy of the uncertainty principle.

Einstein never found a hidden principle of variability. The Danish theoretical physicist Niels Bohr and his Copenhagen school rejected hidden variables as qualities that could not be measured. After rejecting quantum theory at the fifth international conference on electrons and photons at Solvay, Brussels, in October 1927 (which marked the triumph of quantum mechanics), in 1930 Einstein had a bruising meeting with Bohr which ended in his grudgingly accepting quantum mechanics.[8] Einstein worked on quantum gravity and proposed hypothetical, unevidenced "gravitons", gravity particles which may also act as gravity waves. His general theory of relativity had predicted black holes from whose intense gravitational fields light cannot escape, and also massless gravitons travelling like photons at the speed of light.[9] If gravitons are proved to exist, the gravitational force could be described in terms of an exchange mechanism. It would operate like the electromagnetic force of repulsion between two electrons due to the exchange of a virtual photon between them, like the weak force in which the decay of a neutron is caused by the exchange of a virtual W-particle and like the strong force between quarks, due to the exchange of a gluon. There would then be a unified theory of all force, a huge step to a Theory of Everything.

Einstein was sure the universe was ordered rather than a random accident, and he hated the random uncertainty principle of quantum mechanics which had overtaken his own work. Bohr and Einstein talked throughout the 1930s but eventually had no more to say to each other.[10] The split between them was reflected in a split in physics that has lasted to this day.

Einstein spent the last thirty years of his life working unsuccessfully, trying to re-integrate gravity with the other three forces, attempting a unified field theory in which the strong and weak forces, electromagnetism and gravity would be united in an underlying field.[11]

But because Einstein never found a hidden principle of variability does not mean that it cannot be found. The Universalist philosopher detects within the hitherto undiscovered hidden variability workings of a universal principle of order that is behind the apparent randomness of all nuclei and the uncertainty principle. May this hidden principle also include the workings of Newton's expanding force? And may it be associated with metaphysical Light?

David Bohm, an American theoretical physicist and expert on quantum mechanics and Einstein's *protégé*, supported hidden variability. He had discussions with Einstein and improved the Einstein-Podolsky-Rosen thought experiment. Bohm wrote a book on quantum theory in 1951 and in January 1952 took Einstein's side against other physicists in an article in *Physical Review*, two papers published together entitled 'A Suggested Interpretation of the Quantum Theory in Terms of "Hidden" Variables.'

Bohm said that in quantum theory a system physically functions as waves that can only be measured in terms of probable results. "Hidden" variables would determine the precise behaviour of a system. Bohm admitted the internal consistency of quantum theory, but regarded its present form as incomplete as hidden variability is missing.[12] Roger Penrose has said that there *can* be hidden variability if it is non-local (meaning that the hidden parameters must be able to affect parts of the system in arbitrarily distant regions instantaneously),[13] a view he confirmed to me in conversation in 1991.

The Quantum Vacuum and Infinite Order

Bohm saw the quantum vacuum as a reservoir of order. To Bohm, quantum theory has certain basic assumptions although it has been formulated differently by Heisenberg, Schrödinger, Dirac, von Neumann and Bohr. These assumptions are that the laws of quantum theory are to be expressed in terms of a wave function, that the physical results are to be calculated with the aid of "observables" which are sharply defined in certain circumstances but fluctuate at random (lawlessly) in others. As we have just seen, Bohm countered Heisenberg's indeterminacy principle with his theory of hypothetical hidden variables, which he believed explained essential features of quantum mechanics.

He proposed a hidden order beneath the apparent chaos and lack of continuity of the particles of matter of quantum mechanics. He saw a hidden dimension behind the surface, an "implicate order" which is the source of all visible, "explicate" matter in our space-time universe.[14] This implicate order has "infinite depth".[15] The term is another way of describing infinity.

Bohm saw our three-dimensional world as an explicate order of objects, space and time. He believed we live in a multidimensional world[16] – an aspect of the multiverse – and that the implicate order gives rise to our physical, psychological and spiritual experience and is in a fifth dimension of which we are largely unaware. (Time is the fourth dimension). A devotee of Krishnamurti, he held that there is also a superimplicate order, a more subtle sixth dimension which is the source of our spiritual experiences.

Bohm does not appear in physics textbooks because he introduces hypothetical, unevidenced new dimensions, and textbooks rightly restrict themselves to space-time. However, his view is in keeping with the philosopher's view of the infinite. In his implicate order, the order is unfolded like a radio wave containing the sound or like a television wave containing the picture, and it needs an explicate order to unfold it (like a radio or TV set).[17] If we leave to one side Bohm's view of the "implicate" as a folding process, like folding over a piece of paper, and also his hypothetical multidimensional view, for which there is no evidence, then his order can be restated in terms of an infinite order from which an energy surges into chemical composition (via DNA) and plant growth, as Newton proposed.

To Bohm every individual unit contains within itself the totality of existence. Individual units of matter (nuclei) contain the totality within themselves, and the same is true of individual human beings. To Bohm Reality is an unending process of movement. It is not static, but a wholeness or totality from which the thinking observer and his or her ego cannot be separated.[18]

To Bohm the rules of quantum theory applied to gravity result in a gravitational field comprising wave-particles, each having zero-point energy (the energy of the totality), and so the gravitational field cannot be completely defined.[19] So-called empty space contains energy, and he held that matter is a small quantized excitation on top of a sea of energy, like a tiny ripple in a vast sea. He held that we calculate the difference between the energy of empty space and the energy of space with matter in it, and we miss the energy in empty space.[20] This may be the dark energy which was hypothetically postulated in 1998, six years after Bohm died in 1992. Space is full rather than empty, as Parmenides and Zeno held and as Democritus opposed.[21]

In quantum theory atoms affect each other simultaneously in non-local ways when on opposite sides of the universe and display order across space. To Bohm, the apparent randomness, probabilities and uncertainties of quantum theory have order behind them, as the tides order the sea. Bohm saw matter and consciousness as "not distinct" in the sense that we are in the same sea as physical objects.[22] The universe is a quantum vacuum from which we appear like particles and to which we return like particles.

To Bohm the Big Bang was a little ripple in the middle of an ocean. Its ripples spread outwards, Bohm maintained, constituting our expanding universe.[23] This vast sea of cosmic energy has to be on a scale larger than the critical length of 10^{-33} cms.[24] Most physicists agree that there is some kind of granularity on a scale of 10^{-33} cms, which is 10^{20} smaller than an atomic nucleus,[25] and Bohm's sea would therefore be composed of the tiniest conceivable, ether-like, granular structures. The Leibnizian philosopher Geoffrey Read has proposed that myriads of tiny particles arise from an underlying reality below quantum mechanical waves and particles in space, and has attempted to describe this level of reality in terms of a "hidden variable" theory.[26] In conversation with me in 1991, Bohm said: "The Big Bang was not a big event, it was a local excitation within the infinite totality, like a ripple or wave in a sea." He said that one cannot have a vision of the Whole as it is infinite and our minds are finite. But, as I pointed out, mystics hold that there is a part of our consciousness that is universal and capable of cosmic or "unity" consciousness, which is able to see Reality beyond space-time. (Whitehead put it slightly differently: "Our minds are finite, and yet even in the circumstances of finitude we are surrounded by possibilities that are infinite, and the purpose of life is to grasp as much as we can of that infinitude.") Of his theory of wholeness and the implicate order, Bohm said to me with disarming modesty, "These are just ideas." He meant, "There is no evidence." But Bohm is of interest as he was an expert in quantum mechanics continuing Einstein's tradition, having had discussions with Einstein, and he was proposing that there is an infinite order behind the uncertainty principle of quantum mechanics and within the quantum vacuum.

The energy within the quantum vacuum has applications for the emerging universe, an angle that has been taken up by a Belgian theoretical physicist in Brussels. Edgar Gunzig[27] of the Belgian school of physicists has done work on how a universe emerges from the quantum vacuum by quantum processes through virtual particles. Real nothing is emptiness. By principle, a vacuum is non-empty. It has an underground life but is the most possible emptiness that can be imagined. It is as close to an absence of particles and radiation as can be imagined, and is the lowest possible energy state. Virtual particles do not exist, they are potentialities of existence. A vacuum with zero energy trembles due to the gravitation of the curvature of space, which is a reservoir of energy that can be given to virtual particles. Virtual particles are expelled from the vacuum in pairs, live for a split second and then return to the quantum vacuum.

If one virtual particle of a pair attracts and picks up energy from the geometric background of the curvature of space and enters space-time, then it becomes a real particle which could be a first proton and could turn into photons which cannot then be undone. This happened in different parts of the quantum vacuum and there were numerous real particles with positive energy created by negative energy from the curvature of space, emerging from zero energy. Through quantum processes a quantum vacuum emerged and as virtual particles became real, so did space-time. Out of this came a hot beginning and inflation as the universe expanded from zero energy. On this view there did not need to be a Big Bang, just a hot beginning for a pre-quantum vacuum could have pre-existed the Big Bang. Inflation would do the rest, expanding seeds in the microworld to the galaxies of our universe.

Gunzig's team were given the Gravity Research Foundation International Award for mathematics in this field and only in 1991, after the prize was given, did they realise that their mathematical solution explained inflation.

This view can be reconciled to the classical general theory of relativity which, as Penrose and Hawking showed, requires a beginning of time in a point of infinite density and infinite curvature of space-time where all known laws of science break down. In other words, time began in infinity. Our universe began as a superdense concentration of mass smaller than a proton and containing zero energy, for the positive energy of mass is balanced by negative gravitational energy. It expanded through inflation – perhaps through Newton's expanding force of light which he believed was mixed with "ether" and "spirit" and which in some traditions is seen beginning in Fire.

Looking back over 20th-century physics, we can see the incompatibility between classical and quantum physics. Classical physics, meaning pre-quantum physics and thus including Einstein's special theory of relativity, has become universally accepted and its layers of theory are well-established. Quantum physics, however, has been in flux with a number of conceptual frameworks and alternative methods of calculation, the overlapping bits of which agree. Whereas classical physics works through observable phenomena (for example, iron filings are observed moving towards a magnet), quantum physics have no such observable phenomena, the phenomena of quantum theory having a merely mathematical reality. In fact, quantum mechanics is not a physical theory but a mathematical theory. Quantum theories correspond to the theories of classical physics (for example, electromagnetism, gravity) and focus on the physics of the very small (at the atomic and subatomic scales). Quantum theory is dominated by the uncertainty principle and allows entanglement.

The shortcomings of our theory of gravity have been illustrated by the gravitational force exerted by a mass when it rotates: gravitomagnetism. This force was dramatically employed in the successful first use of gravitational slingshot to speed up and propel Mariner 10 past Venus on its way to Mercury in 1974. According to Einstein's general theory of relativity, a rotating mass such as a planet twists space-time, and a spaceship near it is dragged round by the vortex and speeded up.[28] Gravitomagnetism was detected (in the form of a signal) in a laboratory in Austria in 2006 when Martin Tajmar's team found the gravitomagnetic force to be trillions of times stronger than they had expected it to be.[29] There is no explanation in either the general theory of relativity or quantum mechanics, and a new quantum theory of gravity that includes this force is urgently needed.

There needs to be a complete theory of gravity that includes quantum mechanics. This has not been found. Calculations on quantum gravity lead to mathematical infinities. Much of recent physics has involved a largely unsuccessful attempt to bring together classical relativity and the new quantum theory, to integrate the very large and very small. Yet enough work has been done to show that the effects of quantum mechanics were crucial during the first 10^{-43} seconds after the Big Bang, when the universe had a density of 10^{93} grams per cubic centimetre. The era until 10^{-43} seconds after the Big Bang, the Planck era, is often also called the "quantum era", when the entire tiny universe would have been full of uncertainties and fluctuations, with matter and energy appearing and disappearing out of a quantum vacuum in the split second at the very beginning.

It must be pointed out that there is some uncertainty about zero-energy as there are disputed findings from different methods of calculation. Quantum zero-energy emerged from the kinetic theory of an ideal gas. (At absolute zero, -273.16C or 0 K the molecular kinetic energy of an ideal gas is zero, whereas substances have a minimum value called the zero-point energy, suggesting that not all molecular motion stops at absolute zero, 0 K.) The possibility – indeed, likelihood – remains that when zero-point energy is applied to a quantum vacuum, quantum corrections to familiar forces raise the Casimir effect. This was named after Hendrik Casimir, the Dutch physicist who discussed an experiment in this area with Bohr: the force of attraction towards each other of two conducting metal plates placed close to each other. A lowering of the zero-point energy between the plates appears to prove the existence of zero-point energy and of the zero-point energy field. However, the zero-point energies of frequencies produce an infinite total energy, which is meaningless from a mathematical point of view. (Strictly speaking, the calculated energy "diverges to infinity" and so no meaningful value is produced and it is deemed "formally infinite".)[30] Depending on which mathematical approach is taken, zero-point energy may refer to a quality that is infinite or zero, if it really exists. Despite these doubts, the possibility remains that a vacuum has zero-point energy, and this has been accepted by the Belgian team, as we have seen.

Physicists are attempting a Grand Unified Theory by working towards the reunification of three of the four forces (excluding gravity), but the temperature at which they diverged, and might therefore presumably reunite (somewhere between 10^{14} and 10^{27}K, and perhaps 10^{28}K, the temperature for the first 10^{-35} seconds after the Big Bang), is too high to replicate. The particle accelerator chambers would have to extend far into space to achieve such a temperature, and this looks a forlorn effort.

In an article in *The Daily Telegraph*[31], Hawking claimed: "By colliding particles at the kinds of energies that would have been around at the Big Bang, we have unified three of the four forces: electromagnetics, and the weak and strong nuclear forces." The article was a transcript of a two-part television programme shown later, and Hawking repeated the claim word for word in the second programme. He should have said "theoretically and potentially unified" or "in low-energy consequences". It is understandable that, being disabled and only able to communicate at three words a minute, should use words sparingly, but the unification of the three forces has only been solved in Hawking's mind. In *A Brief History of Time* he writes: "The strong nuclear force gets weaker at high energies. On the other hand, the electromagnetic and weak forces, which are not asymptotically free, get stronger at high energies. At some very high energy, called the grand unification energy, these three forces would all have the same strength and so could just be different aspects of a single force....A machine that was powerful enough to accelerate particles to the grand unification energy would have to be as big as the solar system....Thus it is impossible to test grand unified theories directly in the laboratory."

CERN have confirmed to me that the average temperature its particle chamber can reach is 10^{17}K. On 5 November 2007 a CERN expert wrote of the Large Hadron Collider (LHC): "The average collision energy at the LHC (with beams of protons) will be around 10TeV which correspond to about 10^{17}K. We don't know if and where the unification of forces will happen but the LHC is certainly the first-ever-instrument to provide such a high energy." The temperature at which grand unification takes place may be 10^{28}K. Collisions were eventually to be ramped up to their maximum energy of 14TeV. Although CERN have preferred to remain silent when I have twice written to challenge Hawking's claim, we can take it that there has been no practical grand unification and that Hawking was either talking in principle or overclaiming in the article and television programme, which many will have read or seen.

From September 2008 the LHC conducted the world's largest scientific experiment in the 17-mile-long circular tunnel 330 feet under the Alpine foothills near Geneva. Particles in proton beams raced round the route of pipes which contained a vacuum 11,245 times a second at 670 million miles per hour to smash headlong into each other at temperatures hotter than 100,000 times hotter than the centre of the sun (which is 13,600,000K) [32], i.e. 10^{12}, to reveal the first particles that existed in the split second after the birth of the universe in the Big Bang. The largest of four detectors, Atlas, which is bigger than Canterbury Cathedral, looked for the Higgs boson. It is hypothesised that this only exists at very high energies. Immediately after the Big Bang all particles are thought to have had no mass, and as the temperature cooled, it is suggested a field of theoretical bosons consisting of mass and little else came into existence which stuck to them, making them heavy. Theoretical bosons were proposed by Professor Peter Higgs, a particle physicist at Edinburgh University, in 1964. The stakes were high. If they found the Higgs boson, it would become the standard model of particle physics. It would establish that there is a field like *aither*, which may be the ether that was not discovered in the 19th century, which may explain variability as well as mass. If they did not find it, then Nature had another way of giving particles mass – which would turn science on its head. It may be that there is an *aither* or metaphysical Light which is an ordering principle that gives particles their mass, operating like Higgs bosons, and that it has varying intensities round different particles and in different localities, embodying hidden variability. The huge experiment was expected to create 15 million gigabytes of data a year (equivalent to 21.4 million CDs), and scientists have had to create a new form of the internet to cope with it. Many of the findings might therefore not be known for some while. (See Postscript on p368.)

Physicists hope that all four forces can be reunified to include gravity and create a Theory of Everything, but the temperature to be replicated, 10^{32}K, would be even higher than the temperature to reunify three forces, and the particle accelerators even larger. Nevertheless, if the technology could be devised, all four forces *can* be reunified as they were unified at the beginning of their life shortly before 10^{-43} seconds after the Big Bang.

A Theory of Everything must go beyond both quantum physics, which successfully explains much of the particle, and the general theory of relativity, which deals with the universe and gravity. Both these theories work well, and so a Theory of Everything must include both. All particle theories before string theory lead to infinities when gravity is included. These infinities cannot be renormalised and cause the equations to break down. The five main candidates for a Theory of Everything emerged from hypothetical, unevidenced string theory in the 1980s and from an underlying hypothetical, unevidenced "M" theory (M for Mystery) that has not yet been found. The main candidate predicts that the universe has 10 hypothetical dimensions instead of the three actual spatial dimensions, and the other seven have not been found.[33] Universalists leave multidimensional theories to one side as mathematical speculations that cannot be proved.

The search for a Theory of Everything, putting together the very large (relativity) and the very small (quantum mechanics) to explain quantum gravity, has progressed in my lifetime. As we have seen (see p60), Oppenheimer and Snyder introduced black holes in 1939. Penrose established that a universe could collapse into a singularity, a black hole of infinite density, and reversing this, running the film backward, Penrose and Hawking established that a universe could come out of a singularity (now a white hole). Later Hawking established that a black hole (if it exists) must have a glow of radiation round it as pairs of particles drawn towards it split. One, the positive particle, escapes, giving off radiation, while the other, the negative particle, disappears into the black hole. Black holes never lose their heat and so have the potentiality to explode. We can see how a Big Bang could come out of a singularity, but how did gravity behave in the singularity before the universe inflated? It must have held back the expanding force of light or radiation within the singularity until it could hold it back no longer. A Theory of Everything requires an expanding force of light or radiation within the singularity, against which quantum gravity acts as a contracting force. These two opposites must both have been present in the first singularity before the Big Bang. A Theory of Everything cannot be established until it is shown what was behind these two opposites, light and gravity, and what lay outside the singularity before the Big Bang: the infinite.

Three of the four forces are of roughly the same strength, but gravity is much weaker. (A magnet can pick up a chunk of iron on a mountain in defiance of the entire earth's gravitational pull.) The original symmetry when there was one superforce within the dot-sized universe shattered soon after the Big Bang, and whereas the fragments of the electromagnetic, weak and strong nuclear forces can be put together theoretically by weakening the strong nuclear force at high energies and strengthening the electromagnetic and weak forces at the same high energies,[34] gravity seems irreconcilable as it is so weak. I am convinced that the most likely path to a Theory of Everything is not by speculating regarding superstrings, M theory and ten or eleven dimensions and seeking to put together the very large and the very small in black holes – pinning hopes on a multiverse – but by revisiting the idea Newton worked on: that there is an expanding force in light which counteracts the contracting force of gravity. In other words, gravity is weak because of the push of light against it. If physicists move away from string and M theory and analyse the pushing photons of light in relation to the pulling of gravity, I believe they will find a Theory of Everything – but it must include the infinite from which the pre-Big-Bang point emerged. To assert with Hawking[35] that the earth had no beginning or singularity or boundary but has not existed forever, having somehow emerged like a bubble, is a colossal logical and scientific fudge.

For decades cosmologists have accepted the notion that the universe of space-time began as a singularity, whose temperature, density and everything else about it were infinite. If a Theory of Everything marries quantum and gravity, the marriage will have to permit actual infinity which fills the point or singularity from which the universe began.[36]

At a conceptual level, our survey of quantum theory has made evident zero-energy's links with the infinite. The quantum vacuum contains the latent potentialities of existence: virtual particles. It is finite within the inflated balloon of the universe, but it emerged from the infinite singularity in the first second after the Big Bang and has its origin in the infinite. In this sense it is at one with the limitless, boundless infinity outside the balloon of the expanded singularity. Having emerged from the Void that preceded the Big Bang, its emptier, pre-space-time form, the quantum vacuum now contains more virtual particles than when there was a Void or empty nothingness.

Quantum theory plays down the significance of when time along with space began at the moment of the Big Bang, and of when time will end if the universe ends in a Big Crunch (as is thought not to be the case). However, before time began, underlying time and continuing after time ends was, is and will be the infinite timelessness that pre-existed the universe and will survive it: the Void from which the hot beginning and inflation happened and which has yielded, since the Big Bang, a seething quantum vacuum.

Overlaid on this underlying timelessness are spatial events as we saw on p78. Time is a succession of 10^{22} (10 thousand million million million) spatial events every second. That colossal number of events has happened every second since the earth cooled 4.55 billion years ago. In this Leibnizian sense, time has no reality of its own but is an artificial measurement of spatial events which succeed each other. Einstein, discussing time with his friend Besso on a May evening in 1905, saw time as relative, like H.A. Lorentz's "local time", and he linked time with space and matter, the curvature of space distorting time. Space to Einstein did not have a primary independent existence as it did for Newton, and time was also secondary rather than primary.[37] Within Einstein's mathematical theory, clock time is an artificial measurement of spatial events. I would say that the underlying timelessness is primary and absolute, and the measuring of spatial events by time is as relative as the finite is within the infinite. Time is not moving towards disorder (a point accepted by most modern cosmologists and theoretical physicists including Einstein, Hubble and Bohm) as the universe appears to be expanding forever and as it is overlaid on timelessness and infinity, from which it emerged.

Philosophical Summary

What can the Universalist philosopher take from physics for the new philosophy about the universe? There was a timeless Void. The quantum vacuum as a reservoir of order beneath apparent randomness is a very important point, and its operation as an infinite/finite sea of energy. Also, Newton's expanding force of light that counteracts gravity and which is linked to photons and virtual photons, and to undetected "ether", a network of tiny undetected particles. It is also linked to the cosmological constant. Universalists are very interested in manifesting metaphysical Light and light's powers to stimulate chemical composition and biological growth.

To sum up, philosophy can take from physics:

the timeless Void;
the quantum vacuum as a reservoir of order;
a sea of energy;
order within the vacuum;
order beneath apparent randomness and uncertainty principle;
photons and virtual photons;
the expanding force of light that counteracts gravity;
the cosmological constant;
undetected ether-like, dark-energy-like network of tiny particles;
manifesting metaphysical Light;
light stimulating chemical composition and biological growth.

We have been holding order at the back of our minds. The evidence for the universal principle of order, or order principle, in physics, then, is the order within the quantum vacuum which may have produced our universe from zero energy and may counterbalance gravity with an expanding force of light which may control chemical composition and growth. This order is reflected in Einstein's constant and in Bohm's order. This hidden order may be found to control the apparent randomness of the uncertainty principle by determining which nuclei decay through hidden variability. It may explain how the fall-out from the Big Bang created such perfect conditions for life to form.

We have been biding our time on the order in the universe. The Universalist philosopher must now determine the model of the universe's structure that best accords with our findings so far, and see if science can detect a self-organising system, life principle and principle of order, or order principle, in the bio-friendly operations of biocosmology, which covers cosmology, astrophysics and physics in their bio-friendly capacities.

5

Biocosmology: The Bio-Friendly Universe and Order

I have already formed some conclusions about the structure, context and future of our "boundless"-born, shuttlecock-shaped universe. The scientific evidence so far suggests that the expanding universe may not end in disorder, in a Big Crunch. It is between expanding forever and being a flat universe balanced between gravitation and electromagnetic forces which may eventually bring the expansion to a halt. At present it looks as if we are expanding forever provided the hypothetical, unevidenced dark matter keeps us accelerating and does not slow us down.

We have seen on (p69) that what lies spatially ahead of our accelerating shuttlecock-shaped universe is an "edge", beyond which there is a Void which our astronaut surfer breasts as he rushes forward into it. Universalists hold that it contains within itself latent properties that include virtual pre-particles, one of which could have surfaced as a singularity and begun our universe.

The universe, space and time are finite in the sense that they are bounded by extremes: the curvature round the sides of our universe resembling the curve round a shuttlecock's feathers; and the hot beginning (the Big Bang) in our past and the conceptual end of the universe and Nature in the future. Time's arrow is between the singularity at its beginning and the singularity at its end. Time, we have just said, is a succession of material spatial events that began when spatial events began. Time is overlaid on timelessness, the Void or infinity which preceded the Big Bang to which the accelerating universe is rushing, and which would continue after a Big Crunch, the end of the moving universe when all motion ceases in infinity at a temperature of absolute zero or $-273.16C$. This Big Crunch would only happen if the pull of dark matter caused the universe to decelerate, and by then there would be no life on earth as the sun will burn up in five billion years' time.

Although the laws of Nature break down when being sucked back into a singularity or black hole, they emerge from a singularity as a working unity, having "revved up" from inactivity in a reverse of "breakdown". My view of the Big Bang has suggested that our universe came out of an infinite timelessness as one point or singularity in which the laws of Nature and four forces were united, to separate within a second in structures conducive to order; and that our universe co-exists with the infinite timelessness from which it emerged until it eventually returns to it. The shuttlecock-shaped universe is within finite boundaries and is surrounded by the boundless infinite into which the spacesuited surfer is continuously borne.

Six Models for the Structure of the Universe

Physicists' theories about the structure of the universe identify six possible models:[1]

(1) the absurd, accidental, random universe that cannot be explained, it just is and intelligence is an accident;

(2) the unique physical universe unified by a Theory of Everything (perhaps string/M theory);

(3) the multiverse (an infinite series of universes);

(4) intelligent design by God or Aristotle's cause, a theological explanation;

(5) a bio-friendly life principle that works with the four fundamental forces to drive the universe towards life and intelligence – a more metaphysical model than (2);

(6) the self-explaining or participatory universe (Wheeler's "boot-strap" universe in which intelligent life can create the universe that gives rise to it by backward causation, a reversal of time, mind collapsing matter backwards in time to make the universe more bio-friendly forwards in time).

I omit obviously bizarre and freakish models such as one that asserts that the universe is a fake, a virtual-reality simulation in which we, too, must also be simulations.

Even before our investigation has begun, the Universalist can reject the absurd, accidental model (model 1) for enshrining randomness and ignoring order. The Universalist can also reject backward causation (model 6) for its bizarre idea that we are now creating the past universe. Suggesting that the present gives rise to the past – for example, that we are now creating the Second World War – is incredible both to philosophical reason and to common sense. Universalism focuses exclusively on *this* universe and as we saw on p76 is not interested in speculative other universes of an unproven multiverse (model 3). (The multiverse model shifts the problem of how the universe began into the multiverse – but the problem then becomes, how did the multiverse begin?) Our unique physical universe (model 2) is a disguised variation of model 3. It rejects the infinite and puts its trust in string theory and many dimensions, which are hypothetical and unevidenced. We have seen that a Theory of Everything has eluded theoreticians so far, and that there is no prospect of finding a purely physical solution as the temperatures required to reunite the four forces experimentally are so high. Universalism can reject the unique physical universe.

That leaves us focusing on a model created by an intelligent designer/God (model 4), which we shall consider shortly; and a bio-friendly life principle (model 5).

The Bio-friendly Model

I will now focus on the bio-friendly model (model 5). One of the principal reasons for rejecting the idea of a chance Big Bang and a haphazard universe condensing, cooling and in a random way putting up layered atmosphere is that the conditions left by the Big Bang turned out to be ideal for human life and any tweaking of these conditions would make human life impossible. In the words of Fred Hoyle, the universe looks like "a put-up job".[2]

Here I must distance Universalism from the anthropic principle while incorporating some of the data it cites. The anthropic principle arguably began with Alfred Russel Wallace, who proposed a theory of the origin of species through natural selection independently of Darwin, in 1903: "Such a vast and complex universe as that which we know exists around us, may have been absolutely required…in order to produce a world that should be precisely adapted in every detail for the orderly development culminating in man."[3] The principle was shaped by the US theoretical and observational cosmologist Robert Dicke in 1957: "The age of the universe 'now' is not random but conditioned by biological factors…. (Changes in the values of the fundamental constants of physics) would preclude the existence of man to consider the problem."[4]

The anthropic principle was formed when Dicke wrote a paper in the British journal *Nature* in 1961[5] taking up British physicist Paul Dirac's observation that many of the constants of Nature (the speed of light and the mass of an electron, for example), when multiplied and divided in a certain way, equalled the then suspected age of the universe (10 billion years), which he found an extraordinary coincidence.[6] The age of the universe is now 13.7 billion years, so the detail (as opposed to the principle) of Dirac's point looks unreliable today.

Dicke explained that the numbers were not a coincidence but necessary for humankind's existence, a statement of the weak anthropic principle.

The anthropic principle was fully stated in 1973 when Brandon Carter, the British theoretical astrophysicist, speaking in Krakow on Copernicus' 500th birthday, reacted to the Copernican principle which states that humans do not occupy a special position. Copernicus held that as the earth revolved round the sun it had become a tiny corner in a vast universe and humankind ceased to be special.

Carter articulated the anthropic principle and identified weak and strong versions.[7] Of the weak version he said, "We must be prepared to take account of the fact that our location in the universe is necessarily privileged to the extent of being compatible with our existence as observers."[8] "Location" for Carter was a space-time position. His version states that humans are privileged to be observers and to be located in such a good place in the universe. Of the strong version he said, "The universe (and hence the fundamental parameters on which it depends) must be such as to admit the creation of observers within it at some stage."[9] This version states that the parameters of physics must favour observers.

John Barrow and Frank Tipler restated the principle in different terms in their book *The Anthropic Cosmological Principle* in 1986. Of the weak principle they wrote, "The observed values of all physical and cosmological quantities are... restricted by the requirement that there exist sites where carbon-based life can evolve and by the requirements that the universe be old enough for it to have already done so."[10] They emphasized carbon-based life rather than observers. Of the strong principle they wrote, "The universe must have those properties which allow life to develop within it at some stage in its history."[11] They interpreted this last statement in three ways: "There exists one possible universe 'designed' with the goal of generating and sustaining 'observers'" (an interpretation welcomed by Intelligent Design); "Observers are necessary to bring the universe into being" (evoking John Archibald Wheeler's "participatory universe"); and "An ensemble of other different universes is necessary for the existence of our universe"[12] (an interpretation favouring the multiverse, which Universalists leave to one side along with the "participatory universe", model 6).

Of the weak version Penrose wrote, "The argument can be used to explain why the conditions happened to be just right for the existence of (intelligent) life on the earth at the present time. For if they were not just right, then we should not have found ourselves to be here now.... This principle was used very effectively by Brandon Carter and Robert Dicke to resolve an issue that had puzzled physicists for a good many years. The issue concerned various striking numerical relations that are observed to hold between the physical constants (the gravitational constant, the mass of the proton, the age of the universe, etc). A puzzling aspect of this was that some of the relations hold only at the present epoch in the earth's history, so we appear, coincidentally, to be living at a very special time."[13]

Followers of Carter saw the anthropic principle in terms of the multiverse and string theory (which, as we have seen, predicts a large number of possible universes). Barrow and Tipler saw its antecedents in the notions of Intelligent Design and in the philosophies of Fichte, Hegel, Bergson and Alfred North Whitehead, and also in the Omega point of Teilhard de Chardin.[14]

Since Barrow and Tipler there have been a number of books on the anthropic principle,[15] and the main point in them is that at the microscopic, macroscopic and cosmological levels the universe appears to have been incredibly fine-tuned from the moment it began for the production of intelligent life on earth. The existence of intelligent carbon-based life on earth depends on physical and cosmological quantities, and if one of these quantities is slightly altered the balance would be destroyed and life would not exist. In other words, the universe seems to have been designed with the teleological goal of generating and sustaining observers, and creation is man-focused. An even stronger version of the principle, the final principle, asserts that intelligent information-processing must come into existence in the universe and will then never die out. The anthropic principle can be broadly stated as holding that the laws of Nature must allow for the appearance of living beings capable of studying the laws of Nature.[16]

Universalism, focusing on this universe and not other possible universes, has a different angle on so-called anthropic data. It sees the laws of physics and the initial cosmological conditions as providing a *context* for life that seems orderly rather than fortuitous and random. It would be truer to say that Universalism affirms a Contextual Principle, that the context for life is a system that is too precise to have formed by accident or chance or to be random and which is an Order Principle, a principle of order.

There are obvious indications of the orderly system in the rhythm of the days, months and years. The regular rising and setting of the sun, moon and stars suggest order, along with the regularity of day being following by night, the tides and the seasons. Looking at these regularities we ask ourselves: the laws of Nature seem so orderly, could they have come into being by random accident? It seems inconceivable that they could.

But the system is more extensive than the regularities we are aware of each day suggest.

The main conditions of this system can be extracted from work in classical cosmology, physics and astrophysics, quantum mechanics and biochemistry. There are at least 10 conditions that are just right for human life. I now present them, not in their particular disciplines, although the first 12 happen to come from cosmology, but arranged in relation to the importance and fundamental nature of their contribution to the structure of the universe. I broadly proceed from macrocosm to microcosm, but break up different emphases and avoid slavishly lumping together all the different treatments of gravity and protons while seeking to illuminate through juxtaposition. Thus, a force mentioned in one condition has a carry-over effect to the next condition, assisting understanding. These 40 conditions are:

(1) The Big Bang. At the moment of creation, through a combination of ideal circumstances the Big Bang gave the universe perfect density, without which the cosmos would have recollapsed the moment it began to inflate; the right amount of ripples and temperature variations, without which there would have been no stars, light or warmth; and perfect symmetry of matter in relation to antimatter, without which no matter would have come into being.[17] All these conditions occurred simultaneously within the first second of creation and the beginning of space-time, and without this combination of ideal circumstances there could have been no life.

(2) Gravity and rest-mass energy. The Big Bang contained imprints of the seeds of clusters of galaxies, galaxies, stars and planets, the structure of the universe. They are held together by gravity, which is a proportion of their rest-mass energy ($E=mc^2$), the ratio between the two energies – gravity and rest-energy – (known as Q) being 1/100,000 or 10^{-5}. This is a small number, meaning that gravity is weak in the fabric of the universe. The pull of the earth's gravity is in fact 1,000 billion times weaker than that of a neutron star. If Q were smaller than 10^{-5} and other conditions remained as they are, the universe would be inert and structureless, and the formation of stars would be slow and inefficient for material would be blown out of galaxies rather than being recycled into new stars and planets. If Q were larger than 10^{-5}, gas would not condense into structures bound by gravity and the universe would remain dark. The ripples would be turbulent waves and galaxies would condense early and collapse into black holes (assuming they exist). Those

galaxies that survived would be more tightly bound with stars too near. Life could not have evolved if Q were not 10^{-5}, as it is.[18]

(3) Dimensions. The number of spatial dimensions, known as D, are three. If D were two or four, life could not exist. In a three-dimensional world, forces such as gravity and electricity obey an inverse-square law, which means that a mass or charge is four times weaker if one goes twice as far away, making the orbits of planets stable as gravity (which tends to pull them inward) is balanced by the centrifugal effects of their motion (which weaken gravity). Time, a fourth dimension, is different from the spatial dimensions because the "arrow of time" moves forward towards the future, and strings events out so they do not all happen at once. Two spatial dimensions would send speeded-up planets spiralling out into darkness. In a two-dimensional world, animals would have to climb over each other to pass each other, there could be no circulation of the blood and digestive waste would have to come out of mouths. Four spatial dimensions would give an inverse-cube law, sending slowed-down planets into or away from the sun so life would be burned up or frozen. The sun would be unstable and either collapse into a black hole or fall apart – and would be unable to provide heat and light to earth. Electrical forces would behave like gravitational forces and pull electrons from atoms or make them spiral into the nucleus, so atoms would not exist in a recognisable form as there would be no stable orbits for electrons. The conditions on earth are just right. There is a problem with any universe of more than three spatial dimensions. Space-time has to have three space dimensions that are not curled up small for life to be able to exist.[19]

(4) The rate of expansion. The tension between gravity and expansion energy of the material in the universe (galaxies, gas and hypothetical dark matter, the invisible something which seems to exert a gravitational pull on galaxies and stars) is just right. The initial expansion speed after the Big Bang was just right so galaxies and stars could form without the universe collapsing. If the universe had started expanding too slowly, it would have come to a halt and collapsed back by now, and galaxies would be falling towards each other. The universe expanded neither too fast nor too slow but at just the right rate to allow elements to be cooked in stars. The rate of expansion in the universe seems fine-tuned to avoid too fast an expansion, in which case no galaxies could form, or too slow an expansion, in which case the universe would have collapsed before evolution. The initial conditions of the universe were set 10^{-43} seconds after the Big Bang. At 10^{-43} seconds after the Big Bang the universe was expanding at a speed rate

of H_o, the Hubble constant,[20] with a total density close to the critical value between recollapse and expansion forever. Even a decrease of one part in a hundred thousand million million when the universe was 10^{10}K 10 seconds after the Big Bang would have resulted in the universe's collapse long ago. An increase would have prevented the galaxies from condensing out of expanding matter. As to the future, if our expansion does not continue forever the burning-up of the sun in about 5 billion years' time will cause the unwarmed earth to die with it, and the Andromeda galaxy would crash into the Milky Way about the same time.[21]

(5) Density. The actual density of the universe (the quantity of mass per cubic metre of the matter in the universe) is linked to the scale of the universe, which is known with only 10-20 per cent precision because of the Hubble constant (which is assumed to be 65 kms per second per megaparsec). The fate of the universe depends on the ratio between the actual density of the universe and the "critical density", which is known as Ω (omega) and is just right for life. The critical density is about 6 hydrogen atoms per cubic metre (close to a vacuum). If Ω had exceeded 1, the gravitational pull would have defeated the expansion unless another force had intervened and the universe would already have collapsed. If all the atoms in the universe were spread evenly, there would be 0.2 atoms per cubic metre, 30 times less than the 6 hydrogen atoms per cubic metre needed for gravity to bring cosmic expansion to a halt. If actual density is lower than critical value of 1, the universe will expand forever. Ω seems to be 1/25 or 0.04, but hypothetical dark matter has been found to exert a gravitational pull on stars and galaxies which may fill 23 per cent of the universe. After an adjustment for dark matter and exotic particles such as neutrinos, Ω can be put at 0.3 of the density needed to halt expansion. For Ω to remain between 0.1 and 10.0 today after 13.7 billion years and after the universe has expanded to 10^{30} times its original size, Ω must have had an initial value of $1 - 10^{-59}$ and $1 + 10^{-59}$. The kinetic and gravitational energies had to be initially equal to one part in 10^{59}. What physical process could achieve such a fine balance in the first 10^{43} seconds?[22] One second after the Big Bang Ω must have been close to unity – almost exactly 1, differing from unity by one part in 10^{15} (a million billion) – for the universe to have expanded for 13.7 billion years with a value of Ω, at 0.3 of the density needed to halt expansion if our estimate of the quantity of undetectable dark matter in the universe is correct, and to be still close to unity now. It could be that there is enough undetected dark matter for Ω to equal unity or 1? But even if Ω is slightly less than 1, gravity will not halt the expansion of the universe, which can be expected to last forever. The kinetic and

gravitational energies are not exactly equal today – why have they become unbalanced?[23]

(6) Anti-gravity and acceleration. The expansion of the universe is accelerating. A new force, cosmic "anti-gravity" known as λ (lambda) was hypothesized in 1998 after observations through telescopes with 10-metre mirrors confirmed the acceleration. There was no deceleration as one would expect after a long period of acceleration and the speeding-up of the expansion could only be explained by postulating a force that causes cosmic repulsion and balances gravity even within a vacuum, which resembled Einstein's cosmological constant of 1917, which he called λ. Einstein abandoned his static universe in 1929 when he accepted Hubble's view that our galaxy was one of many and was receding from us, and he abandoned λ at the same time. λ is thought to control the expansion of the universe by dominating gravity. λ is near zero, and so weak it can only compete with the weakened gravity of intergalactic space. If λ were five times stronger it would have destroyed gravity, stopped galaxies and stars from forming and there would have been no evolution. The consequences would have been catastrophic. If λ were smaller it would not have made a great deal of difference. Life required λ not to be too large. It seems likely that if λ is not zero then expansion will continue indefinitely out into extragalactic space beyond our universe's horizon, the infinity that lies beyond our finite universe.[24]

(7) Gravitational-electromagnetic ratio and constant. Gravity, the force that holds the moon and planets to their courses, binds galaxies and pulls stars into black holes (if they exist), and may pull the Andromeda galaxy so it crashes onto the earth, is nevertheless weak, tiny compared with the strength of the electrical forces. The structure of stars and stellar balance are dependent on the exact ratio of gravitational to electromagnetic forces, which allow the sun, a middling star, to exist. The sun in turn heats and lights the earth and makes life possible.[25] Gravity is weaker beside the electrical forces governing the universe by the number N, which measures the strength of the electrical forces divided by the force of gravity between them. N is 10^{36}. Another way of measuring the weakness of gravity is by measuring the gravitational fine-structure constant (on a scale in terms of the mass and electrical charge of a proton). The gravitational fine-structure constant is 10^{-40} while the electric fine-structure constant is less than 10^{-2} (10^{-38} times stronger than the gravitational fine-structure constant). The universe is therefore large and diffuse (billions of light years in extent) and provides enough time for elements inside stars to be cooked and for complexity

to evolve around one star. If N were smaller, for example 10^{30}, and if gravity were stronger, say with a constant of 10^{-30} instead of 10^{-40} (or with a gravitational-electromagnetic ratio of 10^{-26} rather than 10^{-36}), the universe would be miniature and short-lived as objects would be smaller and speeded up. Stars would be packed more closely together and living organisms would have no more than about 10^{20} atoms rather than 10^{28}. The mass of each star would be 10^{-15} (1 millionth of a billionth) of the sun's mass and in this smaller universe a year would have about 2 million days.[26] If N were less than 10^{12} there would be no human beings. If N were large enough for gravity to compete with the electrical forces, stars would be around a billion times less massive and would become a million times faster. A star would only live 10,000 years and there would be no time for evolution. No creature would be larger than an insect. As all stars would be small – white dwarfs – and as carbon and the heavy elements are only produced in massive stars, the planets and earth could not exist. If N were larger still, all stars would be large – blue giants – and our sun would be too small to exist and so there would be no earth or life. The law of gravity is an inverse-square law described by Newton in the 1680s, in which the force between two objects is proportional to 1 over the square of the distance between them. If the law of gravity were an inverse-cube the orbits of the planets would be unstable, and a planet moving towards the sun would fall inwards permanently while those that moved away from the sun would recede forever, making life impossible to sustain. The ratio of gravitational to electromagnetic forces and their fine-structure constants allowed by the structure of the universe are just right to allow the sun and earth to exist in conditions conducive to evolving life.[27]

(8) Nuclear binding. Atomic nuclei, atoms, the power from the sun and how stars transmute hydrogen into the atoms of the periodic table (which places 92 atoms in relation to the number of their protons) are all controlled by the binding of atomic nuclei, known as E (epsilon), whose value is 0.007. Thus the hydrogen gas that powers the sun in its core converts 0.007 of its mass into energy when it fuses into helium. If E were 0.006, hydrogen would be less efficient and the sun and stars would not live as long. (We have seen that the sun has at least 5 billion years to go.) The nuclear glue would be weaker, hydrogen would not have fused into deuterium, the first step to forming helium, and in a universe of hydrogen no life could take place. If E were 0.008 no hydrogen would have survived from the Big Bang. For the hydrogen would have transformed itself into deuterium and helium too readily and would have disappeared,

leaving no hydrogen to fuel the stars and no chance of life in the universe. Either way, at 0.006 or 0.008 there would have been no carbon-based biosphere and humans could not exist. A universe with complex chemistry requires E to be between 0.006 and 0.008.[28]

(9) The heat of the universe. The NASA WMAP (Wilkinson Microwave Anisotropy Probe) satellite which mapped the sky in 2003 using heat, produced a thermal map of the universe about 380,000 years after the Big Bang, about 13.32 billion years ago, a snapshot of what the universe was like less than 0.003 per cent of its present age. It showed temperature variations. The over-dense regions are the "seeds" around which clusters of galaxies formed. (See (1).) Without the temperature variations no clusters of galaxies, galaxies, stars or planets would have come into existence, and there would have been no human life.[29]

(10) Mathematical laws. Mathematical laws underpin the atoms, stars, galaxies and people in the universe. The laws of physics are associated with the principles of symmetry. Einstein was amazed that the laws of physics applied to the earth and remote galaxies, and can be understood by human minds. The laws are bio-friendly for without them there would be no human life. They are just right for life.[30]

(11) Flatness. We saw in (4) that in a flat universe expansion and gravitational pull are balanced so eventually expansion stops without collapsing back, making the relation between distance and angles the same as in Euclidean space. Matter is contained in the universe in such a density that it expands at the right speed to allow carbon-based life forms and human life to form. We have seen that our universe can be open (with continued expansion forever) or closed (with expansion until it collapses back) or balanced between these two and flat, with a critical density of matter in space of $\Omega = 1$, a universe which expands at an ever-slowing rate and comes to a stop at zero velocity. We saw in (5) that today Ω has a value of 0.3 (after an adjustment for dark matter), and the solar system is expanding so fast that eventually it might fly apart. Today the density of the universe is within a factor of ten (between one-tenth and ten times) of the critical value corresponding to a flat universe. As the universe is now, 13.7 billion years later, only a factor of ten away from flatness, one second after the Big Bang the universe must have been flat to within a factor of 10^{15} (one part in a million billion) and so the amount by which the density differed from the critical value was 0.000000000000001. Between time zero (the Big Bang) and 10^{-43} seconds after the Big Bang when the gravitational force separated, soon after which at 10^{-36} seconds after the Big Bang the universe

began expanding, the flatness of the universe was precise to within one part in 10^{60} – which precisely gives all the conditions at the very outset that allow stars, galaxies and human life to form.[31]

(12) Carbon. The universe was and is almost precisely flat and the amount of baryonic matter in the universe (protons and neutrons, 25 per cent of which are hydrogen and helium) is close to the amount of hypothetical dark matter, which exceeds it by 10 to 1. In other words, the ratio of invisible dark matter to baryons is 10:1. This allows for the existence of carbon and heavier elements, without which human life would be impossible. Carbon (with 6 protons and 6 neutrons in its nucleus) is made by combining three helium nuclei. This happens after two helium nuclei combine to form beryllium – which is unstable until a third helium nucleus can be included, which may never happen. Fred Hoyle realised that carbon is made inside stars. His reasoning was in effect, "Since we exist, carbon must have an energy level at 7.65 MeV," or an upward shift of 4 per cent above the combined energy of helium-4 and beryllium-8, a bio-friendly prediction. When the experiments were carried out, carbon was found to have an extra 4 per cent. The ratio of strong force to electromagnetism makes it possible for carbon to achieve an excited state of 7.65 MeV. The carbon nucleus was found to have a "resonance" with a special energy that advances the prospects of beryllium capturing another helium nucleus. Thus carbon-12 has an energy 4 per cent above the combined energy of helium-4 and beryllium-8 and allows it to fuse with them and produce carbon nuclei. The combination is just right for carbon-12 to form carbon nuclei, but it is wrong for oxygen-16, which has 1 per cent less energy than helium-4 plus carbon-12. A downward shift of 4 per cent to the combined energy of helium-4 and beryllium-8 would deplete the amount of carbon that could be made, thus reducing the chance of human life happening.[32]

(13) Molecules of life inside stars. Besides carbon, oxygen, nitrogen and phosphorus, the essential molecules of life, were manufactured by thermonuclear processes inside stars after the Big Bang. This is because as we have seen gravity, expansion, density and flatness combine to make stars just right for the production of these materials. Exploding supernovae blew these molecules out into galactic space, where they helped develop life forms including evolving human beings, who are made of such stardust.[33]

(14) Temperature. About three minutes after the Big Bang the universe had cooled from 10^{32}K to 1 billion degrees (10^9K). If the universe had remained at this temperature for a long time or if nuclear reactions had happened faster, all

the atoms would have been processed into iron. The expansion was fast enough to quench nuclear reactions before they converted more than 23 per cent of the hydrogen into helium. This meant that the atoms did not become iron. The Big-Bang theory predicts that there should be temperature fluctuations at 10^{-5} seconds to seed galaxy formation, and in 1992 the COBE satellite found such fluctuations.[34]

(15) Weak and electromagnetic forces. The weak nuclear force that governs radioactivity, radioactive decay and neutrinos has been found to be linked to the electric and magnetic forces which had been unified by Maxwell. The linking of weak and electromagnetic forces was done by the US physicists Sheldon Glashow and Steven Weinberg, by the Dutch theoretical physicist Gerard 't Hooft and Pakistani theoretical physicist Abdus Salam, who predicted new particles that were found in CERN Laboratory simulations and won the 1979 Nobel prize for their work. The electroweak and strong forces were united at the beginning of the universe and separated only when the universe cooled below 10^{15}K when it was 10^{-12} seconds old. The weak and electromagnetic forces separated from the electroweak forces at 10^{-4} seconds after the Big Bang. The unification involving the weak force lasted long enough to give the conditions in which the galaxies, stars and life later evolved.[35]

(16) The strong nuclear force. The strong nuclear force, the dominant force in the microworld, holds the protons in helium and heavy nuclei together in fusion. This was strong enough to prolong the warmth of the sun for without the strong force's nuclear energy the sun would deflate within 10 million years. A 1-per-cent stronger strong nuclear force would burn all carbon into oxygen. A 2-per-cent stronger strong nuclear force would have prevented the formation of protons and yielded a universe without atoms. A 3-per-cent stronger strong nuclear force in relation to the electromagnetic force would allow none of the known chemical elements to form. A 1-per-cent weaker strong nuclear force would make carbon atoms unstable. Hydrogen would be the only stable element and no other elements could exist, including carbon and oxygen. There would therefore be no carbon-based life. Other tiny variations in the strong nuclear force's relation to the electromagnetic force would have given a universe that was 100 per cent helium or one in which no supernovae explosions could occur and therefore no chemicals could be spread from the stars to create life. A 5-per-cent weaker strong nuclear force would have unbound deuteron and would have given a universe without stars. The strong nuclear force acts at short range and is not effective on nuclei heavier

than iron, which become less tightly fused, not more so.[36] The reunion of the strong force and the electromagnetic and weak forces (a GUT or Grand Unified Theory) must in theory be possible as they were originally united at the beginning of the universe until the strong force separated at 10^{28} and the electroweak union separated 10^{-12} seconds after the Big Bang at a temperature of 10^{15}K, a temperature which the CERN Laboratory can reach. However, if as has been thought unification does not happen until there is a temperature of 10^{28}K degrees (the temperature for the first 10^{-35} seconds after the Big Bang), a million million times higher than experiments such as those at the CERN Laboratory can reach, 10^{17}, then at the present level of technology unification would require an accelerator bigger than our solar system. The reunification of these three forces may therefore never be replicated on earth at the present level of technology, and therefore may never be proved outside mathematical theory. The unification of the three forces lasted long enough after the Big Bang to determine the conditions in which the galaxies, stars and life evolved.[37]

(17) Gravitons. Einstein's theory of curved-space time required the existence of gravity waves, "ripples" in space-time's fabric which, he predicted, would be made of – it has to be said, so far undetected – gravitons (particles with zero mass and spin 2). According to hypothetical, unevidenced string theory, which details strings of particles and is an aspect of the multiverse we have left to one side, there could be lines of strings of gravitons. String theory says that every particle associated with a force (such as a graviton) must have a partner (a gravitino). Physicists have conjectured that supersymmetry plus string theory gives superstrings, from which a Theory of Everything may come. The reunion of gravity with the other three forces must in theory be possible as all four were unified at the beginning of the universe, gravity separating at 10^{-43} seconds, the shortest time interval that can be measured, when the temperature was 10^{32}K. If 10^{28}K can only be reached at the present level of technology by using an accelerator bigger than the solar system (see (16)), 10^{32}K will require one proportionately larger, and at the present level of technology this can never be replicated on earth and therefore can never be proved outside mathematical theory. The unification, though the minutest part of a split second in duration, lasted long enough to determine the conditions in which the galaxies, stars and life evolved. If gravitons exist, uniting gravity and quantum gravity, they have provided just the right conditions for the evolution of galaxies, stars and life.[38]

(18) Neutrinos. Some stars end by exploding violently as supernovae,

spreading carbon-based heavier elements in clouds of material and gas that reached earth. In such a star, burning hydrogen is converted into helium which in turn burns to produce carbon. When the hydrogen is exhausted helium takes over. When helium is exhausted, the star collapses and cools into a white dwarf star which, when silicon burning is completed, is an iron ball that may have the mass of 25 suns within the size of our sun. The inner regions of the star are squeezed inwards, electrons and protons are forced to merge, forming neutrinos, and the star is transformed into a neutron star. The star bounces back, the squeeze taken off it, and releases a flood of neutrinos produced by the collapsing-inwards in a shock wave which blows the star apart, releasing the heavy elements needed for life. The shock wave alone cannot do this. The neutrinos achieve the explosion by weakly interacting with baryons (protons and neutrons). If the weak interaction were weaker, the neutrinos would escape from the star without pushing apart the outer layers of the star. If the weak interaction were stronger, the neutrinos would be caught up in reactions in the core of the star and would not reach the region of the outer layers or escape. The weak interaction is just right for neutrinos to drive large quantities of gas filled with heavy elements into space whence they could reach the planets and earth so that human life could take place.[39]

(19) Weak force and helium. The same weak interacting force (see (18)) determined how much hydrogen was processed into helium during the Big Bang. The weak nuclear force is 10^{28} times stronger than gravity.[40] If the weak force had been weaker, all the hydrogen in the universe would have been turned to helium, making water impossible. As all the baryons (protons and neutrons) would have been converted into helium in the Big Bang, the neutrinos could not explode stars and scatter material (see (18)), and as all the stars would have been made of helium they would have burnt out more quickly. If the weak nuclear force had been slightly stronger, no helium would have been produced and all the stars would have been made of hydrogen. Either way life could not take place in its present form. The weak nuclear force is of just the right weakness to prevent all the original hydrogen from being converted into helium and to allow exploding supernovae.[41]

(20) The nuclear forces, di-protons and stars. The nuclear force of attraction between two protons, which creates a di-proton, an atomic nucleus, is overwhelmed by the stronger electrical force of repulsion between two protons, which wipes them out unless uncharged neutrons hold them in balance. As a result, stars can gain energy by fusing protons and neutrons into such balanced nuclei. If the nuclear

forces were slightly stronger and could fuse pairs of protons into di-protons, then stars would not be able to fuse protons and neutrons and would be different. If the nuclear forces were slightly weaker, the universe would consist of hydrogen protons and electrons, and there would be no nuclei for stars to fuse.[42]

(21) Proton electron ratio. The ratio of the mass of a proton to an electron, 1836.15, nearly 2000, could not be 2 or 2 million because these figures would produce a physics, chemistry and biology incompatible with life.[43]

(22) Vacuum energy. During inflation, around 10^{-38} seconds after the Big Bang, the energy scale of the vacuum must have had an energy index of $\omega = 16$, whereas the energy in the universe today is $\omega = -11.5$, three trillion quadrillion (1 quadrillion being 10^{27}) times smaller than that of inflation. The unevidenced dark energy today is therefore not the same today as the dark energy that scientists assume led to inflation, and is just right for life.[44]

(23) The size of the universe. As we saw on p68 there are at least 10^{23} stars in the observable universe, as many as the grains of sand on earth, and 10 million billion stars will be similar to our sun. There are 10^{24} planets; 10^{17} hypothetical black holes; 10^{11} galaxies; 10^{7} clusters.[45] There are 10^{20} planets suitable for life forms[46] of which one, discovered in 2007, orbits the star 41 light years from earth known as 55 Cancri,[47] and another, also discovered in 2007, orbits a small red star 20.5 light years from earth known as Gliese 581. (Currently 228 extrasolar planets are known to orbit stars, some of which may be able to sustain life.) [48] It may seem a hugely wasteful scheme to have so many suns in the universe with life happening only on the earth. Yet those billions of potential homes for life may have to exist so that life can happen on earth. The linear size of the universe may be about 13.7 billion light years (a figure based on the beginning of the universe 13.7 billion years ago) from the base point to the rim of our expanding "shuttlecock", assuming inflation is only going in one direction. We can in theory see as far as light has had time to travel during the last 13.7 billion years although we cannot see beyond the observable horizon, which is taken to be about 10 billion light years, the farthest limit current telescopes can see. It takes a few billion years for a galaxy to form, for the first stars in it to process hydrogen and helium into heavy elements and for neutrinos to explode stars to scatter materials into space. It takes more time for new stars and planets to be formed from the débris and for life to evolve.[49] If the universe had just one galaxy the size of our Milky Way containing 200-400 billion stars, it would have expanded for only a month, according to mathematical formulae, and would not

have been able to produce life.[50] The process from start to finish cannot be completed short of 13.7 billion years as it requires a good supply of débris-providers, and the billions of stars and billions of light years may be necessary if life is to be produced on just one star, our planet.

(24) The emptiness of the universe. The universe has vast reaches of empty space between stars. If it were smaller and lacked this empty space there would be collisions between stars. Near misses would detach planets from orbit around their suns and pull them into interstellar space where they would cool to hundreds of degrees below zero. Either way, the universe would be uninhabitable. The empty spaces in space help evolving life.[51]

(25) Proton-electron charges. The charges of a proton and an electron have been measured and found to be equal and opposite, in balance, every plus cancelled by a minus leaving the sum total as zero. If an electron had more of a charge than a proton, the charges would no longer total zero and every object in the universe would be charged negatively and repel other objects. A table would repel the lamp on it, the tree would repel a nearby parked car, and so on. A 1-per-cent difference between the charge of an electron and the charge of a proton would rip the arms and legs off a human or animal body and send them hurtling into space. A difference of 1 part in 100 billion (10^{11}) would cause stones and people to fly apart. If there had been an imbalance between the charges of a proton and an electron, the cosmos would have been filled with a uniform substance like air, every object in the universe – human bodies, trees, planets, suns – would explode violently. Larger structures like the sun require a perfect balance of one part in a trillion or billion billion (10^{12}). The equality of the charges of protons and electrons is exactly right for human life.[52]

(26) Electron-neutron mass ratio. The mass of an electron (9.108×10^{-28} gms)[53] is 1/1837th of the mass of a neutron. The difference between the two masses, known as B, ensures that nuclei have well-defined, relatively invariant locations. The small difference of B enables well-ordered structures to take place. If atoms stray from their locations under heat, then materials melt. If molecules stray, their materials will dissociate. The size of the B difference guarantees that there are large chain molecules of the right size and kinds to make biological phenomena – organisms – possible. Any change to the B difference would always change the size and length of the rings in the DNA double helix, which would then not be able to replicate itself, making life impossible.[54]

(27) Quantization. Quantum theory claims to explain the basis for stability in

Nature. The quantization principle Niels Bohr proposed in 1913 is essential for the existence and stability of atomic systems. It restricted the energy of orbiting electrons to certain discrete (individually distinct) values which are multiples of a universal energy quantum fixed by Planck's constant. If an electron is added to a proton there is only one orbital radius available to it in quantum theory and so all hydrogen atoms are identical. Environmental disturbances do not upset the structure of atoms as an entire quantum of energy must be added before the electrons' orbit is altered. If this were not the case and the rules governing non-quantum atoms applied to Nature, electrons could possess all possible energies and reside in any orbital radius if there were enough velocity to establish an equilibrium between centrifugal and Coulomb forces. (Coulomb forces are named after the French physicist Charles-Augustin de Coulomb, who in 1785 described the attraction or repulsion of particles or objects because of their electrical charge.) There would then be a continuous buffeting of electrons by photons which would cause a steady change in electron orbit. The ensuing instability would make for an unstable Nature in which structures would be in permanent change, and there would be no stable conditions in which life could evolve. The functioning of quanta help make Nature just right for life.[55]

(28) Freedom of particles. We have seen that from around 10^{-35} seconds after the Big Bang leptons and quarks were formed. At high energy elementary particle interactions became asymptotically free (like a line approaching but not meeting a curve). Had this not been the case, the early universe would have been in an intractable, strongly interacting state in which interaction times would be too short to obey gas laws, and in primordial nucleosynthesis the number of neutrino species would not have been smaller than or equal to 4. The universe would consequently have been burnt to helium during the primordial nucleosynthesis and it would now have no hydrogen, water – or life.[56]

(29) A neutron-proton mass differential. If the difference in mass between a proton and a neutron were not exactly as it is (about twice the mass of an electron) then all neutrons would have been photons and vice versa. A neutron outweighs a proton by a tenth of 1 per cent even though both are constructed of smaller particles, quarks. (In fact, a neutron outweighs a proton and an electron combined.) As a result, protons are stable and neutrons decay. If a proton outweighed a neutron, the position would be reversed: neutrons would be stable and protons would decay into neutrons. As hydrogen's atomic nucleus is just a single proton (on whose stability the element of hydrogen depends), hydrogen would not exist and

in consequence water (H_2O) would not exist. The sun is made of hydrogen, and the sun too would have ceased to exist after 100 years as its protons would have decayed like neutrons and it would have slowly run out of fuel and gone out before life could form on earth.[57] The proton-based structures of all stars and planets would decay into neutrons and collapse into neutron stars or black holes. Indeed, the protons (75 per cent of the universe) would have decayed into neutrons, in which case no atoms would have formed and no life could have taken place. In fact, the neutron-proton mass differential is just right, and the "coincidence" that allows protons to take part in nuclear reactions in the early universe prevents them from decaying as a result of weak interactions.[58]

(30) Sunlight and chlorophyll. The sun radiates ultraviolet light, which may have caused chemical reactions in the primordial soup and produced the first cell, although this has not been proved. Sunlight (along with carbon dioxide and water) is trapped in the pigment chlorophyll, the molecule that accomplishes photosynthesis in plants. Chlorophyll in tiny discoid chloroplasts is converted into the chemical nutritional energy of specific organic compounds (sugars, starch, amino acids and lipids), producing oxygen as waste. Chlorophyll is programmed to absorb light (photons) of a heated brilliant yellow colour and energy and so can absorb sunlight which is of the right temperature. There is a matching between the sun's temperature and chlorophyll's absorption. It acts like a TV receiver tuned to a transmitter. If there were no sun, but only nearby cooler stars, the chlorophyll could not receive the duller-coloured light from stars made of hydrogen at a lower temperature and no other molecule could replace it to receive such starlight. The absorption of light is accomplished by the excitation of electrons in molecules to higher energy states. If the earth had had no sun nearby and had looked to a cooler star it would have had oceans, rainclouds and warm breezes but no plants and therefore no animals. If the earth had looked to a hotter star, the heat would have torn molecules apart and dissociated them. A high-temperature star would have emitted dangerous ultraviolet radiation, and still hotter stars would have emitted X-rays and more dangerous radiation. In fact, the sun-chlorophyll balance is just right for life.[59]

(31) Bonding elements. In biochemistry the essential elements which make up 1 per cent of all living organisms are oxygen, carbon, nitrogen, hydrogen, phosphorus and sulphur. These form multiple bonds. Carbon is especially versatile at forming bonds. Nitrogen is essential for life as nearly 80 per cent of the earth's atmosphere consists of nitrogen whose compounds comprise the

building blocks of living organisms. Phosphorus and sulphur have additional bond-forming ability. They form more open and usually weaker, wide-spaced bonds than nitrogen and oxygen and, adding electron pairs under certain conditions, induce an instability that promotes exchange reactions and plays a role in the transfer of energy between groups of molecules. Sulphur forms three types of molecules with high-energy bonds for biochemical reactions and biological compounds, and therefore for life. These elements act as agents that transfer energy between groups of molecules in chemical reactions. They create three classes of sulphur compounds and one class of phosphorus compounds which together form all the known categories of biological high-energy compounds, and are just right. If phosphorus and sulphur had stronger bonds then biochemical compounds could not happen in the right way and the conditions for life would not be right.[60]

(32) Atmosphere. The earth's atmosphere requires gravity to pull strongly enough to prevent if from evaporating into space, as would happen on the moon if it had an atmosphere, for gravity is weaker there. The atmosphere requires the presence of water, and the earth's collision with an asteroid gave it a shield from solar wind and created the moon which, by gravity, pulls the oceans towards it, creating tides that assist evaporation into the atmosphere and aid life. The earth must not be too hot like Venus or too cold, or too near or too far from a long-living stable star such as the sun. Otherwise there would be no life. The earth is made of iron and during volcanic eruptions its molten core releases carbon dioxide, which regulates the earth's temperature. The earth's orbit must be stable. The atmosphere must be rich in oxygen, which means it must be transformed by bacteria in its early history.[61] The present atmosphere is 21-per-cent oxygen, the upper limit for safety of life as the risk of forest fires being started by lightning increases by 70 per cent for every 1-per-cent increase in the atmosphere's oxygen. If the oxygen were 25 per cent, all land vegetation would be at risk from being incinerated from lightning-strike fires. Oxygen forms ozone in the upper atmosphere, which screens out far ultraviolet radiation which would otherwise destroy carbon-based life. Until plants produced sufficient oxygen to sustain creatures which crawled out of oceans, life was only possible in water, which filters out far-ultraviolet radiation. The loss of hydrogen gas from the earth's atmosphere should be counterbalanced by an influx of hydrogen from the sun. As nitrogen forms nearly 80 per cent of the earth's atmosphere and the nitrogen in the atmosphere can be converted into ammonia, the source of nitrogen in biological compounds, nitrogen is essential for the building blocks of amino acids and

therefore for the chemistry of life.[62] All these conditions can be found on earth, which make it just right for an atmosphere with air to allow life to form and survive.

(33) Water and rainfall. The earth lies at the correct distance from the sun for liquid water to exist. Water is lighter in its solid form than in its liquid form due to properties in the hydrogen atom: ice floats. If this were not so the oceans would freeze from the bottom up and the earth would be covered with ice. Moisture-laden air is blown on to land from the oceans. Desalinated rain is deposited on ocean-facing slopes of hills and mountains. As air moves on over the land it receives moisture by transpiration or evaporation of water from the aerial parts of plants and trees, by evaporation from lakes, rivers, soil and dew on plants, and by clouds of water vapour. Water prevents the earth's climate from being too severe and is indispensable to every organism. It plays a vital role in photosynthesis and is the source of all oxygen in the atmosphere.[63] The earth has just the right conditions to produce adequate rainfall in temperate zones and nurture life.

(34) Particles-antiparticles. There is a symmetry between the subatomic particles of matter and the subatomic antiparticles of antimatter (the existence of which was first proposed by Dirac in 1928 and which have since been detected in experiments). There are 10^{78} atoms in the matter of the observable universe, which have a life span of about 10^{35} years; but there are not as many antiatoms. There are 10^{87} neutrinos in the universe; 10^{87} photons (radiation particles); and 10^{78} protons[64] If particles and antiparticles had been mixed in equal numbers shortly after the Big Bang all protons would have been annihilated with antiprotons during the dense early stages and the universe would have been full of radiation and dark matter but no atoms, stars or galaxies, and life could not have taken place. According to the Russian developer of the H-bomb, Andrei Sakharov, during the cooling after the Big Bang a small asymmetry favoured particles over antiparticles as some particles could not find an antiparticle to be annihilated with. In 1964 Cronin and Fitch confirmed this view by finding that particles and antiparticles decayed at slightly different rates. Nevertheless there is still a strong symmetry between particles and antiparticles: a positron (a positive electron announced by Dirac and identified from cloud-chamber studies of cosmic rays by Anderson in 1932), is the antiparticle and opposite of an electron and carries a positive instead of a negative charge. The spinning antineutron, the antiparticle of a spinning neutron, has an opposite magnetic polarity. The antineutrino, which spins clockwise if viewed from behind, is the antiparticle of the neutrino, which spins anti-clockwise if viewed from behind

and travels at the speed of light. This symmetry suggests an underlying order, and the asymmetry allowed protons to survive and therefore the universe, and life, to form. Without the existing symmetry and asymmetry of particles and antiparticles, the universe would have been atomless.[65]

(35) Bosons. Bosons, virtual subatomic particles considered carriers of basic physical forces, named after the Indian Bengali physicist Satyendra Bose, include gravitons (carriers of gravity), gluons (which transmit the colour force that holds quarks together), photons (which transmit the electromagnetic force) and intermediate vector bosons or weakons (the W and Z particles). If they exist, they would unite particles and forces. Each has an integral spin (an angular momentum measured in quantum-mechanical units of 0,1 etc.). Electrons, protons, neutrons, muons and neutrinos spin on their axis at a rate that is described as ½. Spins are multiples of ½ and other particles have spins of 1, 1½ or 2. Massive or heavy particles are knows as hadrons, light ones as leptons. Each is massless except for the intermediate vector bosons which have a very large mass 80/90 times that of the mass of a proton. For one particle to influence another through the weak force, bosons have to be created. As a boson's mass is huge in relation to most particles, it appears out of the vacuum (which quantum uncertainty allows), travels a short distance to the other particle, lives a short time and then returns to the quantum vacuum. When there is sufficient energy for bosons to be about, the electromagnetic and weak forces become indistinguishable. During the Big Bang there was perfect symmetry in which all forces and particles were indistinguishable, and as cooling began as we have seen gravity broke away and the symmetry was split into two smaller symmetries, which split again, and yet again, giving the variety of forces and particles we have. With the symmetry between electromagnetic and weak forces broken, bosons ceased to be real particles and became virtual particles. The CERN Laboratory's collider sought the Higgs boson, a theoretical particle that may help explain why gravity is weak in contrast to the other three forces. Finding the Higgs boson may be an important step towards the Grand Unified Theory, which seeks to reunify three of the four fundamental forces, and perhaps towards a Theory of Everything, which seeks to reunify all four fundamental forces without replicating the temperature of 10^{32}K when the gravitational force separated 10^{-43} seconds after the Big Bang. Theory suggests the existence of a supersymmetry in which every boson has a fermion partner (named after the Italian physicist Enrico Fermi, such as gravitinos, photinos and axions which may be the particles of dark matter). There is a

detailed order among the subatomic particles which, despite quantum theory's "random" associations, suggests the workings of several very small systems without which the very large could not exist.[66]

(36) Cold dark matter. Cold dark matter is hypothetically considered to form 23 per cent of the matter in the universe that remains undetected, presumably because it does not emit electromagnetic radiation. In the view of Big-Bang theory, it is needed to explain clusters of galaxies and hold galaxies together by a gravitational pull of between five and ten times more material than we see, but its existence outside mathematics has not yet been proved. I must stress that at the time of writing dark matter has not yet been detected. Cosmologists define the properties dark matter ought to have to explain observed features in the flat universe (i.e. in the expanding, accelerating universe in which gravity and electromagnetic forces are balanced). Dark matter could be present in cold clouds of interstellar dust, MACHOs (massive compact halo objects such as planets, brown dwarfs, black dwarfs and black holes) and neutrino WIMPs (weakly interacting massive particles). WIMPs may not be the answer as for neutrinos without dark matter to form superclusters of galaxies, break down into clusters, then galaxies and then individual stars, would take longer than 13.7 billion years, which would make individual stars very young – yet some of the individual stars seem to be nearly as old as (some say older than) the universe itself.[67] Cosmologists hope that dark matter will prove to be cold, in the sense that its particles have speeds lower than the speed of light, and that it does not disperse on galactic scales as would neutrinos, which are hot dark matter. To be slow the particles of dark matter would have to be heavy compared with an electron. Low-velocity cold dark matter particles would be bound together more easily and form dwarf galaxies and clusters.[68] The ratio of dark-matter particles to baryons – protons and neutrons – is, as we saw in (12), 10:1, suggesting there are billions more dark-matter particles than protons and neutrons, and there seems to be some significance in this ratio which is not understood but which contributes to the holding together of the galaxies. It may be that the halo of our galaxy is dominated by non-baryonic cold dark matter with the lightest supersymmetric partner. Assuming that this hypothesis is right (and the lack of evidence for dark matter is still the most urgent problem in astronomy and physics today), then the universe will be flat (see (11)) with gravity and electromagnetic forces balanced, a condition at the time of the Big Bang. This flatness will guarantee the right amount of energy to be converted into mass (in accordance with Einstein's equation

$E = mc^2$, Energy = mass times the square of the velocity of light). So a flat universe expands and, having produced cold dark matter that exceeds protons and neutrons by 10:1, is still flat at the end, allowing life to form.[69] And dark matter, the dominant stuff of the universe, did not form galaxies, stars and humans. If cold dark matter exists, its ratio to baryons holds the universe together.[70]

(37) Dark energy and zero vacuum. Dark energy, like dark matter, is undetected but is a hypothesis to explain, in terms of the Big Bang model, the powering of the acceleration in the universe's expansion (see (6)). Dark energy has been known as anti-gravity or λ, but also as vacuum energy which is thought to fill 73 per cent of the universe. Empty space is full of particles – the emptiness is really a fullness, a quantum sea of latent particles – and the energy in empty space has the opposite effect of Einstein's equation $E = mc^2$ for whereas atoms, radiation and dark matter slow down expansion, this energy speeds expansion up. This speeding-up is due to pressure on the vacuum – the positive energy in the vacuum is balanced by pressure that is negative, creating a tension – and it pushes rather than pulls with a cosmic repulsion measured at 10^{119} stronger than if it did not exist. If the measured value of dark energy were 10^{120}, as opposed to 10^{119}, less than its natural value, the universe would fly apart too fast for galaxies to form and there would be no life.[71] It is the same force that began inflation at 10^{-35} seconds after the Big Bang, leading to a flattened universe beyond the light-year dimensions of the observable universe (currently over 13 billion of the 13.7 billion light-years since the Big Bang and including about 100 billion galaxies), but the dark-energy acceleration is 50 odd powers of 10 slower than the acceleration at the start of the universe. Nevertheless this slower acceleration is still enough to make the solar system move (if not fly) apart. It has been suggested, as an idea, that the repulsion does not come from the vacuum but from a fluid that is an effect of the expansion, which decays. The universe's expansion may go on forever, but if it comes to rest the distance between the earth and the "edge" of the universe on which our surfer surfs may be a figure with millions of zeros after it. Yet somehow fierce expansion was "switched off" while the galaxies, stars and human life were formed, and has subsequently "come back on" again more slowly with such precision that the dark vacuum energy is (research has established) very near to exactly zero like its net electrical charge. The significance of the vacuum energy's being nearly zero is not known, but the symmetry behind the concept is overwhelming. In classical times there were four elements – earth, air, fire and water – and the heavens were ruled by a fifth essence or "quintessence", whose

role this dark energy or anti-gravity energy or vacuum energy now seems to have assumed – provided it exists. If dark energy exists, it creates the accelerated expansion that is a condition of the current galaxies, stars and life.[72]

(38) Glacials and climate. There are long-term cycles which control the earth's climate. There are cycles of Ice Ages and glacials (warmer periods). Every 413,000 years there is a super-orbit. Every 100,000 years a new orbiting cycle begins in which the earth moves closer to and then farther from the sun, causing changes to the earth's climate. Every 41,000 years the earth's axis tilts between 21.5 and 24.5 degrees, and the extent of the tilt affects the amount of sunlight the earth receives. (See 'Ice Ages' in ch. 6 for further details.)[73] Approximately every 1,470 years there is a cycle linked to solar radiation, the Dansgaard-Oeschger events.[74] These cycles have helped to create conditions in which evolving life can take place.

(39) The speed of light. In the "Grand Unified Theory era" which began at 10^{-43} seconds after the Big Bang the universe expanded by a factor of 10^{54}, faster than the speed of light. Space-time itself expanded, so there was no violation of the special theory of relativity. We do not know what is beyond the horizon of over 13 billion light years, the distance our biggest telescope can see in our "shuttlecock-shaped" universe. (The 1990 Hubble telescope could see to a distance of about 10 billion light years.) Depending on the precise cylindrical contours of our "shuttlecock-shaped" universe, the farthest horizon of the observable universe is beyond all the galaxies between earth and over 13 billion light years away, centred on earth and moving out in all directions, whereas the horizon of the whole universe is strictly 13.7 billion light years in radius. In theory, if the cylindrical "shuttlecock-shape" is large enough, we can see over 13 billion light years in all directions, and so the radius of the observable universe is 13 billion light years. In theory the universe may be longer than 13.7 billion light years as the universe expanded by a factor of 10^{54}, faster than the speed of light, which would have elongated its length in relation to its light-year distance. As the tiny singularity inflated with a whoosh, expanding outwards in all directions from no centre, in theory, if the size of our cylindrical "shuttlecock-shaped" universe permitted, we could look across the horizon to 13.7 billion light years away in all directions, and perhaps more as inflation was faster than the speed of light. In theory the radius of the whole universe is 13.7 billion light years plus the extra distance caused by the expansion of the universe faster than the speed of light. Light is at the correct speed, 186,000 miles per second, for

biological life to take place.[75]

(40) Constants and ratios. See Appendix 1 for a list of the 326 fundamental physical constants in alphabetical order, for a list grouping the most important of these under field headings and for a list of 75 ratios extracted from the 326 constants. All the laws of Nature have constants associated with them: the gravitational constant, the speed of light, the electrical charge, the mass of the electron, Planck's constant for quantum mechanics, the strong force coupling constant, the electromagnetic coupling constant. The range of possible values for these constants is very narrow, between 1 and 5 per cent for the combination of constants. For example, we have seen (see (5)) that critical density is 0.3 of the density needed to halt expansion after an adjustment for dark matter but was originally close to unity (or 1), suggesting that the expansion of inflation was fine-tuned.

The following cosmic densities all have the value of unity or 1:

the density of the universe   in protons per cubic metre;
the density of galaxies   in protons per cubic centimetre
galactic dark matter density   in particles per cubic centimetre
density of stars within galaxies   in grams per cubic centimetre
the mean density of the sun   in grams per cubic centimetre
the typical density of planets   in grams per cubic centimetre
the typical density of life-forms   in grams per cubic centimetre
the density of a white dwarf   in metric tons per cubic centimetre[76]

The contents of the cosmos also total unity or 1 in density, suggesting that inflation did indeed take place:

Component in the cosmos	Density as proportion of cosmos density (Ω x)
Radiation	0.00005
Stars	0.005-0.01
Neutrinos	≥ 0.003
All baryons	0.04
Dark matter	0.25
All matter	0.30
Dark energy	0.70
Total	1.0 [77]

The detail of this table is slightly out of date as dark matter and dark energy are now thought to form 23 per cent and 73 per cent of the contents of the cosmos respectively (i.e. 0.23 and 0.73 as a proportion of 1), but the point is, the proportions that contribute to a density of 1 are broadly right.

Outside this range, human life would be impossible. If the constant of gravity were stronger – 10^{25} times less powerful than the strong nuclear force instead of 10^{38} times weaker – the universe would be small and swift, the average star would have 10^{-12} times the mass of the sun and would exist for only a year and no life would develop. If galaxies were to form, the early universe would have to be broadly homogeneous, meaning the same everywhere. This is only approximately true, but is an excellent approximation when averaging over the large regions of galactic space. The inhomogeneity ratio must not be greater than 10^{-2}, otherwise non-uniformities would condense into black holes before stars formed. It must not be less than 10^{-5}, otherwise inhomogeneities would be insufficient to condense the galaxies.[78] Classical cosmology permits a new parameter, S, the entropy (disorder) per baryon in the universe, which is about 10^9. Unless S is less than 10^{11}, galaxies would not be able to form, and planets and therefore life could not exist. S is a consequence of the baryon asymmetry in the universe, quarks and antiquarks being asymmetrical before 10^{-6} seconds after the Big Bang.[79] In short, no matter what constants or ratios we examine, all are within a correct band that permits galaxies, stars and human life to form. It is my contention that all these 326 constants and 75 ratios within them are evidence that the universe is controlled by the universal principle of order, or order principle.

The above 40 conditions provide a context for life to develop and therefore allow us to exist. If just one of these 40 conditions had been changed slightly, life could not have happened. The same is true of the 326 constants and ratios, each of which could be described in the detail I have drawn on in (7) above. The universe had to be exactly as it is for the forces in the Big Bang to create nuclear furnaces in stars for carbon-based life to exist.

Each of the 40 conditions seems to be such a coincidence that it is hard to believe it could be a chance, random event. All these 40 "coincidences" (including the 326 constants) had to happen *in relation to each other* for life to take place. Put together, the 40 conditions derive cumulative energy and force from each other. For all these conditions to be present in our universe in total is staggering. The probability or likelihood of constructing one short protein molecule of 100 amino acids in length has been calculated as one chance in 10^{130}, and a protein of only 150 amino acids in length as 10^{180},[80] well beyond the most conservative estimates of the universal probability bound. The universal probability bound below which chance could definitely be precluded varies from 10^{150} to 10^{94}, but 10^{120} has been taken as the best indicator as this represents the maximum number of bit-operations the universe could have performed in its entire history.[81]

The probability of all of the contextual conditions happening by chance and accident can therefore be taken as 10^{120}, even though to construct a protein of 150 amino acids in length is 10^{180}. The probability bound of 10^{120} is many more times the total number of all the atoms in the observable universe (10^{80}).[82] After considering all 40 it is impossible to believe that the universe is a chance, random accident that happened to explode into all these parameters and ratios in a haphazard way.

Materialists who assert the primacy of matter have founded their belief on 4 per cent of the stuff of the universe. Universalist philosophy would prefer to base its view on 100 per cent of the stuff of the universe. It is embarrassing for Materialists and reductionists that the detectable matter in the universe comprises only 4 per cent.

It is even more embarrassing for those who assert that the universe is a random accident and have to see each of the 40 interconnected contextual conditions as random accidents that they have to base their belief on a probability rating of 10^{120},[83] meaning that their belief in the accidental origin of the universe is extremely improbable. Universalist philosophy holds that the case for Materialism, Reductionism, Randomism and Accidentism has collapsed before the highly complex, interconnected order of the 40 contextual conditions, the minute amount of detectable matter and the unacceptably high improbability rating.

Universalists leave to one side the anthropic principle in both its weak and strong forms. The view that the parameters of physics show that the universe was created for human observers or that teleologically there was a Creationist plan to create a universe in which human beings can exist is beyond the remit of philosophy. The task of philosophy is to consider the place of human beings in relation to the universe, and that means studying the universe to determine the model that should be reflected in philosophy. Universalism's universe reflects the ordered universe of these 40 conditions.

Returning to our six models of the universe, I have disregarded model 2, string theory. I have not used the argument that an intelligent design needs an intelligent designer, and so I have pushed model 4 aside. I have been working with model 5: a bio-friendly "life principle" in the tradition of Bergson's *élan vital*, working alongside the four fundamental forces as a fifth force, invisible like electricity, an ordered life force known to Vitalism.

I have seen the "life principle" as our universal order principle, which works from the Whole to the individual.

We have kept the thought of a principle of order at the back of our minds, and it is now time to bring it to the foreground of our attention. The evidence for the universal order principle controlling and structuring the development of the universe is: the 40 contextual conditions and the constants and ratios which reveal the universe's underlying order and indicate that the universe will probably last forever, or at the worst be sustained in a self-organising equilibrium like Einstein's cosmological constant.

Philosophical Summary

What, then, does biocosmology offer the "new scientific interpretation of the universe and Nature" which is so important to the new philosophy of Universalism? It seems that the structure of the universe has been shaped by an order principle that works within randomness to set up parameters that are suitable for life: 40 contextual conditions including 326 constants and ratios. The order principle is a life principle. Physical circumstances (gravity, dimensions, the expansion rate of the universe, density, the speed of light, anti-gravity/dark energy, flatness, the size and emptiness of the universe) reveal order. The four forces and many particles and their mass ratios and charges, and all forms of matter show order. The atmosphere, glacials, climate and temperature all show order. In so far as the Big Bang shaped these conditions, the Big Bang also reveals order – and shaped order.

To sum up, philosophy can take from biocosmology:

a bio-friendly universe;
the order principle which may work with four forces as a fifth force;
the order principle as a life principle;
constants and ratios in relation to each other reveal order and are right for
    life;
physical laws in 40 conditions (gravity, dimensions, expansion rate of the
    universe, density, speed of light, anti-gravity/dark energy, flatness, size and

emptiness of the universe) reveal order;

the four forces and many particles and their mass ratios and charges, and all
forms of matter reveal order;

the atmosphere, glacials, climate, temperature reveal order;

the Big Bang reveals order.

The "new scientific interpretation of the universe and Nature" must include
evolution, the biological focus from the origin of life to the rise of higher (or
deeper) consciousness. Now that we have seen that the structure of the universe
improbably offers the right conditions for life, the Universalist philosopher must
see if science can detect a self-organising order principle in evolution as we move
from galaxies to emerging life on the newly-formed earth.

6

BIOLOGY AND GEOLOGY: THE ORIGIN OF LIFE AND EVOLUTION

We have seen that the physical universe came out of one point, a singularity. All biological species came out of the same point, and initially out of one cell. We need to start by looking at the origin of life on earth for how that first cell emerged affects the meaning of man's place in the universe, which is important to the new philosophy of Universalism.

The Origin of Life

The earth, which began about 4.6 or 4.55 billion years ago,[1] is one of the smallest planets grouped in our solar system round a central star (the sun). As all the planets revolve in the same plane round the sun, it is likely that they condensed from a single revolving disc of gaseous matter and that the earth originated as a mass of molten rock with no atmosphere. As it cooled, a crust formed. It was disturbed by volcanic activity (hot gases escaping from the molten interior). When the surface cooled to 100 degrees C, an atmosphere developed that consisted of water, ammonia, carbon dioxide, methane and hydrogen. Geological evidence shows that the early earth was virtually without oxygen – hence iron in reduced form is found in rocks. As the earth continued to cool, water vapour in the atmosphere condensed and fell as rain, forming rivers, lakes and oceans[2] about 4.4 billion years ago[3] while erosion formed rocks.

Reductionism eliminates all features of a complex process that are not essential to its function to isolate its most important features. It then creates a simple model for a process that may otherwise be too complicated to study effectively. Reductionism subdivides complex situations into components which are simple enough to be investigated. According to reductionists, if evolution is a fact of life then the origin of life and the first cell (the first chemical evidence for which was 3.8 billion years ago)[4] should all be explainable within the evolutionary (neo-Darwinist) scheme, as should the origin of eukaryotic cells (cells containing genetic material), the origin of multicellular organisation, the rise of flowering plants, the origin of vertebrates from non-vertebrates and the origin of viruses.

According to the reductionist view, life began when it was formed from

127

earthbound, chemically non-living matter, a process known as abiogenesis.[5] There are a number of biogenetic hypotheses, none of which have yet been proved.

The best known reductionist hypothesis is that life began in a "primordial soup".[6] Charles Darwin had suggested in a letter to Joseph Dalton Hooker on February 1, 1871 that the original spark of life may have begun in a "warm little pond, with all sorts of ammonia and phosphoric salts, lights, heat, electricity, etc. present, that a protein compound was chemically formed ready to undergo still more complex changes".[7] He added that "at the present day such matter would be instantly devoured or absorbed, which would not have been the case before living creatures were formed".

In 1924 the Russian plant physiologist Aleksandr Ivanovich Oparin showed that atmospheric oxygen prevented the synthesis of organic molecules that are the building blocks of life. In 1936 in *The Origin of Life on Earth,* he suggested that life began in the sea, and that a primeval "soup" of organic molecules could be created in an oxygenless atmosphere through the action of sunlight. He argued that these molecules would combine and dissolve into droplets which would fuse with other droplets and reproduce through fission.[8] Around the same time the English geneticist and physiologist J.B.S. Haldane suggested that the earth's pre-biotic oceans would have formed a "hot dilute soup" in which organic compounds would have formed, the building blocks of life.[9] The formation of small molecules (such as amino acids and monosaccharides) would lead to the formation of larger polymers (proteins, lipids, nucleotides and polysaccharides) and in due course from polymers to the formation of organisms (simple prokaryote cells).[10] In 1953, taking his cue from Oparin and Haldane, the American chemist Stanley Miller, a Chicago graduate student working under Harold Urey, carried out an experiment on the primeval soup.[11]

Miller sent an electric spark through a chamber containing methane, ammonia, hydrogen and water (the ingredients of a reducing atmosphere). On condensation the water changed colour and a tar-like substance formed in the flask. Within two weeks it produced organic compounds including 13 of the 20 amino acids that form proteins (which in turn form cells and are therefore the building blocks of life), the organic base adenine and the monosaccharide ribose.[12]

There were problems with the "soup" theory. Amino acids have to become proteins and one protein requires 100 amino acids of 20 varieties which requires 10^{130} combinations of amino acids. The amino acids are building blocks, not the assembled structure, and it would be hard to achieve the right protein by accident.

Furthermore, the soup theory assumes a lack of any presence of oxygen in the earth's primeval atmosphere as that would poison early forms of life. However, geological evidence of oxides in minerals suggests that the early atmosphere included oxygen; oxides in minerals could not have formed without the presence of oxygen. Oxygen aside, the early atmosphere on earth was formed of different gases from those used by Miller and Urey. There was no methane or ammonia in the primeval atmosphere, and experiments with the atmospheric gases that would have been present such as carbon dioxide did not produce plentiful amino acids. Carbon in particular does not break out to make larger organic molecules. What is more, amino acids would have to form protein while the second law of thermodynamics applied. This says that a system becomes less and less organised over time, meaning that amino acids cannot be more and more organised over time and form protein. The primordial soup would have been too diluted to generate proteins and lacked a "mechanism" – a self-organising principle – that could have created proteins.[13]

Repeat experiments based on speculations regarding varying earth conditions in early times have formed more biochemicals (all 20 amino acids, sugars, lipids, polynucleotides and ATP). Despite the experiments and work of Sidney Fox (1950s and 1960s), Juan Orowin (1961), Manfred Eigen and Peter Schuster (early 1970s) Günter Wächtershäuser (1980s) and Tom Scheck, who discovered riboenzyme (enzymes made of RNA, 1986), no one has succeeded in synthesising a protocell using basic components. Most reductionists nevertheless believe that prebiotic synthesis took place on the earth. The "primordial soup" hypothesis therefore remains an unproved hypothesis.[14]

There have been a number of alternative hypotheses. Miller's experiment created the field of exobiology, the study of life beyond the earth. This looked for the origin of life elsewhere in the solar system. The extraterrestrial hypothesis favoured by Sir Fred Hoyle, who criticised abiogenesis on the grounds of improbability, asserts that primitive life may have originally formed in space or on a nearby planet such as Mars or the satellites of Jupiter, or the outer solar systems.[15] Hoyle sees life including that of viruses and insects as coming to earth from space.[16] He was supported in this view by the British biochemists Francis Crick and Leslie Orgel, both of whom thought that life began with spores and bacteria borne on meteorites. It is interesting that Orgel, having spent a lifetime researching into how life began on earth, concluded that molecules preceded RNA, which preceded DNA – which was too complex to be the first repository of genetic information, a view Crick, the discoverer of the structure of DNA, shared – and that these molecules did not arise from a primordial soup but from outer space. Meteors, dust and comets may have supplied a few per cent of organic compounds.[17]

Various hypothetical models have been offered to explain how organic molecules become protocells. There are theories connected with RNA further to the work of Tom Scheck, who discovered that enzymes made of RNA can replicate themselves. There are theories that look at metabolism rather than genes further to Wächtershäuser's iron-sulphur theory. There is a clay theory put forward by Graham Cairns-Smith in 1985 that holds that organic molecules arose from clay in silicate crystals. The same applies to honeycombs and seashells, which also have repetitive order with low information content. There is a view by Thomas Gold in the 1990s that life began some five kilometres below the earth's surface in deep rocks where nanobes (filomental structures that are smaller than bacteria but may contain DNA) have been found. A 2,600-feet-deep volcanic system of towering pinnacles and chimneys in a hydrothermal vent field on the mid-Atlantic ridge produces hydrocarbons – oil and gas and molecules, the building blocks of life – by the chemical interaction of sea water and rocks. Their source is non-living, and they are organic building blocks from a non-biological source. It has been suggested that the floor of the Atlantic Ocean contains the origins of life.[18]

There is a view connected with lipids. Phospholipids spontaneously form bilayers in water, the same structure as in cell membranes. These self-replicating molecules were not present in primeval times but may reveal early information storage which led to RNA and DNA. There are views connected with polyphosphates, ecopoesis and polycyclic aromatic hydrocarbons (PAHs) in nebulae. There is a view popularised by Richard Dawkins, the British ethologist, in his book *The Ancestor's Tale*, citing the Scripps Research Institute in California (2004), that autocatalysis accounts for the origin of life. Autocatalysts are substances that catalyze the production of themselves and therefore are molecular replicators for Dawkins. All these experiments and theories are unproved hypotheses.[19]

The trouble is that for every theory and experiment there are difficulties and counter-arguments. The problem with RNA, for example, is that there are no known chemical pathways for the abiogenic synthesis of the pyromidine nucleobases cytosine and uracil under prebiotic conditions. There is also the difficulty of nucleoside synthesis (from ribose and nucleobase), ligating nucleosides with phosphate to form the RNA backbone and the short lifetime of nucleoside molecules, especially cytosine. There are doubts about the size of an RNA molecule capable of self-replication. Different hypotheses lack complete evidence, and cannot be related to the exact conditions in the prebiotic universe.[20]

So where does this leave the origin of life? In plain English, a Thermos flask with water, a heat source, a flash to simulate lightning, simple molecules and a coil demonstrated that two weeks after this soup was left cooking there was molecular activity. Simple liquids threw up enzymes like bubbles from dishwashing detergent. This suggested that the action of wind, heat and lightning on waves would have a similar effect and throw up molecules that might form a cell. However, an initial cell could not copy itself until the molecules were complicated enough to replicate themselves and become a lipid group with proteins trapped inside along with RNA. RNA could copy itself if a wave hit an enzyme and broke it into two small "bubbles", which became a cell and a printed copy.[21] The chances of our world coming out of this soup by accident are extremely slight.

To the reductionist, chemical life began when carbon combined with five other atoms: hydrogen, oxygen, nitrogen, sulphur and phosphorus. Life, to a reductionist is an electrical gradient or imbalance of charges across a membrane. A membrane is like a sandwich with no filling: two layers of fat (the slices of bread) with water-loving heads of lipids on the outside of one layer and water-avoiding tails of lipids pointing inwards on the outside of the other layer. If three sodium ions are pumped in and two potassium ions are pumped out there is a charge. That charge, to a chemical reductionist, is life.[22]

All life and all species grew from a first cell, and there are now trillions of trillions of cells. It takes more than 10^{15} cell divisions to proceed from one cell to a human being and in each mitosis there is a one-in-a-million (10^{-6}) chance of error, of a mutation being made. For every time a cell divides it copies DNA, and every time there is a copy of DNA there is a one-in-a-million (10^{-6}) chance for an error to be made in human cells. The four building blocks of DNA are A, P, G and C. If a cell copies itself so P comes first, there is a mutation, and the two cells are different. One cell will grow better and take over the population of cells that stems from further mitosis from that cell. We cannot say that one cell is better than the other, just different. It will prove to be better if it survives and will prove not to be better if it is extinguished. If a cell receives too many instructions it commits suicide. If the lipid pump is poisoned, a cell will die.

Mutations that are different and do not die are creative.[23]

There have been a huge number of cells created over the years. Multiply the number of cells in each human being, more than 10^{15}, [24] by the number of human beings in the world at present, 6.7 billion. Making allowances for births and deaths and working backwards, taking account of the fact that there were one billion human beings alive in 1804 and two billion in 1927,[25] we can in theory work back to the first human – or at any rate the estimated number of early humans alive 40,000 years ago – and estimate the number of human beings who have ever lived: somewhere between 45 billion and 125 billion, most probably between 90 and 110 billion.[26] We can then calculate the number of human cells that there have ever been. Bearing in mind that life began on earth between 3.8 billion and 3.5 billion years ago[27] we can in theory calculate an estimate for the number of cells in all mammals, reptiles, birds, fish, insects, plants and bacteria since the very beginning.

All these cells came out of one cell, the first cell or protocell. A long time had to elapse during which cells learned to hold together and exchange information. We thus have a Tree of Life for the period and pattern from the first cell to the inflation of species (see below), which replicates the concept of the period and pattern from the Big Bang to the inflation of galaxies. But whereas the atoms that came into existence with the Big Bang are still in existence and have not died out, cells – and branches of species – *have* died out. Now all human beings have one thing in common: we are all cells.

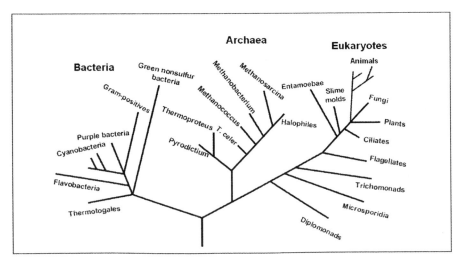

The Tree of Life

We are made of atoms of hydrogen that comprise one-tenth of our weight, which came from the Big Bang, and of stardust, which forms 0.01 per cent of the universe and comprises: atoms of helium from the Big Bang and later stars, iron atoms carrying oxygen from exploding white dwarf stars, oxygen from exploding supernovas and carbon from planetary nebulas.[28]

The reductionist view of the origin of life is a string of hypotheses that are not proved. Reductionism identifies self-replication but has not identified a "mechanism" by which self-replication takes place in a right way that leads from cell to organism. "Mechanism" is a reductionist word whose mechanistic implications the Universalist does not accept, hence the inverted commas. What is missing is the "mechanism" for a self-organising, ordering principle which can take organic molecules to the first cell, and the first cell to the many species of our world without being driven by blind, random accident, the chances of which happening are very highly improbable.

Plate Tectonics

Having established itself, emerging life had to contend with unstable geological circumstances: land masses that moved about, changing environments and creating the conditions in which evolution could take place.

The theory of continental drift was first proposed by the German meteorologist Alfred Wegener in a book published in 1915, *Die Entstehung der Kontinente und Ozeane* (*The Origins of Continents and Oceans*). Watching the drift of ice floes in a fjord in Greenland, Wegener imagined continents drifting across the earth's surface in the same way and it has since been confirmed that the continents drift on their plates.

Plate tectonics[29] developed in 1967-1968[30] and drew together continental drift, oceanography, seismology, vulcanology and mountain building. It supplies a unified theory for the development of continents, oceans, earthquakes and volcanoes. The core of the earth is made of iron and covered with a 3,000-km-thick layer of mantle. Above it is the 30-km-thick crust, the continents, oceans and atmosphere. When asteroids, meteorites and comets came together in collision to form the earth, the iron was pulled down to the centre by gravity. The earth's inner core is cooling and solid iron. The outer core is liquid magma. Rock is solid and its weight causes rock beneath it to behave with plasticity and flow just as a glacier is solid until 40 metres above ground level when the weight of the ice gives it plasticity and causes it to flow. Magma is kept in place by the pressure of rock, and when pressure reduces as a result of cracks in the rock caused by steam, magma flows up. After the earth was formed gas escaped in the form of steam and turned to water, and around 4.4 billion years ago formed the oceans. The earth was a series of floating layers (an isostasy, the rising and falling land and the sea forming an equilibrium of the crust). Continents were passive riders on the plates.

The earth's crust is a series of slowly moving plates 6-30 kms thick that interlock like pieces of a jigsaw puzzle and float on dense semi-molten rock, driven by the convective heat flow deep inside the earth. In horizontal movements, plates pull away from each other, stretching parts of the earth's crust and forming basins, including the mid-Atlantic rift. Plates can slide past each other, causing earthquakes. They can collide, compressing each other, and magma is forced up, creating mountains and also causing earthquakes in mid-ocean ridges and oceanic trenches at the plates' edges. When oceanic plates are pulled under a continent, magma rises from 100 kms (60 miles) down, cracks rocks and cools, and its composition changes into lava. Lava wells up in the mid-ocean ridges, creating new land, causing the oceans to spread and driving continents apart.

The energy behind plate tectonics is the energy in the centre of the earth as heat flow from the molten core drives tectonics, volcanoes and earthquakes. Plate movement caused rifting as in the East African Rift Valley and the Red Sea, where basins filled with water. And as rifting spread the Atlantic Ocean basin was formed.

Some 200 million years ago all lands were one land mass. In 1885 the Austrian geologist Eduard Suess proposed that there was one supercontinent, and that it should be called Gondwana. In 1912 Wegener proposed that the first supercontinent included all lands surrounded by a world ocean extending from North to South Pole and spanning 80 per cent of the earth's circumference, and that it should be called Pangaea (Greek for "all lands"). Geologists now refer to the one continent as Pangaea. An example of what this land mass looked like can be seen in Drygalski fjord, South Georgia in the north Antarctic: rocks speared up by the Scotia plate which connects South America and Antarctica, made of the earth's crust as it was then. By 180 million years ago the land split. In 1937 the South African geologist Alexander Du Toit proposed that Pangaea split into two continents, which he called Gondwana to the south and Laurasia to the north.[31]

Gondwana was one southern supercontinent with Antarctica in its centre, round which were South America, Africa, the Falkland Islands, India and

Australia. It broke up, perhaps due to hot spots in the earth's mantle beneath its surface. Around 150 million years ago a seaway formed between west Gondwana (South America and Africa) and east Gondwana (Antarctica, Australia, India and New Zealand). Around 130 million years ago South America separated from the African-Indian plate, which also separated from Antarctica. Around 100 million years ago Australia and New Zealand separated from Antarctica, which then had luxuriant vegetation. Gondwana was now fragmented into six or more microplates, most of which are now in west Antarctica. One microplate rotated to form the Falkland Islands and South Georgia. Plants and animals, including dinosaurs, had been stranded on different continents, separated by oceans, and remnants of Gondwana can be found in South Africa, Patagonia, Antarctica and Tasmania.[32]

Around 65 million years ago Africa collided with Eurasia and threw up the Alps, and North America and the British Isles separated from Eurasia, perhaps as a result of the shock of the impact. About this same time (65 million years ago) the dinosaurs were wiped out, perhaps as a result of the consequences of the plate collision. The extinction of the dinosaurs at this time has traditionally been attributed to the devastating impact of a giant asteroid which struck Mexico at a site named Chicxulub after a nearby town 65 million years ago. (There is a school of thought that dinosaurs evolved into birds, and some certainly did, such as Archaeopteryx – feather-covered dinosaurs – and pterodactyls.)[33] India had run into Eurasia about 70 million years ago, forming the Himalayas. The Andes were thrown up by a collision between the South-American plate and the Nazca plate. The plates on which South America and Africa rested moved apart, creating the mid-Atlantic ridge. Around 50 million years ago Australia separated. Around 35 million years ago Antarctica separated from South America and developed an ice sheet and circumpolar oceanic current. Around 34 million years ago (or more exactly, 33.7 million years ago)[34] Antarctica began to ice up. Gradually our present world took shape.

The earth's magnetic base is weakening (by 5 per cent per century in recent times). For periods between 145 and 65 million years ago the magnetic field, now centred on the North Pole, has been reversed so that the magnetic field centred on the South Pole. The last long reversal of the earth's magnetic field (called the Brunhes-Matuyama reversal) is thought to have taken place 780,000 years ago, and it may be that the earth's magnetic field reverses every 250,000 or 500,000 years, or more – in which case a reversal is now long overdue. At present the magnetic field maintains a shield between the earth and the sun, and diverts dangerous solar radiation.[35]

The sun's magnetic field reverses every 11 years, and abnormal magnetic effects have been encountered off the coast of South-East Africa and in the south Atlantic. Computer researchers have been unable to detect a pattern within the reverses and have said that it seems to happen randomly, which is another way of saying we do not understand the pattern. It may be that when the switches between North and South Poles of both the earth and the sun are understood we will have an explanation which will make the process appear anything but random.

All these plate movements and continental drifts have contributed to a land-mass context on earth which is right for life and which can sustain evolution.

Ice Ages

Evolving life also had to contend with inclement climatic conditions which suddenly created inhospitable environments but contributed to temperate atmospheres capable of sustaining life. Evidence for these cold periods has been found in rocks.

Evolution is the development of life in geological time, literally the "unrolling" of life as "evolution" comes from the Latin *evolvere*, meaning "to unroll".[36] Evidence for early evolution is found within palaeontology, the study of plants and animals of the geological past through their dead remains preserved as fossils in sedimentary rocks.[37] We have seen (on p127) that the earth was formed about 4.55 billion years ago, and that water condensed into oceans about 4.4 billion years ago.

The geological history of the earth spans nearly 4 billion years: the oldest known rocks have an isotopic age of 3.96-3.9 billion years ago. The earliest fossils may be 3.8 billion years ago.[38] From the time the earth came into existence about 4.55 billion years ago according to geologists and astronomers there are about 650 million years (between 4.55 and 3.9 billion years ago) for which no geological record exists.[39] The Precambrian period stretches from 3.9 billion years ago to 570 million years ago, 75 per cent of the earth's history which included the first photosynthesis about 1.5 billion years ago.[40]

After this period geochronology measures geological time in terms of three eras[41] the Paleozoic Era (from the Greek for "ancient life") which covers marine animals from 570 to 245 million years ago; the Mesozoic Era (from the Greek for "middle life") which covers the ancestors of the major plant and animal groups from 245 to 66.4 million years ago; and the Cenozoic Era (from the Greek for "recent life") which covers from 66.4million years ago to the present.

Each era is subdivided into periods. The Paleozoic Era is subdivided into six periods: Cambrian; Ordovician; Silurian; Devonian; Carboniferous; and Permian. The Mesozoic Era is subdivided into three periods; Triassic; Jurassic and Cretaceous. The Cenozoic Era is subdivided into two periods, the Tertiary Period (66.4million years ago to 1.6million years ago), which is subdivided into five epochs: Palaeocene, Eocene, Oligocene, Miocene, and Pliocene; and the Quaternary Period (1.6 million years ago to the present), which is subdivided into two epochs, the Pleistocene Epoch (traditionally dated by radiocarbon as 1.6 million years ago to 10,000 years ago) and the Holocene Epoch (10,000 years ago to the present). The date of the end of the Pleistocene Epoch and the beginning of the Holocene Epoch is sometimes held to be earlier as a major climatic warming took place about 13,000 years ago, and a compromise date of 11,750-11,500 years ago (about halfway between 13,000 and 10,000) is frequently given – around c.9600BC. Some authorities would extend the Pleistocene Epoch to our present time. Others hold that the human impact on the world has been so comprehensive that a new human-based geological era came into existence c.1800, the Anthropocene era.

A new discipline, palaeoclimatology, has established that the slow cooling of the earth over tens of millions of years cooled the climate of the earth, which underwent at least four and probably five Ice Ages.

An Ice Age is a period when the earth's climate is reduced for a long term, as a result of which continental and polar ice sheets expand, as do mountain glaciers. In glaciological terms, an Ice Age is a period when there are ice sheets in the northern and southern hemispheres.[42]

Our Ice Age began 40 million years ago with the growth of the ice sheet in Antarctica and extended with the icing-up of much of the southern hemisphere 33.7 million years ago.[43] In the middle of the Pliocene Epoch before 3 million years ago the earth's climate was actually quite warm. Forests grew on the Arctic coast of Greenland, ice caps at the poles were thinner – or even absent – and the sea was 35 metres higher than today. The Ice Age intensified during the Late Pliocene period when ice sheets spread in the northern hemisphere 2.67 million years ago. It continued through the sharp cooling 2.5 million years ago and throughout the Pleistocene Epoch, which lasted from 1.65 million years ago to about 11,750/11,500-10,000 years ago. It is sometimes wrongly said that our "ice age" ended at this time.

We are still in an Ice Age because there are still ice sheets in Greenland and Antarctica.[44] (The North Pole is in the Arctic Ocean, about 450 miles north of Greenland.) About 40 per cent of the earth comprises continents, 60 per cent ocean basins. However 71 per cent is now covered by ocean and only 29-percent land, so 11 per cent of the earth is flooded continent. Of the 29-per-cent land, approximately 10 per cent is covered by ice. Approximately 13 per cent of the earth's surface is influenced by sea ice in the course of any year.[45]

Colder periods within an Ice Age are known as glacials. The most recent glacial period ended 11,750/11,500-10,000 years ago. We are now in an interglacial (or warmer period), still within an Ice Age. There is evidence that for the last 20 million years there have been glacial periods lasting 90,000 years followed by interglacials never longer than 12,500 years.[46] On this evidence, we are somewhere between 10,000-11,500/11,750 years into our interglacial and can expect a colder glacial period in which mountain glaciers advance in getting on for 1,000-1,500 years' time, or more, c.AD3000-3500.[47] Global warming may be seen within this long-term cooling context. However, recent research considers that our interglacial has similarities to a previous interglacial that lasted 28,000 years. It could be that we should not expect a new glacial until about 16,500-18,000 years from now.[48]

Evidence for Ice Ages can be geological, chemical and palaeontological. There are problems with all the evidence. Geological evidence of rocks being scratched by ice, glacial moraines, drumlins, the cutting of valleys and the depositing of till or tillites and glacial anomalies can be erased by successive glaciations (as glaciers advance after retreats). Chemical evidence based on isotope dating (the ratios of isotopes in sedimentary rocks, ocean sediment cores and ice cores drilled through the ice caps of Greenland and Antarctica) can be obscured by other factors in isotope ratios. Palaeontological fossils require appropriate temperatures in which to survive, and they have to be found by researchers.[49]

On the basis of such patchy and varying evidence, the known Ice Age – periods when whole continents have been covered by ice sheets in the northern and southern hemispheres – have traditionally been held to be as follows:[50]

(1) 2.7 to 2.3 billion years ago, the Huronian Ice Age during the Proterozoic Era in the Precambrian times. This Ice Age is hypothetical and is not well documented

by geological, chemical and palaeontological evidence. Some authorities date the beginning of the first Ice Age to 2.3 billion years ago.

(2) 850-630 million years ago, the Cryogenian Ice Age. This was the most severe Ice Age in the last billion years and may have resulted in the entire earth's being covered in sheet ice, making the earth resemble a snowball. This is the earliest well-documented Ice Age, and it is possible that its end may be responsible for the Ediacaran and Cambrian explosions which saw progress towards the emergence of *Homo sapiens*.

(3) 460-430 million years ago, the Andean-Saharan Ice Age, during the late Ordovician and Silurian Periods.

(4) 350-260 million years ago (in another view 370-260 million years ago), the Karoo Ice Age during the Carboniferous and Early Permian Periods.

(5) Around or between 40 and 33.7 million years ago to the present, our Ice Age in which Antarctica froze over some 33.7 million years ago and in which we are in the Holocene interglacial period after a glacial that peaked 125,000-115,000 years ago when ice covered a third of the earth's surface. During this time ice sheets or glaciers covered most of North America – for example, 97 per cent of Canada when the ice over Hudson Bay was some 3,300 metres thick – and Eurasia until about 11,750/11,500-10,000 years ago.[51]

There is a consensus that these Ice Ages were caused by a combination of conditions: the composition of the atmosphere, and in particular the concentrations of carbon dioxide, methane, sulphur dioxide and other gases; changes in the earth's orbit around the sun and possibly the sun's orbit around the galaxy; the movement of tectonic plates which affected the continental and oceanic crust; variations in solar energy; the orbits of the earth and moon; the impact of large meteorites; and the eruption of super volcanoes. Of these the most significant were likely to be: the changing of continental positions and the lifting of continental blocks; the reduction of carbon dioxide in the atmosphere; and changes in the earth's orbit. In our own time, greenhouse gases may affect the composition of the earth's atmosphere and bring about climate change.[52] On the other hand, climate change may have been caused by the glacial-interglacial pattern which can affect the earth's atmosphere by changing the rate at which the weather removes carbon dioxide.

Today the scheme of Ice Ages has been updated by work on the advances and retreats of mountain glaciers. In the 1960s and 1970s supercomputers confirmed[53] the work done in the 1920s and 1930s by Milutin Milankovitch, a Serbian mathematician, on the orbit and tilt of the earth. Doing mental calculations, while in prison, he found that:[54]

(1) The earth's orbit round the sun varies from circular to elliptical, causing an eccentric (in the sense of wandering) glacier advance that happens around every 100,000 years (which is really an average of cycles lasting 80,000-120,000 years[55]), with a more severe advance every 413,000 years, that results in less summer snowmelt;

(2) The tilt of the earth's spin axis relative to its orbit round the sun steepens and shallows from about 21.5 to 24.5 degrees, a cycle that happens every 41,000 years;

(3) The earth's spin axis wobbles like a spinning top about to fall over (a phenomenon Milankovitch called precession). This wobbling causes a slight change of direction of the earth's axis of rotation in relation to the fixed stars every 23,000 years (a refinement of Joseph Adhémar's calculation of 26,000 years and another calculation of 19,000 years).[56] In addition, the orbital ellipse precesses in space because it interacts with Jupiter and Saturn, shortening the precession of the equinoxes to every 21,000 years. This affects the equinoxes every 21,000 years.

To expand, during an Ice Age the polar regions are cold, there are great differences between the temperature on the equator and the temperature on the poles, and continental-size glaciers cover much of the earth. Ice sheets or glaciers advance in 100,000-year cycles because of the earth's orbit, with a severe advance every 413,000 years, and later retreat. The earth moves closer to and farther from the sun in these cycles and glaciers advance and retreat in a regular pattern. Within this cycle, first less polar ice melts each summer and ice sheets or glaciers grow larger, then the pattern is reversed and more polar ice melts each summer and ice sheets and glaciers grow smaller, as is happening now. This cyclical pattern has happened throughout our Ice Age.

The tilt cycle involves the tilting of the earth's axis. The earth's tilted axis, spinning like a top, describes a circle every 23,000 years, and the tilt changes between about 21.5 and 24.5 degrees every 41,000 years. Today the axis is tilting at 23.44 degrees. The degree to which the earth tilts affects the amount of sunlight the earth receives during a year.[57]

We now see Ice Ages in terms of the interaction between geology's millions of years of continental drift, astronomy's thousands of the years of cycles and meteorology's annual view of winter snowfall and summer snowmelt. There are no warm waters at the North and South Poles because they have been blocked by continental drift. In the south, cold water circulates round cut-off Antarctica. Cold water produces little evaporation, therefore little snowfall and therefore little snowmelt. In the north, the volcanoes of Central America connected North and South America and blocked the flow of warm water proceeding westwards across the Atlantic, sending it up towards Greenland 2.67 million years ago (a date based on records of sediment on the ocean's floors). Warm water produces plenty of evaporation and therefore a lot of snowfall, which over 2.67 million years led to ice more than 40 metres thick and glaciers in Canada. There was plenty of snowmelt, but not enough to melt glaciers over 40 metres thick. About 20,000 years ago the land and sea between Alaska and the British Isles were covered in sheet ice so that one could walk between the two.[58]

In theory, since 2.67 million years ago there should have been 27 cycles of 100,000 years. However during the first 1.67 million years tilt and wobble dominated, whereas from 1 million years ago eccentric (wandering) orbiting dominated; no one knows why. It can therefore be said that since 2.67 millions years ago, for the first 1.67 million years there were 40 advances of mountain glaciers (i.e. 1,670,000 divided by 41,000), and in the last million years there were 10 advances (i.e. 1,000,000 divided by 100,000), giving a total of 50 advances of mountain glaciers. This figure does not take account of 413,000-year and 100,000-year advances in the first 1.67 million years or 413,000-year and 41,000-year advances in the last 1 million years. There may, therefore, in fact have been more than 50 advances of mountain glaciers during the last 2.67 million years.[59]

It is worth pointing out that a 100,000-year glacier advance peaked about 200,000 years ago when, according to the "out-of-Africa" view, *Homo sapiens* first left Africa, perhaps on account of climate change. The next 100,000-year glacier advance ended about 115,000 years ago. On a 250,000-year-range from a 5-per-cent glacier advance to an over-1-per-cent glacier retreat, 11,750/11,500-10,000 years ago eccentricity was 2 per cent, showing glacier retreat. The last glacier advance ended about 11,750/11,500-10,000 years ago, after which our interglacial began. The end of this glacier advance coincided with the last tilt peak at 24.25 degrees, and since then there has been a strong glacier retreat. About 11,750/11,500-10,000 years ago there was also a peak for less earth-sun distance in June, showing a glacier retreat.[60] All three Milankovitch patterns show a retreat of mountain glaciers around 11,750/11,500-10,000 years ago.

The retreat of mountain glaciers caused by the end of the last 100,000-year advance can be expected to end in 7,000 or 8,000 years' time, when a new advance of mountain glaciers can be expected to begin, culminating in the peak of another 100,000 advance in the distant future (perhaps in 85,000 years' time) when the earth will be "an orbiting ice house".[61] We have seen (on p138) that our interglacial can be expected to end between 1,000-1,500 and 16,500-18,000 years from now, and 7,000-8,000 years' time is towards the middle of this range.

There are claims that the warming of the earth's climate is not due to carbon emissions by humans. It has been pointed out that our solar system seems to be warming up, that unexpected warming has been detected on Neptune's moon Triton, on Jupiter, on Pluto and on Mars's shrinking southern ice cap, which cannot be attributed to humans.[62] However, this planetary heating may be a consequence of seasons – their summer – and experts do not accept that there has been an increase in solar radiation. Milankovitch orbits and tilt may explain planetary heating. It cannot be maintained that global warming is due to rising temperatures throughout our solar system.

Milankovitch's work has been supplemented by the work of three scientists, Willi Dansgaard of Denmark, Hans Oeschger of Switzerland and Claude Lorius of France, who in 1996 won the Tyler Prize (equivalent to a Nobel Prize in environmental science) for discovering that ice cores taken from the ice caps of Greenland and Antarctica indicate that a 1,470-year cycle has dominated the earth's weather since the end of the last glacial 11,750/11,500-10,000 years ago. This cycle has influenced the earth's temperature for 900,000 years. It has been linked to variations in solar activity which are obscured from the earth by atmospheric conditions. In November 2001 a team led by Gerard Bond found that satellite data showed that the earth's climate has been linked to variations of solar activity amplified by cosmic rays for 32,000 years. Holger Braun found from computer studies that sun cycles 87 and 210 years long generated a

1,470-year cycle.[63]

Thus the temperature on earth is determined by the following four long-term factors:

(1) Its long-term position in an Ice Age that began 40 million years ago, its now being 11,750/11,500-10,000 years into an interglacial which may last another 1,000-1,500, 7,000-8,000 or 16,500-18,000 years;[64]

(2) Its position in relation to its 100,000-year and 413,000-year orbiting cycles;

(3) Its position in relation to its cycle of tilting between 21.5 and over 24.5 degrees; and

(4) Its probable susceptibility to variations of solar activity amplified by cosmic rays in cycles of 1,470 years following sun cycles every 87 and 210 years.

The earth is thus now on a long-term cooling – *not* warming – of glacial-interglacial Ice-Age cycles which can be expected to last for a very long time. All these climatic cycles and Ice Ages have contributed to creating temperate conditions which are right for life and which can sustain evolution.

Evolution by Natural Selection

Life, having originated, has evolved despite – or perhaps because of – drifting continents and harsh Ice Ages which have provided extreme environments. These are the background to the theory of evolution which explains the development of all species as they adapted to changing environments and varied, and in particular our development from primitive unicellar jellies to humans with brains that can understand the universe. The theory of evolution was arguably the 19th century's greatest intellectual achievement.

Early biologists believed that species were created by God and had continued without change. Charles Darwin's grandfather Erasmus, a doctor, believed in "organic evolution" (the evolution of the living world) and became the first of five generations of Darwins to become a Fellow of the Royal Society in 1761. He read Buffon, the 18th-century scientist who realised that species were not fixed and unchanging.

Erasmus Darwin came to believe in evolution by variation and

143

"improvement" of species by their own activities. He believed that characteristics acquired by a parent can be transmitted to their offspring. He wrote about the evolution of life from "tiny specks in the primeval sea", which led to the forming of fish, amphibians, reptiles and human beings.[65]

His grandson Charles Darwin read his books. He was supposed to become a doctor but abandoned the idea after attending operations on children when no anaesthetics were used. He also abandoned his second choice of career: to be a clergyman. As he researched into evolution "disbelief crept over me at a very slow rate" (*Autobiography*).[66]

In 1831 he became an unpaid naturalist on board HMS *Beagle*, which was given the task of mapping coastlines during an Admiralty-commissioned voyage to the southern hemisphere. During the voyage, which lasted five years, he developed the idea of organic evolution by natural selection. He accumulated geological and fossil evidence that supported the change of organisms with time, and he studied the *flora* and *fauna* of South America and the Galapagos Islands, four of which he visited from the *Beagle* over a period of five weeks in 1835: Charles, Chatham, James and Albemarle (today known by their Spanish names as Floreana, San Cristobal, Santiago and Isabela). He took specimens.[67]

Darwin was not especially interested in the specimens at the time. He took 40 giant tortoises on board to eat as fresh meat. The crew lay the tortoises on their backs and cut off a leg a day for food until a tortoise had no legs left. Then they killed it and threw the shell and dead body overboard. Sometimes they boiled the tortoises in their shells. Darwin went on to Australia and donated three giant tortoises which had not been eaten, one of which, about 15 years old in 1836, died in 2006 at the ripe old age of at least 184.[68] There were finches on every island. Darwin thought them sparrows, and one variation wrens. He did not label his specimens. John Gould, his companion on ship, labelled specimens of mockingbird.[69]

When he returned to England Darwin handed his specimens to experts in Cambridge and London. Their analysis demonstrated there were 14 separate species of finch (whereas Darwin had thought they were subspecies) – all crossbills, 13 from the Galapagos Islands and one from Cocos Island[70] – and four separate species of mockingbird.[71] The giant tortoise was one species with 14 subspecies.[72] There were also different species of lizard.

Analysis in London demonstrated that the species had adapted to local conditions and undergone variations which were passed on to their own offspring. Darwin did not know about DNA, which was not discovered until 1869 and whose role in genetic inheritance was not known until 1943, but he saw that these variations were transmitted to offspring.[73]

The 14 species of finch were – and are still – mainly distinguished by their bills. These varied in size and shape according to diet. The original seed-eating ancestor or ancestor group of the finches is thought to have come from St Lucia via Costa Rica and may have arrived before other land birds and found no competition. The warbler finch may have diverged first to feed on insects and spiders. The ground finches continued to eat seeds. The cactus finches ate opuntia (cactus) blossoms. The tree finches fed off trees but also on the ground. The woodpecker finches used an opuntia spine to tap the crevices of trees for grubs. The vegetarian finches curled away bark to feed on tree tissue and on buds, leaves and fruits. Sharp-beaked ground finches pecked wounds in seabirds and drank their blood – so they were called "vampire" finches – and they cracked open birds' eggs on rocks. Other species picked parasites from iguanas and giant tortoises.[74] In the second (1845) edition of his private diary of the voyage, the *Journal*, Darwin wrote, "One might really fancy that, from an original paucity of birds in this archipelago, one species had been taken and modified for different ends."[75]

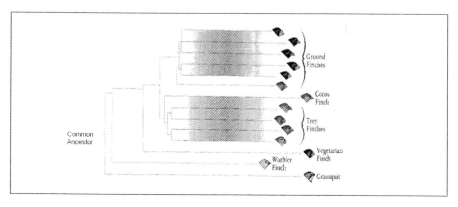

Darwin's 14 finches

Of the mockingbirds, the Galapagos mockingbird was the most common, and the three other species varied slightly in colour, size and bodily proportion: the Hood, Chatham and Charles mockingbirds localised on Espanola, San Cristobal and Floreana).[76]

The surviving subspecies of giant tortoise occur in two main forms. On arid islands where vegetation is sparse tortoises have evolved a saddleback carapace with a raised front that enables them to reach up to leaves. On moister islands where there is vegetation, tortoises have evolved a domed carapace enclosed at the front.[77] The population of one subspecies is now down to one: Lonesome George, who faces extinction when he dies.[78] (His wizened face was used as the face of ET in Spielberg's film.)

What had happened was that once these creatures had come to the Galapagos Islands from South America, their descendants had adapted to local conditions and undergone variations in a struggle to survive. (A similar adaptation can be found on North Ronaldsay in the Orkney Islands, where sheep excluded from fields by a wall have adapted to the beach and eat seaweed – and now fall ill if they eat grass.)

What the analysts told Darwin in London set him thinking. He began compiling notes on evolution in 1837. He had found evidence of adaptive radiation among the birds of the Galapagos Islands – that one ancestor or ancestral group had produced a variety of forms adapted to various environments, whose beaks were adapted to finding different foods. In 1837 he grasped the principle of natural selection. On 28 September 1838 he read Malthus's *Essay on the Principle of Population*, which focused on a human struggle for survival in a contest for food. Darwin saw a similar struggle among all living things. On that same day, Darwin wrote in his *Notebook on Transmutation of Species,* "On an average every species must have same number killed year with year by hawks, by cold, & c....The final cause of all this wedging must be to sort out proper structure....One may say there is a force like a hundred thousand wedges trying to force every kind of adapted structure into the gaps in the oeconomy of nature, or rather forming gaps by thrusting out weaker ones."[79] In other words, Nature overproduces acorns, eggs, seeds and sperms, and selects those offspring best able to cope with life – by gaining food and escaping predators – to survive. The rest perish. In short, Nature selects.

Darwin was not the first to propose natural selection. Others had had similar ideas long before him. A thousand years before Darwin, Abu Uthman al-Jahith (781-869), a scientist of East-African descent working in Baghdad, considered the influence of the environment on species in his *Book of Animals*, which was based on folklore rather than scientific observation: "Animals engage in a struggle for existence; for resources, to avoid being eaten and to breed. Environmental factors influence organisms to develop new characteristics to ensure survival, thus transforming into new species. Animals that survive to breed can pass on their successful characteristics to offspring." This was a theory of natural selection put forward during the 700 years when the international language of science was Arabic and Baghdad, the capital of the Abbasid Empire, was the centre of intellectual and scientific inquiry.[80]

Darwin saw that all species vary with time, and some variations bring advantages in the contest for food and struggle for existence. Organisms with favourable variations pass these characteristics on, and so new species arise from existing forms. As Darwin put in it his *Autobiography*, "Favourable variations would tend to be preserved, and unfavourable ones to be destroyed.... The result of this would be the formation of new species." He added, "Here, then, I had at last got a theory by which to work."[81] In other words, Nature selects which organisms and species survive. Lacking evidence for his theory of natural selection in action except for the adaptive radiation of the birds in the Galapagos Islands, Darwin had to rely on artificial selection through domestication: he bred pigeons. His theory nevertheless offered a "mechanism" for evolution that could be tested.

Darwin had wanted to become close to God and had chosen to study His creation (the species created by God). He had come to realise that Creation did not need the idea of God as species evolved in a process that is still continuing. He realised that when an animal has young there can be mutations of species. We have seen on p144 that as he researched into evolution "disbelief crept over me at a very slow rate" (*Autobiography*). Darwin was reluctant to publish his findings as he expected opposition from the Church. He was forced to publish when he received a letter from Alfred Russel Wallace containing a comparable hypothesis that threatened to rob him of his life's work.

On the Origin of Species by Means of Natural Selection, or the Preservation of Favoured Races in the *Struggle for Life*, now known as *The Origin of the Species*, was published in 1859 and sold out on the first day. The book stated that organisms produce far more offspring than can ever survive to become mature individuals, and that the number of individuals in a species remains constant. There is therefore a high mortality rate.[82]

The individuals in a species are not all identical but show variations in their characteristics. Therefore some variants are more successful in competing for survival than others, and some are less successful. (This deduction has been tested. Herbert Spencer restated it as: "Nature guarantees the survival of the fittest.")[83] The parents for the next generation will be selected from those individuals that have adapted best to the conditions of the environment. Offspring have hereditary resemblance to their parents. Therefore subsequent generations improve on the adaptation of their parents.[84]

The Austrian monk Gregor Mendel identified the "mechanism" for inheritance, which was published in 1900, after Darwin's death. Darwin believed that inheritance was a blend of male and female parents' characteristics. Mendel proved that characteristics were not blended: either the mother's gene or the father's dominates to give a characteristic. An unblended trait can therefore be inherited.

Neo-Darwinism restates the concept of evolution by natural selection in terms of Mendelian and post-Mendelian genetics. Genetic variation involves chromosome and gene mutations, apparently random assortment of maternal and paternal homologous chromosomes (pairs with the same structure) during crossing over and the recombination of segments of parental homologous chromosomes. Genetic variation is expressed in phenotypes (observable characteristics), some of which are better able to survive than others. Natural selection causes changes in the proportions of particular genes that may lead to new species.[85]

In short, natural selection involves:

(1)   a variation in a population;
(2)   an advantage over others depending on local environment (for example the advantage of one bird over other birds); and
(3)   a variation being passed on to offspring.[86]

Selection is in accordance with local conditions. It starts with a small mutation in DNA that makes for diversity which can be physically inherited. Thus a seed-eating finch encountering no seeds on a new island in the Galapagos, but plenty of cacti, adapts and develops a beak that can eat cactus. Local conditions were determined by the weather, which affects population growth. Thus in the Galapagos Islands the population of the large-billed ground finch, which eats seeds with its large beak, is increased in a dry year when seeds are plentiful but reduced in a wet year when seeds become scarce, causing starvation. Bad-weather conditions caused the first seed-eating finches to defect to eating cactus. The population of the warbler finch which eats insects with its tiny beak is increased in a wet year when insects are plentiful but reduced in a dry year when there are few insects about, causing starvation.[87]

In the same way, man is a species that adapted to local conditions and underwent variations in a struggle to survive. He did not mutate into a new species. A species is defined by breeding: it breeds within itself. (The full definition of a species is "a group of individuals of common ancestry that closely resemble each other and are normally capable of interbreeding to produce fertile offspring".) Unlike the 14 Galapagos species of finch there is only one species of man, modern *Homo sapiens*. If confronted with adverse environmental conditions, man – modern *Homo sapiens* – could in theory face the extinction that faces Lonesome George.

Darwin's theory of natural selection suggests that contrary to Aristotle's "Ladder of Nature" or Great Chain of Being (see pp178-179), which had man at the top above descending forms (mammals, whales, reptiles and fish down through insects to lower plants and inanimate matter), species were not individually created in their present forms but evolved into new species. Thus in the marsupials (whose young are in an external pouch in the later stages of pregnancy) the marsupial tree dweller evolved into the marsupial mouse, the rabbit-eared bandicoot and Tasmanian wolf. In eutherian mammals (whose young are in the uterus), the squirrel evolved into the mouse, the rabbit and then the wolf.[88]

In the same way, a fossilised finned fish known as a tiktaalik that lived in the Arctic, had shoulders, elbows, forearms, wrists, hands and fingers similar to those of humans, emerged on to land 365 million years ago and may have evolved into our common ancestor. The theory of natural selection suggests that this common ancestor evolved into apes and humans over a period of 35 million years.[89]

Apes and Humans

In linking all species including humans to the first cell the new philosophy of Universalism affirms the oneness of all living creatures and is open to, and prepared to be convinced by, the most controversial aspect of Darwin's theory of evolution, humans' descent from apes, though in a context of bio-friendly order rather than one of random accident.

A more immediate common ancestor for apes and humans than the tiktaalik may have been the ape/monkey *Aegyptopithecu* (whose fossils date to 35 million years ago, or mya) and were first found in Egypt). It evolved into the true ape *Proconsul* (whose fossils date from 20 million years ago) and the human-like ape *Ramapithecus* (whose fossils date from 10 million years ago first in India and Pakistan, and later in Kenya and the Middle East, long after Africa joined up with Eurasia and ceased to be an island 16 or 17 million years ago).[90] Nearer our time there were more ape-like hominids:[91]

Sahelanthropus tchadensis	7-6 mya, West Africa, Chad
Orrorin tugenensis	6 mya
Ardipithecus kadabba	5.5-4.4 mya
Ar. Ramidus	over 4.4 mya, Africa

There were 22 intermediate extinct species between apes and, or alongside, man (*Homo sapiens sapiens* being the twenty-third). The intermediate levels began with *Australopithecus, a genus* of fossil hominids, particularly *Australopithecus Afarensis* (a family of three of which left fossilised footprints in volcanic ash in Laetoli 3.6 million years ago, proving the species walked upright on two feet). The first partial skeleton of this hominid, named Lucy, discovered in Afar, Ethiopia in 1974, was 3.5 million years old.[92] It is possible that modern *Homo sapiens* descended from her, and that from her evolved *Homo habilis* and *Homo erectus* before *Homo sapiens neanderthalensis* and modern *Homo sapiens* (ourselves) shared a common ancestor about 400,000 years ago.[93]

The full list is as follows:[94]

Australopithecus	4.2-2 mya, fossils beyond 4 mya
1. *Au anamensis*	4.2-3.9 mya, North Kenya
2. *Au afarensis*	3.8-3 mya, East Africa
3. *Au africanus*	3 mya, found in 1924
4. *Au bahrelghazali*	3.5-3 mya, Chad
5. *Au garhi*	2.5 mya, Ethopia
Kenyanthropus	
6. *Kenyanthropus platyops*	3-2.7 mya, Kenya
Paranthropus	3-1.2 mya
7. *P. aethiopicus*	2.5-1.2 mya, East and South Africa
8. *P. boisei*	2.3-1.2 mya, East Africa
9. *P. robustus*	2.0-1.0 mya, South Africa
Species of the *genus Homo*	
10. *Homo habilis*	2.5-1.5 mya, Olduvai, Tanzania/Africa
11. *Homo rudolfensis*	1.9-1.7 mya, Kenya
12. *Homo georgicus*	1.8-1.6 mya, Georgia

One theory holds that archaic ancestors of *Homo sapiens* belonging to the genus *Homo* arose simultaneously in Africa, Europe and Asia some 2 million years ago.[95]

13. *Homo ergaster*	1.9-1.4 mya, East and South Africa
14. *Homo erectus*	2-0.03 mya, Africa, Eurasia, Java, China
15. *Homo cepranensis*	0.8? mya, Italy
16. *Homo antecessor*	previously thought to be 0.8-0.35 mya, now 1.2-1.1 mya, Spain, England
17. *Homo heidelbergensis*	0.6-0.25 mya, Europe, Africa, China
18. *Homo sapiens neanderthalensis* (chronologically descended from *Homo erectus*)	0.35-0.03 mya, Europe, West Asia
19. *Homo rhodesiensis*	0.3-0.12 mya, Zambia

Another theory holds that archaic ancestors of Cro-Magnon *Homo sapiens* arose in Africa about 0.4 million years ago, and certainly from 0.25 or 0.2 million years ago; and that modern *Homo sapiens* emerged from Africa 0.2 million years ago and replaced all other species of *Homo*.[96]

20. *Homo sapiens*, ancestors of *Cro-Magnon man* (The bones at Cro-Magnon, France, date to 35,000-10,000BC.)	0.4/0.25 or 0.2– 0.01 mya, Africa
21. *Homo sapiens idaltu*	0.16-0.15 mya, Ethiopia, France
22. *Homo sapiens sapiens* (i.e. modern *Homo sapiens*)	0.1/0.07/0.05 mya, Africa/ worldwide
23. *Homo floresiensis* (This may in fact be a dwarf version of *Homo sapiens* rather than a new species.)	0.10-0.012 mya, Indonesia

Humans began with the *genus Homo* (a name chosen by the 18th-century Swedish botanist Carolus Linnaeus for his classification scheme) some 2 million years ago, but studies of human evolution include earlier hominins (a specialist term used by palaeontologists to denote members of the subfamily or tribe of *H011111inae* or "humans", i.e. chimpanzees and humans). Within the genus *Homo* of archaic and modern humans there are species, such as *Homo erectus* and *Homo sapiens*.[97]

According to geneticists, modern humans are hybrids created by interbreeding between early hominids (a general term used by archaeologists for the family of ape-men and humans) and chimpanzees/bonobos, which began 6.3 million years ago.[98] Hominids and chimpanzees/bonobos are thought to have diverged from gorillas between 6 and 8 million years ago, perhaps 7.5 million years ago (though the recent discovery of *Chororapithecus abyssinicus* suggests the split might have been 10-11 million years ago). Hominids are thought to have diverged from chimpanzees and bonobos about 5.5-5.3 million years ago, before which they had a common ancestor.[99] (See diagram on p153.) For a million years hominids acquired chromosomes from chimpanzees/bonobos until hominids finally broke free from them 5.3 million years ago.[100] (Chimpanzees and bonobos are our closest living relatives and if they became extinct then gorillas would become our closest living relatives.)[101]

There is considerable uncertainty regarding the line of descent. *A. afarensis* is widely recognised as the ancestor of *A. africanus*, but *A. africanus* is no longer recognised as the ancestral hominid for modern *Homo sapiens*. The descent from *Australopithecus* to *Homo* is now believed to be more indirect.[102] The diagram on p154 allows for alternative evolutionary pathways for the descent.

There are two competing theories as to how modern humans spread across the globe. One, the "multi-regional" idea, holds that modern humans appeared simultaneously in Africa, Europe and Asia, descended from *Homo erectus* who left Africa some 2 million years ago. This theory sees *Homo erectus* as evolving and producing modern *Homo sapiens* in Africa.[103] Recent research suggests that *Homo ergaster*, who lived between 1.9 and 1.4 million years ago rather than *Homo erectus* is the link between apes and humans.[104] Evidence for this came from the discovery of the bones of over 30 people 780,000 years old at Gran Dolina on Atapuerca Hill in Spain in 1994.[105] On this view, *Homo ergaster* passed through *Homo heidelbergensis* and then branched into *Homo erectus* in China

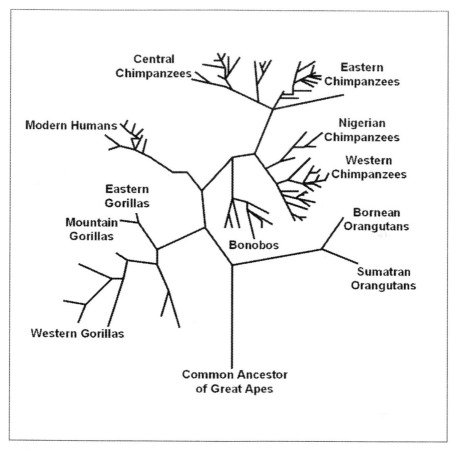

Central Chimpanzees

Eastern Chimpanzees

Modern Humans

Nigerian Chimpanzees

Western Chimpanzees

Eastern Gorillas

Mountain Gorillas

Bonobos

Bornean Orangutans

Sumatran Orangutans

Western Gorillas

Common Ancestor of Great Apes

and Divergence of gorillas, chimpanzees, and bonobos and ancestor humans *Homo neanderthalensis* 0.5 million years ago and of modern *Homo sapiens* 0.4 million years ago.[106] (See map on p156.)

We have seen that a species is defined by breeding and breeds within itself, and that modern *Homo sapiens*, belonging to one species, mates within its species but not outside it. Books say that the predecessors of modern *Homo sapiens*, Neanderthal and Cro-Magnon man, did not interbreed with modern *Homo sapiens* as they were separate species. However, they had a common ancestor, archaic *Homo sapiens*,[107] and there is some evidence that the two siblings did interbreed with modern *Homo sapiens*.[108] When they died out, only modern *Homo sapiens* remained. By 27,000 years ago only one hominid species survived – our own.

The other theory holds that modern humans, known as Cro-Magnons, emerged from Africa some 0.2 million years ago and spread across the globe, replacing all other species of *Homo* including *Homo erectus*.[109] It must be said that there is no

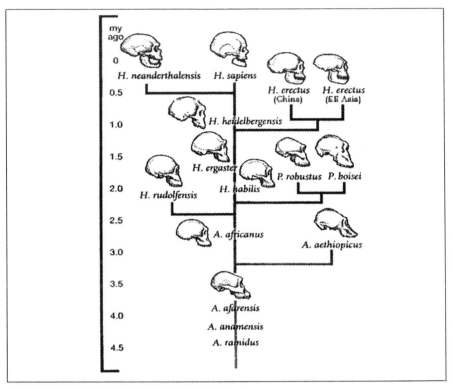

Family tree or bush: descent of modern humans from 4.5 million years ago

clear link between the appearance of modern *Homo sapiens* in Africa 200,000 years ago and older human species.[110] Climate change may have been responsible for modern *Homo sapiens'* move out of Africa, our true "Garden of Eden", into Eurasia. We have seen that a Milankovitch 100,000-year advance of mountain glaciers peaked around 200,000 years ago, and this may have impacted on Africa by bringing colder weather, driving modern *Homo sapiens* northwards. There seems to have been a new migration of modern humans, *Homo sapiens sapiens* (our species), from Africa about 100,000 years ago, and certainly between c.70,000 and c.50,000 years ago (see map on p156), perhaps also on account of climate change. This migration replaced all existing hominid species in Europe and Asia. There is fossil evidence in Israel that modern *Homo sapiens* had reached the near East 100,000 years ago.[111] The "out-of-Africa" idea sees a more recent African origin for modern *Homo sapiens*, and there is more of a consensus for this African siting.

There is a compromise between the multi-regional theory – which proposes

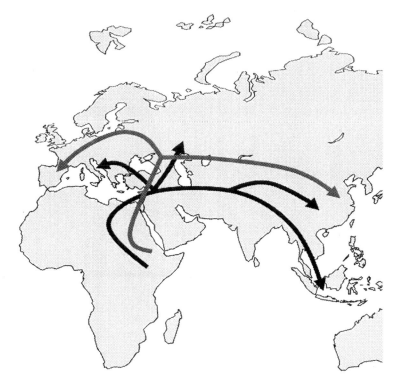

➤ *Homo erectus* expansion beginning about 1.8 million years ago

➤ *Homo heidelbergensis* expansion beginning about 0.6 million years ago

Spread of hominids 1.8 million and 600,000 years ago

that there was no single geographical origin for modern humans and that there were independent transitions from *Homo erectus* (or *Homo ergaster*) to *Homo sapiens* in regional populations – and the "recent African origin" theory that all non-African populations descend from a *Homo sapiens* ancestor who evolved in Africa just over 100,000 years ago. It is the assimilation (or hybridisation) theory that gene flow between the early human populations was not equal. Modern humans could therefore have blended modern characteristics from African populations with local characteristics in archaic Eurasian populations, so that genes are from archaic African *and* non-African populations.[112]

DNA and the Oneness of all Living Things

The Universalist philosopher is open-minded about "the new scientific

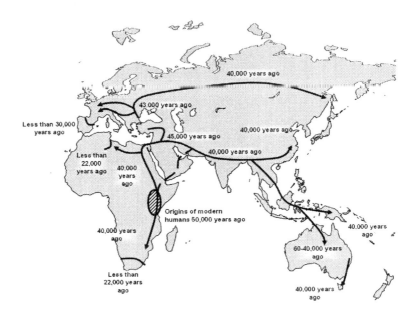

Emergence of modern humans from Africa c.50,000 years ago

interpretation" of the origin of *Homo sapiens*, but it has to be said that there is no widespread agreement on the exact links in the descent from ape to human. Darwin wrote in *The Descent of Man*: "In a series of forms graduating insensibly from some ape-like creatures to man as he now exists it would be impossible to fix on any definite point when the term man ought to be used." Today modern *Homo sapiens sapiens* looks to be less than 100,000, perhaps between 70,000 and 50,000, years old despite 13 previous species of *Homo* and a previous 9 intermediate forms – 22 previous species in all. There is therefore a weakness in Darwin's view that humans are descended from apes.

Nevertheless, modern humans have 98.4 per cent of genes in common with chimpanzees – 99.4 per cent if we omit bases that could be changed without affecting amino acid[113] – and evidence of our descent can be found in anatomical traces of our tails and in the different levels of the human brain (some instinctive, some rational). There are moves to include chimpanzees in the same *genus* rather than the same taxonomic family as humans.[114] As humans share at least 98.4 per cent of their genes with chimpanzees the links are likely to be permutations of the intermediate species in the family tree on p154.

The discovery of DNA, the universal genetic code, has confirmed that all humankind within modern *Homo sapiens* is interrelated.[115] (DNA was first discovered in 1869 but its role in genetic inheritance was not demonstrated until 1943 and its structure was not determined until 1953.)[116]

DNA goes back a long time. It can theoretically survive one million years.[117] The oldest DNA was found in a Greenland ice core in 2007, genetic material taken from pine trees, butterflies and other organisms living 800,000 yeas ago. This surpasses DNA from woolly mammoths, plants and fungi 350,000 years old, found frozen in Siberia in 2003.[118] The oldest DNA for modern *Homo sapiens* was mitochondrial DNA passed down through the maternal line in East Africa (Tanzania and Ethiopia). A genetic study by Dr Sarah Tishkoff's team and others who collected DNA samples from 1,000 Tanzanians between 2001 and 2003 found that the most ancient populations include the Sandawe, Burunge, Gorowaa and Datog people who live in Tanzania. They have links to the female ancestor of every person on earth who is thought to have lived 140,000-160,000 years ago.[119]

In practice, human DNA goes back to "coalescence", the time when any sample of genetic sequences of living things (bacteria, archaea and eukaryotes) can be traced back to a common ancestor in the past. Due to the apparently random elimination of ancient genetic lineages, the most recent common ancestor (as opposed to the most recent common matrilineal, or female-lineage, ancestor for mitochondrial DNA) remains at a constant time-distance from the present, and in our case was living 10,000-11,000 years ago. Human mitochondrial DNA (inherited only from one's mother) shows coalescence at around 140,000-160,000 years ago and Y-chromosome DNA (inherited only from one's father) shows coalescence at around 60,000-70,000 years ago. In other words, all living humans' female-line ancestry can be traced back to a single female (mitochondrial Eve) living around 140,000-160,000 years ago, and all living male-line ancestry can be traced back to a single male (Y-chromosomal Adam) living around 60,000-70,000 years ago. Research on many genes has established coalescence points from 2 million years ago to 60,000 years ago. However, it seems that all living humans' female-line and male-line ancestry can be traced back to a small number of people alive around 60,000-70,000 years ago, when there was a bottleneck of the human population.[120]

Human DNA is thought to go back to around 74,000 years ago at the most. Yet,

TIMELINE
Origin of the Universe and Life on Earth, and the Evolution of *Homo Sapiens*

Billion years ago	Event
Before 13.7	Moving infinite
13.7	Big Bang
13.32	Matter era begins, galaxies and stars formed
5.5	Formation of the sun
4.6-4.55	Formation of the earth
4.4	Liquid water first flowed on earth, first oceans
4.2/3.96/3.9	First rocks
4-3.8	Life begins on earth, first cell
3.8-3.5	First prokaryotic cells
3.7/3.5	First fossils of primitive cyanobacteria
2.7	First chemical evidence of eukaryotes, first stable isotopes
2.7-2.3	Huronian Ice Age
2.6	Bacteria living on land
2.5-2	First multicellular eukaryotic cells
1.9	Oxygen atmosphere develops
1.8	Oldest multicellular fossils
1.5	First photosynthesis

Million years ago	
850-630	Cryogenian Ice Age
800	First multicellular life
700/575	First animals
543	First shelled animals
533-525	Cambrian explosion
520	First land plants
500-450	First fish Insects and other invertebrates move on land First colonisation of earth by plants and animals
460-430	Andean-Saharan Ice Age
400	First vascular and seed plants, and insects
365	Tiktaalik moves onto land
360	Four-limbed vertebrates move on land

355	First reptiles
350-260	Karoo Ice Age
320	First amphibians
290	First conifers
250/225	First mammals and dinosaurs
200	Dinosaurs dominate
160	First birds
135	First beaked birds First flowering plants
65	Large dinosaurs extinct, mammals dominate
60	First primates 40/33.7-present Our Ice Age
35	First apes
15	Apes diverge from other primates
14	Orangutans diverge from other apes
8-6	Gorillas diverge from chimpanzees, bonobos and ancestors of humans
5-5.3	Ancestors of humans diverge from chimpanzees and bonobos
4.43	First hominids
2.5	Bonobos diverge from chimpanzees First stone tools
2.3-2	Early *Homo*
1.8	*Homo erectus*

Thousand years ago

400-250	First archaic *Homo sapiens*
250	First *Homo neanderthalensis*
200	Glacier advance peaks Ancestors of modern Cro-Magnon *Homo sapiens* leave Africa
150-130	Anatomically modern humans (*Homo sapiens sapiens*) arise in Africa
115	Glacier advance peaks
100-50	Modern *Homo sapiens sapiens* leave Africa
40/35-10	Cro-Magnon man in France
29/27	*Homo neanderthalensis* extinct
11.5	Holocene interglacial begins
10	First domestic animals
5	First writing and civilisation

the species of modern *Homo sapiens*, the ancestors of Cro-Magnon man, is about 200,000 years old. So, why does DNA make modern *Homo sapiens sapiens* look younger than he appears from the fossils? Is it because *Homo sapiens sapiens* did not emerge until just over 70,000 years ago, with different DNA from the ancestors of Cro-Magnon man and other *Homo* species, whom he replaced? Perhaps. But the answer may lie in ancient volcanic activity. Some 75,000 years ago, there was a huge eruption of a volcano at Toba in the Indonesian island of Sumatra. This released energy equivalent to about one gigaton of TNT (three thousand times greater than the 1980 eruption of Mount St Helens), and the ash-cloud is thought to have reduced the average global temperature by 5 degrees C for several years and to have intensified the Ice Age. The fall-out and the ensuing climate change are thought to have wiped out most of our species, reducing us to between 1,000 and 10,000 breeding pairs[121] and perhaps damaging our DNA, so that in effect we had to start again. The Toba catastrophe theory is just a theory, but there *is* a dearth of genetic material before 75,000 years ago.

We can all now experience the relatedness of humankind according to DNA. In 1994 Professor Bryan Sykes, a leading world authority on DNA and human evolution, was asked to examine the 5,000-year-old Ice Man, who had been found trapped in glacial ice in northern Italy. By examining the Ice Man's DNA he was able to locate a living relative of the man. His research located a gene which passes undiluted from generation to generation through the maternal line. Having plotted thousands of DNA sequences from all over the world he found that they clustered around 36 groups or clans, of which there are only seven in Europe. Therefore everyone of native European descent can trace their ancestry back to one of seven women. He found that there are 17 paternal clans, five of which are found in Europe. (The Sykes technique traced my maternal line back 10,000 years and my paternal line back 40,000 years.) Thus modern *Homo sapiens* – the 6.7 billion currently alive – can be tracked back to one woman, our "Eve", who lived about 140,000-160,000 years ago as one of only one or two thousand (or as many as ten thousand) individuals whose lines did not survive.[122] This woman is the mother from whom all living human beings are descended, and in terms of her all human beings are literally related as distant cousins many times removed.

We all look back to one "mitochondrial Eve" and "one Y-chromosome Adam", who is thought to have lived some 60,000 years ago. Both, in turn, had a long line behind them. In the "multi-regional" theory, both were at one time thought to be have been ultimately descended from *Homo erectus* via the

Neanderthals. (We have just seen that *Homo ergaster* may be a better link between apes and humans than *Homo erectus*). Fossil and genetic evidence suggest that anatomically modern humans emerged from Africa around 50,000 years ago or at the most 100,000 years ago (as distinct from the migration of their predecessors that began 200,000 million years ago),[123] and in a few thousand years replaced all other species of humans across the Old World.

It has been suggested that our "Eve" might have been Lucy, but she lived 3.5 million, not 140,000-150,000, years ago. In the "out-of-Africa" theory, both Eve and Lucy were descended from the early humans who may have left Africa 200,000 years ago and certainly did leave 50,000 years ago. We have seen that the compromise assimilation theory sees humans as blending African and non-African genes.

In seeing all human beings of the modern *Homo sapiens* species as coming from 36 verifiable clans with one clan mother and clan father, DNA research dramatically reveals the fundamental oneness of the modern *Homo sapiens* species and humankind.[124] The evidence is even more startling,[125] for it just as dramatically reveals that all living things have DNA and their genetic material has a universal genetic code which can be read. Biochemical analysis suggests a common origin for all living things. Thus the evidence suggests that just as biology came out of one point, *all* biological creation is ultimately one and interconnected via measurable DNA, and that all species came out of one cell. DNA brings us face to face with the fundamental interconnectedness between every living thing and with the hypothesis or theory we have just considered, that, as Darwin believed, life itself – the first cell – evolved from non-living, inorganic resources of the earth. (See Timeline on pp158-159.)

Darwinism: Evidence and Doubts

The Universalist philosopher affirms the oneness of living things and is open to the theory of evolution but dubious that our complexity could have evolved by random accident.

The evidence for evolution has accumulated since Darwin's day, when the adaptive radiation of the Galapagos Islands' birds was the strongest evidence for natural selection. There is now considerable evidence from palaeontology – notably from fossils and rock-dating, but also pollen and spore microfossil analysis. (Spores date back more than 450 million years ago, and the oldest pollen to 330 million years ago.)[126] Although no exact links have been found in the descent from ape to human, comparative biochemistry has now established that a chimpanzee is at least 98.4 per cent related to humans in terms of its DNA. Immunology detects differences in proteins and also establishes that chimpanzees and gorillas (African apes) are 97-98 per cent related to humans. More evidence for evolution has been provided by geographical distribution, comparative anatomy, comparative embryology, artificial selection and systematics (the classification of living things within a phylogenetic scheme).[127]

There is a problem with focusing on evolution in the Galapagos Islands as the oldest of these (to the east) are only 4-5 million years old while those to the west are a mere 120,000 years old, having risen from the sea during volcanic action. As dinosaurs died 65 million years ago and it took 35 million years for apes to become humans, and giant tortoises are thought to be 4 million years old, there does not seem to be enough time for evolution to have taken time in the Galapagos Islands. In fact, they are on the north of the Nazca tectonic plate that is passing slowly from west to east under Latin America at the rate of 1-2 inches (5cms) a year. It is thought that there may have been islands in the Galapagos region for 70 or 80 million years, not merely five, and that the old islands have long since passed under Latin America. This would explain how evolution was able to take place in the Galapagos, with some creatures migrating from old islands to newer ones before they slid under the sea with the moving plate.[128]

Reductionist Darwinists, looking at Darwin's entry in his *Notebook on Transmutation of Species* on 28 September 1838 (see p146), maintain that natural selection is a chance, random process. They maintain that weather conditions are chance random events. Individuals, they argue, are of no consequence compared to the species. The first finch to cause a mutation was important, but the succeeding individuals are unimportant. The blue-footed booby lays two eggs. The second and last chick dies in 98 per cent of cases. Darwinists argue that it is only born to guarantee the survival of the species should the first chick be killed by a frigate bird. So it is with humans. The fate of the individual is relatively unimportant to Darwinists, compared with the survival of the species. Darwinists hold that life throws up chance, random species which are successful if they survive.

However, the chance, random setting of life contrasts sharply with the order of the double helix of the DNA cascades, through which the apparently chance and random mutations are transmitted. For the two strands of DNA's double helix wind in opposite directions and mirror each other exactly, suggesting an orderly transmission of an apparently chance or random event. If Darwin's theory of evolution is correct, chromosomes should evolve and not remain unchanged for tens of thousands of years.

There are many doubts about Darwinism.

In biology, Darwin's theory requires complex structures to work immediately if they are to be preserved in the struggle for survival. How are such structures built up without being extinguished? How can a stick insect evolve safely from a structure that does not look like a stick?[129]

In palaeontology, new species develop suddenly. In the Cambrian explosion, biology's Big Bang in Burgess Shale, Canada 530-525 million years ago, at least 19 and perhaps as many as 35 out of 40 phyla first appeared on earth. (Phyla are the highest biological categories in the animal kingdom, for example corals and jellyfish, squids and shellfish, crustaceans, insects and trilobites, sea stars and sea urchins.)[130] (See Appendix 2.) There was also a Big Bang of birds and an explosion of flowering plants.[131] Darwin thought new species would develop gradually, not suddenly.

The earth was thought to be no more than 100 million years old in Darwin's time.[132] We have seen that it is now thought to be 4.55 billion years old, and there have been at least 3.5 billion years of the evolutionary history of life on earth.[133] Life forms are vastly more complex than Darwin thought and 100 million years was not enough for such complexity to develop. 4.55 billion years is not enough when the first 3.5 billion produced only bacteria, leaving too little time for such complex organs as the human eye to evolve. Can random fluctuations in accordance with the theory of natural selection produce such complex and organised patterns which seem to be purposive?

Some species of hydra hunt and protect themselves with poisonous guns, tiny stinging cells that fire a poisoned hair. The flatworm generally will not touch a hydra, but occasionally it seeks one out and swallows the poison guns without digesting them and then places the guns on its own body for its own protection and fires them at attackers. When it runs low on poison ammunition it seeks out another hydra and swallows its poison guns.[134] According to Darwinism chance, random natural selection and the instinct to survive account for this purposive and intelligent, co-ordinated behaviour. This sense of purpose seems to be present in even the lowest forms of life, contrary to Darwin's theory.

Darwin's theory of sexual selection ignores the fact that some species (for example, the whiptail lizard whose young are produced from unfertilized eggs) are only females; that some animals change sex to maintain their numbers and that over 300 species indulge in genetically useless homosexual behaviour, contrary to Darwin's theory.[135] If apes, fish and dinosaurs evolved or mutated into humans, mammals and birds, why have no convincing and undisputed intermediate forms been found, either in fossils or bones?

Other doubts include:[136]

(1) Species that evolve the least last longest.

(2) Antibiotic resistance comes from an orderly, organising "mechanism" (a reductionist word Universalists dislike) within an organism when evolution and natural selection are supposed to be chance, random, blind.

(3) Co-operation is important in evolution, contrary to Darwin's survival of the fittest.

(4) Nature has patterns that exceed chance – for example, marsupial mammals occupy the same or similar niches as eutherian mammals. (See p149) The two diverged in the Cretaceous period and show parallel adaptive radiation.

(5) New species appear without a long succession of ancestors whereas Darwinism requires gradual changes from one species to another.

(6) Some creatures (such as cockroaches and coelacanths) survive without evolving, even though they mutate.

(7) A bird's lung, a rock lobster's eye and a flagellum of bacteria are too complex to be produced by chance.

(8) There is little evidence for large changes as a result of natural selection.

(9) There is not enough time for random evolution, as we have seen, let alone random evolution that is continually impeded by natural selection.

(10) There is no universally accepted descent from apes to humans, as we have seen, and although there are 22 intermediate extinct species, the descent is

subject to academic dispute.

Doubters often put forward their doubts in considered good faith while admiring some aspects of Darwin's generalising theory. Such people have thought deeply and would be shocked by the tone adopted by some Christian dissenters, one of whose websites[137] claims: "I believe that Darwin was dishonest and lied to help get his theory accepted." He claims that the Galapagos finches had almost nothing to do with the formulation of Darwin's theory; that Darwin's claims about the role of the finches in evolution were retrospective in the second (1845) edition of his *Journal*; that Darwin only belatedly came to believe that "one species had been taken and modified for different ends"; and that he revised his *Journal* in 1845 to inject into his pre-1839 text what he had pieced together between 1839 and 1845. "Dishonesty" and "lies" are grave charges to be levelled at Darwin and do not do justice to the process of the evolution of the idea of natural selection in Darwin's mind in the 1830s and 1840s, a slow musing familiar to all creative people and involving revisiting and reinterpreting half-formed ideas.

Darwin's theory of natural selection is a hypothesis, just as much of a theory as the origin of life. It is not finally provable because mutations over millennia cannot be replicated in a laboratory. Has Darwinism erred in being too focused on chance and what is random? Has it missed a design in the universe which suggests an intelligence behind natural phenomena?

Intelligent Design?

The Universalist philosopher is aware of the bio-friendly structure of the universe and its order principle, and while striving to be balanced and neutral, and to observe strictly philosophical categories rather than theological ones, he must take note of any "new scientific interpretation" of design in the universe.

Opposing the Darwinists is the Intelligent Design movement. William Paley was an 18th-century Anglican theologian who in *Natural Theology* (1802) argued that the existence of God can be demonstrated from the design of the universe just as the existence of a maker can be demonstrated from the design of a pocket watch.[138] Looking back to Paley, a group of American intellectuals[139] met in the 1980s to consider the complexity of information theory in relation to Darwinism.

Information theory stems from the work of an American engineer, Claude Shannon, in 1948. It is a special branch of mathematics that measures information in terms of the number of instructions required to describe it.[140] The more complex the structure, the more instructions needed to describe it. Simple order, they held, can occur without design, whereas complex order probably cannot. Every cell has living molecules which act as machines and perform a variety of complex tasks. The information in DNA is far more complex than the repetitive patterns in non-living salt crystals or the random structure of pebbles lying on a beach. Complex information in a book is organised by intelligence, and we should not be looking for how it was organised by chance.[141] The absolute maximum of the number of particle operations since the universe began has been calculated as 10^{150}.[142] The probability of generating a protein of only 150 amino acids in length by chance, at random, is less than one chance in 10^{180}, well beyond the most conservative estimate of acceptable probability.[143]

The Intelligent Design movement took account of a number of post-war developments in assessing complexity. It considered fractals, a class of complex geometric shapes that are self-similar, which extended information theory to the symmetry in apparently irregular natural phenomena. In *The Fractal Geometry of Nature* (1977) the Polish-born Benoit Mandelbrot had seen mathematical similarities in many natural shapes which suggested an underlying order, which he called fractals. The word "fractal" comes from the Latin *frangere*, to break, so a fractal is something broken or shattered that still retains symmetry and a degree of order. Examples of fractal systems include: snowflakes; tree barks;[144] ragged terrains of mountain ranges seen from the air; the branch systems of trees; the curly shell of the nautilus mollusc; overlapping plates of a pineapple skin; the coastline of Britain; the scattering of galaxies; clusters and superclusters across the universe; music in the key of 1/f; the formation of fern leaves; rainfall patterns; fjords; DNA; branching blood vessels in our lungs; the structure of leaves; frost on glass; waves created by the combination of wind, water and tides; the structure of cauliflowers; and the spiral of plants whose whorls all turn the same way. In mathematics, fractals' irregular details or patterns are reiterated.[145]

Similar to fractals was the comparison of people's features in terms of their proportion. The science of proportion has focused on "the Golden Ratio", the number τ or tau (or in the early 20th century φ or phi named after the Greek sculptor Pheidias's perfect proportions). This never-ending, never-repeating number has been known since very ancient times[146] and was used in Penrose's tiles[147] in 1974. The Moorish builders of the Alhambra Palace in Granada used it in producing mosaic tiles with 17 different symmetries, unaware that 17 is the maximum number of symmetries that such tiles can display.[148] The secret of the Golden Ratio, which has been passed on from artist to artist down the centuries, is that beauty has a proportional ratio of 1.6180339887. Thus a perfect nose to a face is 1:1.61803.[149] The same ratio can be found in other parts of the body. The widespread nature of this proportion suggests that the body's limbs are the result of order and design rather than of random chance, which might not have achieved this mathematical figure in so many variations so consistently.

The antiquity of the Golden Ratio can be judged from a tablet discovered in Susa, Iran, in 1936 which says: "1 40, the constant of the five-sided figure." The Babylonians used the sexagesimal system that is based on 60, so the measurement was 1 and 40/60 or 0.666, giving a total of 1.666.[150] In Pheidias's Parthenon the proportion of the columns from the top of the façade to the bottom of the pedestal is also 1:1.61803.[151]

Intelligent Design also considered general systems theory, the study of human life and social organisation in terms of systems and subsystems. The theory had been developed by the biologist Ludwig von Bertalanffy, author of *Perspectives on General System Theory*, in 1936 in the hope that it would identify laws and principles that apply to many systems and make sense of wholeness, differentiation, progress and order.[152] Nine levels of complexity have been identified: the structural framework; clockwork; a cybernetic device such as a thermostat; the cell; the plant system; the animal system including ecological systems; humans; the social system; and finally the transcendental. The social system includes health care, family, body, economic, information, banking, political and international systems. All systems have common elements: input; output; throughput or process; feedback; control; environment; and goal. The main rule is that a system is greater than the sum of its parts.[153]

In the 1960s "universality" came to be used of systems of interacting units, including avalanches, earthquakes, heart attacks and stock market meltdowns. These are made up of thousands of interacting units or events such as faults in plate boundaries and the buying-and-selling transactions by thousands of traders. A system at a critical point can be described by statistical mechanics, one of the oldest branches of natural philosophy. Universality's many systems involve patterns not unlike those of fractals. It reduces all complex systems to simple models and connects them.[154]

Out of systems theory in the 1940s and 1950s grew cybernetics (the science of communications and automatic control systems). This was followed in the

1970s by Catastrophe Theory (a branch of geometry focusing on changes in systems' behaviour such as collapsing bridges or mob violence)[155] and in the 1980s by Chaos Theory (which developed from a paper by Edward Lorenz in 1963, how apparently random or unpredictable behaviour in systems is governed by deterministic laws, for example irregularities in heartbeat, the weather and in the motions of clusters of stars where the randomness may be more apparent than real).

All these systems appear chaotic but have an underlying order. In Chaos Theory, disorder and chaotic systems have measurable characteristics. Universality is the opposite of chaos for systems at critical points are measured by the ubiquity of order, not chaos.[156]

Having considered all these interconnected studies of complexity, supporters of Intelligent Design assume that complex artefacts are not the products of chance or the *theory* of natural selection but have been *designed* (planned, shaped and produced) by an intelligence that came first. (Compare a computer program which is designed by a programmer.) Darwinists assume that life forms are not designed but have developed by chance through the *theory* of natural selection, and see intelligence as an accidental development that evolved by chance some 100,000 years ago as a byproduct of accidental life. T.H. Huxley is credited with claiming that "six monkeys set to strum unintelligently on typewriters for millions of years would be bound in time to write all the books in the British Museum", but these words should probably be attributed to Sir Arthur Eddington in 1927.[157] (Sir Fred Hoyle, countering, said of Darwin's belief that life in the universe happened by accident, that the chances that life just occurred are about as unlikely as a whirlwind blowing through "a junkyard" and reconstructing a dismembered Boeing 747.)[158]

Supporters of Intelligent Design see Nature as comprising law, chance and detectible design, for some features of Nature illustrate design. ("Irreducibly complex", Michael Behe's words, meant "having complexity not susceptible to being reduced by reductionists".) They believe that evolution occurred not solely through law and chance, and that studying design helps us understand Nature better.[159]

Supporters of Intelligent Design are not Creationists, who assert the fundamentalist Christian teaching that the universe was created in six days by God in accordance with *Genesis* 1-6 and that evolution did not occur. Universalist philosophy must leave this Christian teaching to one side for it is non-detectible and untestable, and in the category and realm of theological belief, not science. (Arguably evolution by law and chance without design is also non-detectible and untestable, and in the category and realm of belief, not science – though many Darwinists and neo-Darwinists would be unhappy with this interpretation of their hypothesis.)

Opponents of Intelligent Design maintain that it involves observation and analysis of natural phenomena from astrophysics to micro-organisms and subatomic chemistry, and the discovery that a structure is "irreducibly complex" and could only have been created by an "intelligent designer" who or which is deemed to be the researcher's preferred deity or purposive force. They claim that the irreducible is always becoming reducible. They claim that Intelligent Design is "not science" even though it is empirical, testable and disprovable.[160]

Opponents also claim that the natural selection process is not random and involves selection of the fittest. Variations from one generation to the next may be random. A bird's beak fits into a flower not because it is carved by a designer but because it adapts to its surroundings, like a river shaping itself to a river bed. They evolve together.[161] However, unintelligent selection can be expected to produce an unintelligent result. The Darwinian selections are intelligent. So where did the intelligence come from? Where is it located?

Darwin's theory of natural selection and Dawkins' theory of selfish genes are currently being challenged by Edward O. Wilson, the US evolutionary psychologist or sociobiologist who champions group selection rather than kin selection. He focuses on the organized behaviour of ants, "superorganisms" that (as Virgil knew in *The Georgics*, book IV) act altruistically for the group and their colony rather than selfishly. Ants live and die for the good of the colony, and in this there are parallels with humans. Group selection is not about the working of the Whole, but it does require us to see evolution in a new way. Arguably (and it is not an argument I would press), ants are more important than humans as their species weigh more than the human species: though an ant weighs one-millionth of a human, the 10,000 trillion ants in the world weigh more than the 6.7 billion humans.[162]

Another example of altruistic behaviour for a group is the pied babblers in the Kalahari desert in southern Africa, who forage for food (small snakes and scorpions) in groups. One altruistically stands sentry, watches for predators, sings a "watchman's song" and makes alarm calls when it spots a predator: gabar goshawks, pale chanting goshawks, sparrowhawks, slender and yellow mongooses,

Cape cobras and puff adders. The sentry allows the group to concentrate on finding food without having to look out for predators. Clearly the altruistic sentries co-operate rather than compete to survive, and interact socially.[163]

There is another new way of seeing evolution. Perhaps all the species developed because they *wanted* to develop, and they used willpower to secure their evolutionary development. The drive for evolution may be a matter of inner will, a persistent drive for self-improvement. Each organism may have a deep aspiration to better itself, to realise its possibilities, to transcend itself and take itself to its next stage. Just as a manual labourer thinks, aspires and plans a better future than digging roads and carries the image of a better life persistently in his imagination, so every organism may have an imaginative image to realise better possibilities in relation to its environment. Every organism may follow a deep subconscious drive that is reflected by willpower, which in fact acts out and implements coded instructions inherited through DNA in its cells. In this sense evolution may be a matter of effort, willpower and inner drive of imaginative will – which is contained within the workings of the order principle.

This is not proved of modern *Homo sapiens* and it is difficult to see how it could be proved of all the other species of insects, fish, birds, reptiles and mammals. This view will therefore have to remain a theory, but it is attractive as it suggests that humans owe their present exalted position to their past efforts and striving, which have enabled them to transcend their selves and *become* their image of their future selves.

New thinking in biology has thrown up morphogenetic fields and the morphic resonance of Rupert Sheldrake, which have not yet been proved and remain theory. Shunned by academic biology, Sheldrake has focused on the instinctive fields surrounding animals and all living beings, his proposal being that their sixth sense and memory can be found in their surrounding fields.[164]

Intelligent Design has a place in raising doubts about Darwinism but it has not defeated Darwinism, which remains, for all its shortcomings, the only workable theory of evolution – despite offering an incomplete view of what drives evolution.

Expansionism: "Mechanism" of Order

The Universalist philosopher thus says "Yes" to Darwinism as an imperfect but unbettered description of a process, what happens in evolution, but "No" to Darwinism when it claims that why it happens is always a chance, random accident.

To reductionist cell biologists, biochemists or Darwinists, there is no reason why we are alive beyond accidental combinations of atoms into a cell, and accidental mitoses and DNA copies that have led to mutations and new species. To reductionists the Darwinist process of adaptation, variation and the struggle for survival by accident and chance explain the drive from the first cell to all the species in our universe.

But there is another way of thinking, the Expansionist (or integrationist) way. Whereas reductionism reduces from the very complex to the most simple and from the Whole, which it takes to bits, to its parts, on whose functions it dwells, Expansionism expands from very complex individual cells, individual structures, organisms and species to the whole universe and sees humans as products of the whole universe, the One. Whereas reductionism explores the functions of parts broken off from the Whole and reduced or simplified, Expansionism explores the functions of parts in relation to the Whole, integrates and acts integrationistically,[165] putting back together what has been broken off.

Such thinking is what Coleridge meant by the "esemplastic power of the imagination",[166] "esemplastic" coming from the Greek "*eis en plattein*", "to make into One". It is the opposite to the faculty of the reason which analyses, makes distinctions and reduces into bits. It puts the bits back together into a whole. It is like finding a shattered Roman vase in potsherds in the ground and sticking it together to restore its original shape.

Such thinking is also Universalist thinking. It has found fragments and attempted to unite them. In science it has found mechanism and vitalism, reductionism and Intelligent Design, and has attempted to put them together. Whereas reductionism works by simplifying – some say, oversimplifying – and leaving out crucial aspects such as order, Expansionist Universalism relates the functions of broken-off parts back to the function of the Whole from which they were fragmented, and through this relation brings newer meaning to the parts.

The question now is how to put all atoms and all cells back together into an Expansionist view of the universe that, besides uniting relativity and quantum theory in cosmology in an underlying unified field, includes both the thinking of reductionism and of Intelligent Design.

What is the Expansionist view of the origin of life? It acknowledges reductionist Darwinism but adds a sense that we are products of the universe, of the 40 "coincidences" we looked at in ch. 5. It brings to Darwinism a sense of the complex order in the universe that made life possible: a self-organising drive. It proposes a "mechanism" within cells that has driven evolution by natural selection so that all species have an instinctive place in a vast ecosystem, the whole bio-friendly universe in which, through 40 "coincidences", humankind can sustain itself. It proposes that, as Newton surmised, sunlight has a role in driving the evolution of plants whose seeds have chlorophyll and carotene pigments that trap light energy, and also of insects, fish, birds, reptiles, mammals and humans. It studies the nutrients in leaf and skin. In short, Universalism reforms Darwinism by introducing a "mechanism" of self-organising order which is unfortunately not biology's pressing need to identify.

Biological holism has tried to establish such a "mechanism". Holism is a 1920s idea that originated with Jan Smuts, the South African philosopher who later became prime minister of South Africa from 1939 to 1948. In *Holism and Evolution* (1926) he argued that everywhere we look at nature we see wholes which are hierarchical, whole fields at ascending levels. Smuts held that the universe is dynamic and produces ever higher-and-higher-level wholes which are ever more organised and complex. Evolution is the driver to ever higher unities, and Expansionism proposes that a principle of order drives evolution. The holistic view of organisms holds that irreducible wholes are the determining factor in biology, and that control-and-regulation "mechanisms" operating at molecular level make these organism systems cybernetic or automatic-control systems which follow self-determining and self-organising principles.

The concept of holism has become contaminated by vague, New-Age accreted meanings. One of these is the anti-reductionist Gaia hypothesis put forward by the scientist James Lovelock, who took up a suggestion by his friend William Golding, the Nobel prize-winning novelist, that the holistic universe should be named after the Greek goddess Gaia. The Gaia hypothesis regards the earth, the atmosphere, the oceans and organisms as one living process, a single evolutionary entity whose environment regulates itself as a whole and maintains its self-regulation by homeostasis so that the processes of life remain comfortable.

That the earth is an equilibrium regulating itself by homeostasis is an idea that is being taken increasingly seriously. The idea that the earth is alive as a single entity is as old as the human race and is in tune with Wordsworth's view of the Lake District's mountains. However, it has not convinced scientists, and including a goddess in a holistic label has not helped the scientific acceptance of the holistic view.

The holistic principle is under attack by rationalist reductionists[167] who claim that holists assert that the whole is more than the arithmetical sum of its interconnected (or associated) parts, and that it is a falsity to abstract the whole from the parts and then reify this abstraction, implying that the whole possesses a kind of existence over and above that of its parts in their actual state of interconnectedness. There is also a circularity as it explains the parts in relation to the whole, which is explained in terms of the parts.

Rational Universalists can counter this. It is known that plants have sensory apparatuses as they grope for water and move in the direction of the sun, but can one see the atmosphere, oceans and mountains as being sensory and self-regulatory? Clearly a principle present within atmosphere, oceans and mountains is doing the regulating, and I shall have more to say on this in chapter 9 (see p266). Without this principle within the Whole, holism is essentially an intuitional insight, a perception. In the last analysis, Gaia's form of holism without a principle within the Whole remains a theory.

To Expansionism, the survival of the fittest plants and animals is not controlled solely by external forces of natural selection but by internal forces as well, which may be triggered by sunlight. Mutations are driven by an organism's response to the environment. Thus plants set apart from each other produce more seeds. Snails grow thicker shells if crabs are in the vicinity. They select thicker shells before the crabs appear. In this they display intelligence and are ahead of natural selection. Barnacles grow thicker shells when predatory snails are about. Survival is not a matter of the weeding out of thin-shelled forms by natural selection. Changes in the phenotype ("observable characteristics") are purposeful rather than a random mutation. Evolution takes place at the level of an organism rather than the population. Plants, snails and barnacles are evidence of a Non-Random Evolutionary Hypothesis (NREH)[168]. This hypothesis states that the capacity of organisms to adapt to a new environment is built into their adaptive genes. As the capacity to adapt is within each organism, each adaptation can be made purposefully, raising notions of meaning and design. Cell biologists speak of "molecular vitalism", which evokes Goethe's view of plants.

I return to the idea that the drive for evolution may be a matter of inner will, a persistent drive for self-improvement, an inner inspiration to self-betterment which leads to self-transcendence (see p170). Evolution may be a thrust to realise better possibilities in relation to the environment, a persistent drive in the imagination to fulfil an image of the organism's next stage of development. This may be contained in coded instructions in the cells via DNA, and suggest the workings of the order principle – and also of Expansionism.

Universalism, seeking to reconcile conflicting views, would prefer to steer clear of the term "holism" and start afresh with "Expansionism". Expansionism covers the ground of biological holism and order's purposive drive to ever more complex hierarchical wholes. In this respect it looks back to the philosopher Whitehead's "philosophy of organism" in *Process and Reality*, which was delivered as lectures in 1927-8 and published in 1929. Whitehead, lecturing a year after Smuts' book, held that there is a definite property of wholeness that enables organisms to develop. Universalism, taking the baton from biological holism and its philosophical cue from Whitehead, identifies a biological "mechanism" within the neo-Darwinist model which spearheads an organism's teleological drive at molecular level to greater and greater wholes. In 1953 Crick and Watson elucidated the structure of the DNA molecule and found that strings of nucleotides, precisely sequenced chemicals in DNA, store and transmit assembly instructions for building protein molecules the cells needs to survive. The human genome of one human being consists of six billion "letters" of DNA[169] within 10^{360} possible strand combinations within 10^{23} bits of information[170] and, given the complexity of one individual's genetic make-up – the DNA recipe book that he or she inherited from his or her parents – it is quite possible that among the six billion letters there are coded instructions for the organism to implement the principle of order and to be purposive in driving towards more and more complex hierarchical wholes. The instructions are in four-character digital code-like letters in a written language or symbols in a computer code. Richard Dawkins, the British ethologist (or behavioural zoologist) has pointed out, "The machine code of the genes is uncannily computer-like." The letters appear to have been designed and no theory of undirected chemical evolution has explained the design in these letters. There is too much information in the cell to be explained by chance.

Perhaps this "mechanism" is within the coded information within the complex order of the double helix of DNA which instructs cells to make specific proteins – that perhaps contain purposive drives to more and more complex wholes? Perhaps this "mechanism" is stimulated by sunlight, which also stimulates chlorophyll in plants through photosynthesis, as Newton believed?

There is even more ignorance about the first principles of biology than there is about the first principles of cosmology. Biology is a younger science than cosmology, which looked back over 400 years to Copernicus, Galileo, Newton and later Einstein to establish its first principles. Biology's modern reign only goes back 150 years to Darwin and it has been sustained by the recent discoveries of Crick and his team. Biology has more revolutionary discoveries waiting to be made. The first principles of biology are still not fully understood, and we are ignorant of aspects of the instinctive behaviour of all creatures.

Neo-Darwinists, believing that Darwin explained everything, do not accept that there is still ignorance. Dawkins writes: "Natural selection is the blind watchmaker, blind because it does not see ahead, does not plan consequences, has no purpose in view Yet the living results of natural selection...impress us with the illusion of design and planning....The whole book has been dominated by the idea of chance, by the astronomically long odds against the spontaneous arising of order, complexity and apparent design. We have sought a way of taming chance....To 'tame' chance means to break down the very improbable into less improbable small components arranged in series....Slow, gradual, cumulative natural selection is the ultimate explanation for our existence."[171] However, making purposeless chance palatable and spinning design as an "illusion" cannot conceal the ignorance on key points and the huge chinks in the neo-Darwinist argument. Because of this ignorance, neo-Darwinists such as Dawkins who cite Darwin as if he were an Old-Testament prophet and believe he can do no wrong do not have an open-and-shut case.

Scientific evidence does not show an open-and-shut case for the Materialist, reductionist version of the universe. Looking at Nature through reductionist eyes (through the eyes of Darwin and Dawkins) exposes what is unsatisfactory in the reductionist, scientific view. Successful though it has been as a method in explaining parts, there is no overwhelming evidence for the reductionist, Materialist outlook.

Cosmology has found reasons to suggest that the Big Bang is bio-friendly. The same bio-friendly considerations apply to the emergence of the first cell from the soup, which may have been helped by interstellar bacteria. (See numbers (2), (30), (31), (32), (33), (38), (39) in ch. 5.) The bio-friendly universe and first cell suggest a life principle and principle of order. Where do they come from? If they did not just emerge from the soup by accident in a random way, a view I have rejected, then they developed from a predisposition to order with laws that contain within them the drive to develop towards self-improvement in a certain direction. The action of ultraviolet light on the cosmic "soup" may have been an aspect of this drive to order.

We have seen that there is a return to Einstein's cosmological constant that avoids a Big Crunch, the eventual collapse of the universe. Universalism links the universal predisposition to order to this constant and to the development into higher consciousness of modern *Homo sapiens* which is a self-organising drive. This higher consciousness has become higher and higher over millions of years and can appreciate Beethoven and Shakespeare. It has been able to create sophisticated religions far beyond the powers of the ape.

The stage modern *Homo sapiens* has reached today does not represent the final stage in evolution. We are an intermediary stage, and beyond us is a far more conscious being than we are, a modern self-improved and self-developed *Homo sapiens* with a far higher consciousness that will be able to open to forces that we cannot open to, just as in relation to us the ape could not appreciate beautiful art or open to the mystic Light. For *Homo sapiens* higher and higher levels of consciousness are ahead.

Philosophical Summary

What does biology offer the "new scientific interpretation of the universe and Nature" which is so important to Universalism? First, there is no evidence that the first cell arose by accident, it is not a foregone conclusion that man's place in the universe is a pointless accident. There must be a "mechanism" for a self-organising, ordering principle within the origin of life and evolution. Plate movements and Ice Ages have contributed to the right conditions that sustain life. (They can be added to the 40 bio-friendly conditions.) Evolution by natural selection as a process is right, but it has not developed advanced organs by random accident. There is no agreement on the descent of man from apes but DNA has revealed the oneness of all living things. Darwinism is imperfect but unbettered. There is an ordering thrust within the process of evolution.

To sum up, Universalist philosophy can take from biology and geology:

no evidence for the "soup theory";
no evidence that the first cell arose by accident;
no evidence that man's place in the universe is a pointless accident;
a "mechanism" for self-organising ordering has driven the origin of life and
   evolution;
land mass and climatic conditions have been right for sustaining life;
evolution by natural selection is the process of development, but little is

known about the ordering thrust of evolution;

evolution may be a matter of inner will, a drive for self-improvement, a persistent desire for self-betterment in relation to the environment, to realise the next stage of development;

no evidence on man's descent from apes but it is likely;

DNA has revealed the oneness of all living things;

Darwinism is imperfect but unbettered;

Expansionist, ordering thrust within evolution via DNA and perhaps sunlight.

The "new scientific interpretation of the universe" must include Nature's elaborate and complex system. The Universalist philosopher must now see if science can detect a self-organising system and principle of order within the practical side of biology: Nature's system of ecology.

7

ECOLOGY: NATURE'S SELF-RUNNING, SELF-ORGANISING SYSTEM

A very important aspect of the Universalist philosopher's "new scientific interpretation" of the universe is how all the species of Nature relate to each other here on earth, for the way they live and the function they have are evidence of the workings of the order principle's self-running and self-organising and are anything but random or accidental. They must have a profound impact on the philosopher's description of the universe.

In Nature there is a vast system that works day in and day out, quietly thrumming in the humming grass, open spaces and woods under the sky. Going about our daily business, listening to news and keeping appointments, we may be largely unaware of the vast ecosystem[1] under our noses, a natural unit of interacting living and non-living organisms and components through which energy flows and nutrients cycle to form and maintain a stable system. Individual organisms live in their habitat within one of the many local ecosystems within this vast ecosystem, the study of which is ecology, "a branch of biology dealing with the relation of organisms to one another and to their physical surroundings" (*Concise Oxford Dictionary*).

Until the 18th century philosophers did not doubt that this vast system was "the great chain of being" which began in the demiurge of Plato's *Timaeus* and the biological writings of Aristotle and was systematised by the Neoplatonist Plotinus. It saw all creation and natural beings as an ascending hierarchical order from the most simple to the most complex, from mineral to vegetable to animal to man and eventually divine beings.[2] Christianity incorporated this structure of an ascending gradation of the forms of life on the "*scala naturae*", Aristotle's Ladder of Nature or ladder of life. From the tiniest insect to godhead, all creation was linked and the universe was seen as an organic Whole ruled by order. The idea was especially strong in the Tudor time[3] and in the works of Shakespeare, where disorder at one level in the hierarchical chain (for example, in the mind of a king) is reflected in corresponding levels of the chain (for example, in strange roarings of lions). In its scheme of order all species had an allotted place. The idea was revived in the 20th century by the American historian of ideas Arthur Oncken Lovejoy in *The Great Chain of Being: A Study of the History of an Idea*.

1579 drawing of the great chain of being from Didacus Valades, Rhetorica
Christiana

Out of this "great chain of being" grew a classification system, which imposes a
general plan on the diversity of living things. The plan can be the binominal
system[4] (two names, one in Latin, e.g. common oak – *quercus robur*) or a
hierarchical scheme of seven taxa (kingdom, phylum, class, order, family, genus,
species). The order within the classification may seem to be imposed by the brain,
but the classification reveals family relatedness between species, and the
interconnectedness of species reveals the workings of the underlying universal
principle of order. There is a tree of all living things showing their
interconnectedness. (See p180.)

The system operates round a vast number of species. A world effort begun in

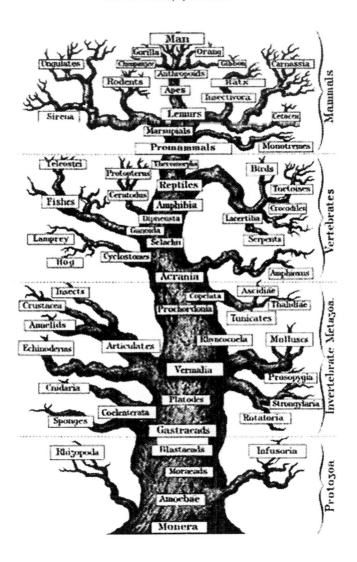

1874 genealogical Tree of Humanity according to the German biologist,
Ernst Haeckel, The Evolution of Man.

2000 to log every species had reached 1.4 million by 2008 and researchers expected to list 1.75 million species by the completion of the programme in 2011.[5] New species are being discovered all the time. In Antarctica, the Larsen Ice Shelf is breaking up, and in 2007 researchers made three visits with remote-controlled vehicles and discovered over 1,000 new species under the ice. These included new sea plants, barnacles, sea cucumbers, crustaceans, sea spiders and a new species of octopus.[6] In central Vietnam eleven plant and animal species were discovered in 2007, including two new butterflies, five orchids and a new snake species named the white keelback.[7]

A recent study has estimated that there may be as many as 2 million species on land. It has been claimed there are between 2 million and 30 million species of animals and plants on the planet, and that about 2 billion have evolved, struggled and become extinct since 540 million years ago. The World Conservation Union estimates that there are between 10 and 100 million species in the sea, 15 million being the most widely accepted figure, most deep down on the deep-sea floor, many the size of an aspirin.[8] The sheer wealth of sea species can be grasped by recognizing that there are 400 species of squid, 875 species of mollusc, 650 species of polychaets, 470 species of anthropods ("cockroaches of the sea"), 299 species of isopods and 310 species of bryozoans (which cling to rocks, for example seaweeds and sponges). There are also 8,800 species of flightless birds.[9]

Instinct: Inherited Reflex or Transmitted Order?

Universalist philosophy, seeking the "mechanism" that thrusts evolution forward to self-improvement and self-betterment, needs to take account of the instincts of all the species on which Darwin focused. Their behaviour patterns are amazingly ordered from the moment of their birth, and they go to elaborate lengths to renew their species and keep the system running. All the species are born with instinctive knowledge of how they are to operate. It seems that this instinctive knowledge has been inherited.

From about 1800 Jean Baptiste de Monet de Lamarck, the French biologist, discussed the idea that species were not "fixed" (or "immutable" in Darwin's word, unchangeable). Lamarck's mechanism of change was his "law of use and disuse" in which organs used a lot become well-developed whereas organs falling into disuse become atrophied – a change that he claimed was transmitted to offspring over several generations or even within a single generation as the inheritance of acquired characteristics. Thus the giraffe, faced with too little vegetation at ground level, stretched its neck to feed from trees and after stretching its neck for several generations acquired the characteristic of a long neck which it passed on to its offspring. Some biologists scoff at the concept of neck-stretching being passed on. August Weisman was largely responsible for the rejection of Larmarckian inheritance of acquired characteristics. He claimed that the gametes (cells able to unite in reproduction) passed from parents to offspring were not changed by the surrounding body cells. The "mechanism" of organic evolution was then thought to be natural selection, which is followed by neo-Darwinism.[10]

DNA has had a profound effect on how we regard inheritance of characteristics. To reductionists, inheritance of such acquired characteristics seems to be bipolar: it moves between a biological tendency (or substrate) passed into DNA by inheritance, and an environment to which it reacts.[11]

Thus, a young bird that has undergone a variation, such as the Galapagos cactus finch, has a general tendency to look for something prickly. The young bird searching for its food has to be guided by its parents to an actual cactus in the environment, which it soon remembers through its high-order memory system. Particular cultural or behavioural inheritance then comes into play via the persistence of memory.

A young robin is born with a tendency to pull at string or grass. It sees a parent pulling a worm, and learns to apply its tendency to particular worms in the environment. Again, a young cat is born with a tendency to tug string. Its sees a parent pull at a mouse's tail, and learns to apply its tendency to particular mice in the environment. In the same way a sea lion's pup is born with a tendency to suck and has to be guided to its mother's nipples in the environment, which it drains as the mother may be absent for five or six days fishing.[12] A similar tendency can be observed in young human babies who mouth a sucking action within minutes of being born, and have to be guided to their mother's nipples in the environment.

Reductionist biologists explain this instinct in terms of skills developed at random by the struggle to survive. To reductionists instinctive behaviour is more often innate (a tendency) than learned, a complex system of conditioned reflex – of stimulus and reflex to survive – learned in the struggle to survive and passed on genetically into DNA by inheritance.

The discovery of the genetic and structural aspects of DNA in 1943 and 1953 has focused anew on the inheritance of acquired characteristics, but with a time span of several generations rather than a single generation. Thus Darwin's cactus-eating finch transmitted the tendency to look for something prickly over several generations. The first cactus finch to transmit this knowledge created a mutation in the species that has continued to be transmitted through DNA.

It should be pointed out that sportsmen cannot transmit their sporting prowess, though they can pass on their physical athleticism through DNA and teach their children or grandchildren to emulate their achievements through cultural and behavioural example. It should be stressed that this is not automatic. Einstein's children did not become geniuses even though they had the tendency to study and were in contact with the cultural and behavioural example of Einstein.

We can see this interpretation if we look at animal behaviour. The observation of animals advanced through the work of Karl von Fritsch (on bees), Konrad Lorenz (on the fixating of jackdaws and greylag geese, which he found to be innate or instinctive, and genetically determined) and Nikolaas Timbergen (on gull chicks pecking a red spot on the side of their parents' bill to make them regurgitate food). They collectively won the Nobel Prize for physiology or medicine in 1973 for establishing ethology (the study of zoological behaviour which is related to ecology, genetics and physiology) as a modern science. Reductionists assert that we must avoid attributing human feelings (anthropomorphism) and planning (teleological outcomes) to animals, whose behaviour is determined by their need to seek out a favourable environment in which to survive. They do this by receiving sense data by using feedback, the "mechanisms" for response and co-ordination in the nervous system and the effecter organs. Their behaviour is therefore controlled by homeostasis.[13]

Instinctive behaviour can be a reflex response to danger or pain (for example, the earthworm's escape) or orientation (as when woodlice move quickly in dry conditions but slow down in greater humidity). Darwin defined instinctive behaviour as a series of complex reflex actions that are inherited – today we would say "genetically determined" – and therefore subject to natural selection. Nest building, courtship and food selection must be completed in the right order if the animal is to survive, and offspring without parental care have to learn to know how to survive very early on. Ready-made behaviour patterns can be found in animals that have no time for trial-and-error learning, like the new-born turtles.[14]

But there is another view, that instinct conveys the principle of order through a law that is transmitted in the DNA of all organisms (the same mechanism reductionists cite). To a Universalist the universal principle of order through DNA transmits innate tendencies or instinctive behaviour to organisms so they can integrate themselves into the limited ecosystem which orders their lives. The many limited ecosystems make up one gigantic whole ecosystem, within which there is a predisposition to order.

The Universalist explanation for this innate, instinctive behaviour – that it is genetically determined through DNA by the universal principle of order which is driving all members of all species to survive and play their part within an orderly system – can be seen to operate in many species. Thus, a cactus finch is driven by the principle of order within its DNA to look for something prickly and eat the fruit of a cactus, for in doing so it takes its place in the order of the ecosystem of the Whole. A young robin, a kitten and a sea lion's pup are driven by the "order principle" transmitted within their DNA to eat a worm, a mouse and to suck nipples.

In this connection, one of the mysteries that puzzles scientists is the "group mind" or collective behaviour of flocks of birds, herds, schools and packs. All birds wheel at the same time without losing formation. The same applies to swarms of bees and locusts. Sixty thousand bats can leave the narrow mouth of a cave in ten seconds without colliding, so complex are their individual radars which simultaneously measure each bat's own space in terms of 60,000 other darting bats. Instinctive co-ordination adds a dimension to cell complexity. A flock of thirty blue-footed boobies flying in the Galapagos Islands in early morning will at the same instant all dive straight down in formation and hit the sea at the same time. They have spotted a shoal of fish and each booby goes for a different fish. They never collide in mid-air or under the water although almost touching as they dive, and each always comes up with a different fish. No two birds target the same fish. The complexity of their communication system fills one with awe and leaves one wondering whether fields are responsible, and how it could possibly be the result of a random development, by chance.[15]

Is this co-ordination a tendency they have inherited and a characteristic they have acquired from the environment as a conditioned reflex, either by example or by inheritance – because they have been innately programmed by DNA to learn from the example of their immediate ancestors? The co-ordination that each bird or bat instinctively possesses can just as easily be seen as the working of an innate "order principle", a tendency inherited in DNA and awakened by information that may be transmitted by photons (see pp242-246) that (as in photosynthesis in the case of plants) drives each creature to develop the co-ordinating skills it needs to take up its rightful place in the environment's ecosystem and in the universe. The birds flock because the "order principle" keeps them together to be safe from predators. The blue-footed boobies dive for fish without hitting each other because the "order principle" dictates. The "order principle" directs bats, bees and locusts into their rightful place in the universe.

Land birds' migrating skills reveal awesome order. Both swallows and swifts migrate and return over thousands of miles to the same nest a year later.

Cuckoos know to where they have to migrate, which is a different country from the one frequented by the bird that has brought them up. The mobility of birds when they migrate is also stunning. As many as seven million starlings can form one flock, and as they fly in formation they form patterns against the sky, including one that resembles a perfect tower, in which millions of birds form a precise, straight-lined outlined shape. Recent research claims to have found that each starling keeps track of only seven other starlings, which is why they can stick together in this way to avoid being attacked by predators.[16] Perhaps this is the explanation for the "group mind", that all creatures including bats monitor no more than seven other creatures when flying in formation. Starlings have bat-like spatial awareness when flying in a flock and do not collide, and the same is true of many species of bird.

Sea birds' sense of direction and the instinctive knowledge of their chicks is equally stunning. All tube-nosed birds – albatrosses and petrels – lay only one egg per season. The young chick can be heard in the egg, calling to get out before it spends two days cutting a hole in its shell. On hatching it instinctively knows how to cross its bill with the inside of its parent's open bill to take regurgitated fish and squid. Both parents of the waved albatross in the Galapagos and the wandering, royal, black-browed, grey-headed and light-matted sooty albatrosses in the Southern Ocean leave their chicks for a month to find food, detecting it by their sense of smell. They may fly 600 miles in a day, and the wandering albatross can be on the wing for a week without resting – and they can live to be 75. Waved albatrosses mate for life and migrate. The male birds arrive back from their migration first, make a nest and sit on it for up to two months waiting for a specific female to arrive from overseas. The females always arrive at the right nest at the right time, returning to their nests with an unerring sense of direction and regurgitating the krill, fish, squid, crustaceans and carrion they have been carrying from their craw. At fledging time the chick instinctively knows that feeding must stop so it can empty its stomach and be light enough to take its first flight. It instinctively knows that it must climb all day to reach the best spot where it can take off successfully with the wind.[17]

Were these migratory and direction-finding skills tendencies they inherited or characteristics they acquired from the environment as a conditioned reflex, either by example or by inheritance – because they have been innately programmed by DNA to learn from the example of their immediate ancestors? These migratory skills can just as easily be seen as the working of an innate "order principle", a tendency inherited in DNA and awakened by information that may be transmitted by photons (see pp242-246) that (as in photosynthesis) drives each creature and directs birds to flee a coming winter and return well-fed the following spring. Birds migrate because the "order principle" dictates that that is what they have to do in relation to the ecosystem.

Equally baffling and bemusing is the way birds scent prey. In Antarctica, in the Southern Ocean between South Georgia and Elephant Island, I have seen at close quarters a pod or school of orca killer whales and calves attack and kill by drowning a minke whale and whereas there were only a dozen albatrosses and petrels of various species following our wake shortly before the attack, with no other birds visible from horizon to horizon, a few minutes after the attack began two hundred birds of different species arrived from far and wide, drawn by blood in the sea, and settled on the water in a great flock, hoping to share in the kill. They knew to come because their sense of smell is so acute that they can detect the odour of the kill in the atmosphere and fly across the horizon straight to the action.[18] There may be a parallel with Komodo dragons, which can detect the smell of a rotting carcass eleven kilometres away, by using sense organs on their forked tongues which pinpoint the direction of the carcass.

Just as amazing are some creatures' auditory powers. Bats have such an acute sense of hearing that they hear the squeaks of other bats over a great distance. The 86 known species of whales and dolphins, which have nostrils in the top of their heads, also have acute hearing. They have large brains and can tell the difference between a harpooner about to attack and a sympathetic scientist, a sign of great intelligence, and they form emotional ties and a social base like humans. Adult whales 200 to 250 years old, which go down one and a half miles and eat 3,000-4,000 pounds of krill there each day, and calf whales, born with an instinct to rise on their own to take a first breath at the surface, communicate with shrill squeaks which they can hear with their intense hearing. We can pick up the squeaks of bats and whales on radios tuned to certain frequencies.[19]

Were these creatures' acute senses of smell and hearing tendencies they inherited and characteristics they acquired from the environment as conditioned reflexes, by example or by inheritance? They can just as easily be seen as the working of an innate "order principle", a tendency inherited in DNA and awakened by information that may be transmitted by photons (see pp242-246). It drives each creature and directs birds to smell and hear what they need to smell and hear in order to fulfil their role in the ecosystem.

Quite bizarre is the Galapagos giant tortoises' immunity from the unpleasant side-effects of manzanillo fruit. The manzanillo is a tree that develops near the ground. Its milky resins irritate human skin, and it bears fruit that resemble a small apple but are highly toxic and may burn human skin and eyes. The fruit is poisonous to humans and if eaten causes very painful stomach cramps. Yet giant tortoises love the "apples" and are done no harm by them. It now looks as though the manzanillos are reserved for tortoises to keep their species alive, and as though they deter humans by giving them unpleasant side-effects.[20]

Did manzanillos originally burn tortoises' mouths and give them violent indigestion, and did tortoises, when there was no alternative food, develop immunity to the unpleasant side-effects and pass on the immunity to succeeding generations as a tendency to be inherited and a characteristic to be acquired as a conditioned reflex, while parents guided their offspring to manzanillo trees? Manzanillo-eating can just as easily be seen as the working of the innate "order principle", which directed giant tortoises to eat manzanillo fruit as their ecosystem now demanded.

One of the great mysteries regarding inheritance of acquired characteristics is the hatching of turtles. Baby turtles hatch under soft sand and burrow upwards with their flippers to reach the surface of a beach. But if it is day and the sun is hot, they stop, for their instinct tells them to wait. They all do this even though there is no parent present to instruct them. They have to cross the beach in order to reach the sea, their new home, and frigate birds circle in the air looking for newly-hatched turtles. They eat 100,000 baby Galapagos turtles a year.

So why do the baby turtles wait? On one view, they have inherited an instinct, a tendency, which tells them to wait until the sand is cool, and do not know or realise that darkness will give them cover from the frigate birds. On another view, they are born with the inherited knowledge of a predator, and perhaps even of the concept of a frigate bird – a view I heard at the Darwin Centre in the Galapagos Islands. Discussion with several trained naturalists and experts gives different answers. Some say their inherited knowledge is confined to hot/cold, others say their instinct is telling them that there is danger from predators.[21]

The danger-avoidance of the turtles can just as easily be seen as the working of an innate "order principle" transmitted in DNA, an inherited tendency guiding each turtle to avoid danger in the interests of its growth to self-transcendence as it makes its way to the sea to take its place in the ecosystem.

Another example of such innate behaviour can be found in sand or digger wasps, which burrow several shafts in sand and seal each. They hover over each one, learning and memorising its position. The females then find caterpillars, sting them to immobilise them but leave their hearts beating. They put the paralysed caterpillar in a burrow, lay an egg beside it and seal the nest. The hatched larvae feed on the fresh food of the immobilised caterpillars. The young emerge from their nests and replicate this behaviour without having met their parents. But like their parents they have to learn to remember where the nests are, a "mechanism" which is not innate.[22]

The principle of order can also be detected in the mating ritual of sticklebacks. Sticklebacks' courtship releases hormones which make the male develop a red throat and belly and blue eyes, and the female swell with eggs. The fish move to shallow water and there is then a ritual: the male builds a nest, a female appears, the male does a zigzag dance, the female shows her belly, the male swims to the nest, shows the female the entrance and prods her belly with his snout. Each response of the male stimulates the next response of the female and vice versa. The female lays her eggs and swims away, the male empties the nest and spreads sperm on the nest, chases females away, then fans his pectoral fins to keep a current of fresh water on the eggs so that by the eighth day there are high oxygen and carbon-dioxide levels in the water round the eggs, which then hatch. The male guards the young fish. The instinctive behaviour of the sticklebacks and their ritual are again innate.[23]

It can be argued that the sand or digger wasps and the sticklebacks are guided by the "order principle", a tendency inherited in DNA and awakened by information that may be transmitted by photons (see pp242-246) to make provision for the hatching of their young so that they too can take their place within the ecosystem.

The birdsong of yellowhammers and canaries, and of cuckoos, is innate. They sing without having heard their parents or being in contact with other birds of their own species. Chaffinches and blackbirds, however, combine innate knowledge of their song with a learned element. The "order principle", it can be argued, guides all these birds to sing as the ecosystem requires.[24]

Similarly, a herring-gull chick that is motivated by hunger pecks at the red spot on the herring-gull's yellow bill because it primed to do so by the universal principle of order acting in the DNA. The chick has a predisposition towards order to do this as it knows the action will result in regurgitation. The universal principle of order acting in the DNA has programmed all chicks to go for the red spot on the adults' bills to achieve the food it needs to survive and evolve, and to play their roles in the self-organising Whole. In tests, a chick pecks a red spot on a yellow bill in a hundred per cent of cases; a black, blue or white spot on bills in 86 per cent, 85 per cent and 71 per cent of the time respectively; and no spot just 30 per cent of the time. The chicks are driven by the "order principle" to do what they need to do in order to take up their rightful place in the ecosystem.[25]

The instinctive behaviour of penguins[26] strongly suggests a principle of order. There are 17 species of penguin, which are all restricted to the southern hemisphere except for the Galapagos penguin. Fossils show that penguins have been breeding in particular areas for between 40 million and 50 million years, and perhaps 100 million years. Rockhopper, chinstrap, gentoo, macaroni and king penguins are all found in the south Atlantic, but only two species – the emperor and Adélie penguins – breed south of the Antarctic Convergence, inside which water temperature suddenly drops five degrees.

In the Antarctic, near the South Pole, male emperor penguins, the largest penguins, feed for three months in the sea and, with their bellies full, shuffle in line – in a kind of caravan – to the breeding ground where they were born, which may be up to 100 miles away. It is not known whether they are guided by the moon and stars, or by an inner compass. Many "caravans", including lines of females, converge on the breeding ground and arrive on the same day, generally at exactly the same time. Here, on the thick ice where they will be safe until next year, the colony splits into pairs. These penguins are monogamous in the sense that they have only one partner per year. In due course the females deliver one egg, which they shelter between their legs. Eventually they roll it to the egg's father. If the egg cracks and the chick's life is ended, the relationship between the pair is over. If the transfer happens successfully, the exhausted, now starving mother leaves the colony to feed at sea. She has lost a third of her body weight and has a long trek of up to 100 miles back to the sea. Female penguins can hold their breath for 15 minutes and dive to a great depth to search for krill and squid, but some mothers are eaten by leopard seals. The father looks after the egg until it hatches by shielding it from the freezing winds beneath a flap of skin on his belly for two months. In temperatures that can reach −80C all the fathers huddle together to protect their single eggs. They do not feed for over four months and, exhausted, remain alive by eating snow. It is now dark all day and the southern lights flicker in the sky above them.

The fathers' chicks hatch, and each father shelters the chick between its feet.

The plight of all the fathers is now desperate. Not having fed for four months, they are starving and have no food for their chicks who are near death. They cough up a milky substance held in reserve at the back of their throats to keep their chicks alive for a couple of days. At this point the mothers return through the dark, shuffling up to 100 miles, incredibly finding their direction across wastes of frozen ice. The mothers hurry and arrive a few days after their chicks were born. Both mothers and fathers trumpet into the air, and the mothers identify their mates by their individual trumpeting sound and retake possession of the chick between their legs. The mothers regurgitate food into the mouths of the chicks.

The starving fathers sing to the chicks, which sing back, so both know the other's voice. Then the starving fathers leave to shuffle up to 100 miles to the sea to take their first food for over four months. They have lost half their body weight. The mothers walk with their chicks between their feet. Sometimes a chick dies, and its mother groans with grief. She attempts to steal another chick but is prevented by the group. Thus penguins display social interaction and awareness, at a level. In due course the fathers return and identify their mates and chicks by their trumpeting sound. Eventually the family of three return to the sea and split up, each swimming off in different directions. They will never see each other again but have continued the species. The same cycle will begin anew at the next breeding time.

The order in the breeding of these penguins is awesome. The random winds, freezing cold and danger of leopard seals, giant petrels and skuas have been coped with by innate instinct genetically determined through the DNA. All the emperor penguins know instinctively what they have to do.

King penguins breed in colonies and there are organised crèches for the downy chicks (which are known as "oakum boys"). I have stood among 80,000-100,000 breeding pairs on Salisbury Plain in South Georgia and have seen the chicks standing together with adults standing guard nearby like teachers, keeping an eye on predatory southern giant petrels and skua birds. The king penguins have an innate instinct to organise their chicks in an orderly way.

Chinstrap penguins make nests of pebbles. They return to the same breeding ground year after year and reuse the previous year's nests of pebbles and reinforce them by finding more. Each male chooses its mate by depositing a pebble at her feet. The breeding ground chinstraps visit is the same breeding ground where they themselves were born. The newly born fledge, leave the breeding ground after two months and do not return until three years later. They travel a vast distance, often through brash ice and fog (difficult conditions that would deter a Zodiac) to reach the colony where they were born. A two-month old chinstrap has the instinctive memory and GP (Global Positioning) system to find its way back after three years. This principle applies to all penguins and is the case with the 8,000 breeding pairs on Penguin Island off King George's Island, one of the peri-Antarctic islands. It can be argued that the chinstrap penguins are guided by the "order principle" to do what they have to do.

Elephant, crabeater, leopard, Weddell and fur seals all leave the sea for the breeding ground where they were born. There they spend a month. The males become beachmasters, chasing all rivals away with a quick, loping, writhing forward-propulsion of their massive bodies. The beachmaster elephant seal can have up to 100 females to make sure its genes are passed on in mating, and it fights competitors. After the pup is born both parents go to sea to feed and the pup is left alone for 23 days. Depending on its species, its mother then returns and regurgitates food into its mouth and makes several more return visits. Its father returns in due course to moult. The pup will one day return to breed at the same place. It can be argued that the seals are guided by the "order principle" to take their place in Nature's ecosystem.[27]

Many species return to the breeding ground where they themselves were born.[28] Salmon are well known for swimming upstream against the current and leaping up waterfalls to return to their breeding ground. Coho salmon also return to their source. Trout have been known to leap three feet from a lake into a round water-pipe and swim against flowing water to reach another wetland where they were born. Krill, despite being shrimp-like with a minute brain, after five years at the bottom of the ocean return to the exact spot near the sea's surface where they themselves hatched. Lobsters process in a single line to the breeding ground where they were born. Crocodiles also return to the breeding ground where they were born, as do sea-horses, eels, turtles, seals, hammerhead sharks and whales. All birds return to where they were born, often the same tree. All albatrosses return to the colonies where they were born. The same is true of many insects, such as cicadas, butterflies and moths, and also monitor lizards (including the largest of them, the Komodo dragon). Humans are drawn to return to their childhood haunts but hospital birth and migratory patterns have meant that humans do not generally breed where they were born.

In theory one of these creatures can persuade a female to be its mate and go to a safe place with a supply of food that is nearer its habitat than the breeding ground where it was born. But until it adapts, its instinctive drive to return to its birthplace is too strong to be overruled. Quite simply, these creatures have been programmed to return to their birthplace.

Why do so many species return to their own beginnings? The colony that produced them is always a safe place, close to food and with reasonable protection from predators. Although mortality rates are still high, a safe place offers the best available chance of protecting the newly born. It is imperative that the good places should not be visited by the majority of the entire species. By instinctively returning to where they were born, the members of a species prevent competition within that species for the best sites and manage breeding so that the available food is distributed among all and there is therefore enough food to go round. Each member of each species quite literally has – and knows – its place, and by returning there it contributes to an orderly system for all. The vast ecosystem of Nature functions smoothly and harmoniously when each member of each species knows its place and where to go, just as a school functions smoothly when each member of the school knows which classroom to go to at a certain time without the entire school trying to cram into one classroom and fighting for space.

In short, each member of each species is driven by an instinct – the "order principle" – that must be inherited within its DNA to travel to a particular place so that in the vast ecosystem of Nature up to 2 million land species and 10-100 million (in fact, perhaps 15 million) sea species (see p181), each with its own ecosystem, can all breed and reproduce without depleting the food stocks for the rest. Nature's ecosystem is amazingly organised and reveals the hidden principle of order most strongly in organisms' collective returns to their own breeding locations.

In all these aspects of breeding behaviour, the universal principle of order may be driving the cycle of stimulus-response. If so, the instinct is not a learned reflex following skills developed at random and genetically determined, but a response to a principle of order that guarantees the place of the species in the local ecosystem and the whole ecosystem of the universe.

Food Chains, Food Webs and Order in Nature's Ecosystems

Further evidence for an order principle can be found in the food chains which form part of wider food webs. Every creature eats or is eaten in such proportions that by and large the balance of every species' population in relation to that of all other species is preserved in a harmonious way, so that there is enough food to go round for all. In the eating habits which all creatures instinctively know from birth the Universalist philosopher detects a self-running, self-organising, self-perpetuating system of bewildering and amazing complexity.

An ecosystem contains many plants and animals living together in a community of organisms and different species. Ecology is the study of interactions among organisms and between organisms and their environment, and plants are fundamental as they derive energy from sunlight.[29] On land and in water, the two life-sustaining processes are photosynthesis as sunlight combines with water to sustain plants, and respiration – as oxygen, the waste product of photosynthesis – adds to the oxygen in the atmosphere. Animals need nutrients which they eat and digest. These nutrients are dependent on plant nutrition.

In every ecosystem there is a food chain during which matter and energy are transferred from organism to organism as food.

Plants convert solar energy to food by photosynthesis. In a predator chain the plants are then eaten by an animal, which is then eaten by a larger animal. In a parasite chain a small organism eats part of a larger host and is then fed on by smaller parasite organisms. In a saprophytic chain, micro-organisms feed on dead or decayed organic matter. As energy in the form of heat is lost at each step, there are usually only four or five eating. When humans eat cereal grains directly without eating animals that have fed on cereal grains, they take in increased energy.

There are four principal ecosystems featuring ponds, woodlands, lakes and oceans. In each of these, there are herbivores, which eat plants; carnivores, which eat animals; insectivores, which eat insects; decomposers, which eat the surface of mud; and detritivores, which eat dead or decayed organic matter and the detritus below the mud. In the food chains of all ecosystems herbivores eat plants and then are eaten by carnivores. There are some 10^{18} insects on earth on which insectivores can feed. In the many food chains there are producers (green plants, leaves, algae, seaweed) and consumers (fish, birds).[30]

Below are the four ecosystems and some food chains:

(1) In and around a pond ecosystem the food chain is plant-fish-bird: green plants are eaten by fish which are eaten by herons. Plants are eaten by water snails and water fleas, which are eaten by carnivores such as sticklebacks or omnivores such as caddis-fly larvae that eat both plants and smaller animals. Sticklebacks are eaten by herons. Earthworms, detritivores, feed on dead leaves and the waste

matter of animals. Decomposers such as bacteria and fungi eat and accelerate the decay and dispersal of dead organic matter, which is eaten by detritivores such as small fish.[31]

(2) In a woodland ecosystem, the food chain is tree-insect-bird. The woodland ecosystem has a four-step food chain of producers and consumers: the oak leaf (a primary producer) is eaten by a leaf-feeding caterpillar (a primary consumer), which is eaten by an insectivorous bird (secondary consumer), which is eaten by a bird of prey (tertiary consumer). Thus, in the grazing food chain living oak leaves are eaten by green tortrix moths' caterpillars, which are eaten by blue and great tits, which are eaten by sparrowhawks. In the decomposer or detrital food chain dead oak leaves are eaten by decomposers such as earthworms, which are eaten by great tits or shrews, both of which are eaten by sparrowhawks. To state it in wider terms, photosynthetic green leaves (producers) are eaten by leaf-eating insects (herbivorous consumers), which are eaten by insect-eating birds (carnivorous consumers such as blackbirds, thrushes, robins and starlings), which are eaten by sparrowhawks (carnivorous consumers). In Africa, grass is eaten by zebras, which are eaten by a lions; and trees are eaten by giraffes, which are also eaten by lions. (Too few trees means there will be too many giraffes that will starve and die; too many trees means there will be too few giraffes that will starve and die, and some lions will starve for lack of dead meat.)[32]

(3) In and around a lake ecosystem, the food chain is algae-fish-bird. Algae and phytoplankton are eaten by zooplankton, which are eaten by herbivores and insectivores such as freshwater shrimps, which are eaten by carnivores such as bleak, which are eaten by perch, which are eaten by northern pike which are eaten by ospreys. Also, plants are eaten by caterpillars, which are eaten by lizards, which are eaten by snakes. Also, dead animals or carrion are fed on by blowfly larvae and carrion beetles, which are eaten by centipedes and ravens. Also, earthworms, nematodes and bacteria decompose organic matter in the soil while dead wood and rotting plants are eaten by detritivores such as pill bugs and fungi. The dead material in a lake exceeds the living material, and detritivores and decomposers thrive.[33]

(4) In and around a sea ecosystem, or more widely a marine-littoral ecosystem, the food chain is sea plankton-fish-bird. Sea plankton is eaten by sea snails,

jellyfish, shrimps and sea stars, which are eaten by tuna and mackerel, which are eaten by dolphins and sharks, which are eaten by man. Similarly, mineral salts, dead plants and algae are eaten by bacteria, which are eaten by dung, which is eaten by kittiwakes, guillemots, fulmar petrel, little auks and puffins, which are eaten by Arctic foxes. Also, phytoplankton are eaten by zooplankton, which are eaten by fish, which are eaten by pelicans. In the Antarctica there are 60 million unicellar organisms (plankton, plants) living in each litre of sea ice. They are incorporated in sea ice by wave movement while ice crystals are being formed and when the ice cracks they are eaten by amphipod crustaceans, which are eaten by krill (shrimp-like crustaceans 6 cms long), which are eaten by whales, seals, penguins and many sea birds.[34]

Food chains are linked together to form a more complex food web in which a "biomass" of organisms are present, which can be found at every stage of a food chain. This is because most organisms consume more than one type of plant or animal: they do not eat the same food every meal.

Below are some food webs of the four ecosystems (which include the food chains for each ecosystem):

(1) In the food webs of ponds and wetlands, microscopic plants are eaten by daphnia/water fleas, which are eaten by sticklebacks, dragonfly larvae and great diving beetle larvae, which are eaten by kingfishers, which are eaten by sparrowhawks. Canadian pondweed is eaten by mayfly larvae, which are also eaten by sticklebacks, dragonfly larvae and great diving beetle larvae. Detritus is eaten by common pond snails which are eaten by dragonfly larvae and by lesser water-boatmen, which are eaten by greater water-boatmen, dragonfly larvae and great diving beetle larvae. Also, algae produced through the sun's energy by photosynthesis are eaten by mosquito larvae, which are eaten by dragonfly larvae, which are eaten by perch, which are eaten by man.[35]

(2) In a woodland's food web, green leaves are eaten by leaf-eating insects (herbivorous consumers), which are eaten by insect-eating birds such as starlings, blackbirds, thrushes and robins (carnivorous consumers), which are eaten by sparrowhawks (carnivorous consumers higher up the food chain). Alternatively grass is eaten by grasshoppers, which are eaten by frogs and toads, which are eaten by snakes, which are eaten by hawks. Again, decomposers such

as bacteria and fungi are eaten by phosphates and nitrates in soil, which are eaten by pines, which are eaten by pine borer insects, which are eaten by golden-crowned kinglets, which are eaten by red-tailed hawks. Pine borers are also eaten by salamanders, which are eaten by snakes, which are eaten by golden-crowned kinglets, which are eaten by red-tailed hawks. Phosphates and nitrates are also eaten by oak trees, whose acorns are eaten by mice, which are also eaten by snakes and red-tailed hawks. Alternatively, acorns and insects are eaten by skunks, opposums, weasels and mice, which are eaten by hawks, owls and foxes. Again, dead plants are eaten by collembola insects, diptera and mites, which are eaten by snow bunting, purple sandpipers, ptarmigan and spiders, which are eaten by Arctic foxes. Also in a woodland food web, grass and seeds are eaten by rabbits, which are eaten by foxes and owls; and grass seeds and acorns are eaten by mice, which are eaten by owls. Also, on an African savannah grasses and trees are eaten by giraffes, gazelles, baboons, zebras and wildebeest, which are eaten by cheetahs, leopards, lions and wild dogs.[36]

(3) In the food webs of lakes such as Lake Michigan, phytoplankton such as algae, diatoms and flagellates are eaten by zooplankton such as water fleas and copepods, which are eaten by amphipods, opossum shrimps and molluscs, which are eaten by forage fish such as lake whitefish, deepwater and slimy sculpins, yellow perch, bloaters, alewives and rainbow smelt, which are eaten by piscivores (fish-eaters) such as salmon, trout and turbot (on which sea lampreys are parasites), which are eaten by man.[37]

(4) In the sea or marine-littoral zone food web, large seaweeds are eaten by herbivorous fish such as grey mullet, which are eaten by carnivorous fish such as pollack. Large seaweeds are also eaten by flat winkle, which are eaten by edible crabs, which are also eaten by pollacks or herring gulls. Tiny seaweeds are eaten by limpets and by sea urchin, and all are eaten by edible crabs, and sea urchins and edible crabs are both eaten by herring gulls. Limpets are also eaten by lobsters and dog whelks. Plankton is eaten by common mussels, barnacles and common prawns. Mussels are eaten by lobsters, and prawns by herring gulls and seals, which are eaten by polar bears. Also, in the marine food web phytoplankton is eaten by sardines and Atlantic herrings, which are eaten by man. And benthic (i.e. bottom) detritovores' organisms are eaten by algae, which are eaten by phytoplankton, which are eaten by zooplankton, which are eaten by

planktivorous fish such as shad, which are eaten by piscivorous game fish. The large predators at the top of the marine food web include tuna, seals and some species of whales.[38]

The complexity of food webs within an ecosystem displays a stunning degree of order. They are maintained in a very sophisticated way. The natural populations are kept stable because any increases or densities return to an equilibrium. There is an amazing system of checks and balances. As soon as one member of the food chains begins to dominate, its population is reduced to bring the equilibrium back into balance.

Population densities in food webs are controlled by both predators and by food limitation. Plant quantity limits the density of herbivores such as grasshoppers, which limits the number of predators. On the other hand, plant densities are limited by the herbivores, which are limited by predators. There is in fact a negative feedback control of population. An increase in numbers from the optimum population produces increased competition and raised environmental resistance, which produces a fall in numbers, and as a result the optimum population is sustained. Alternatively, a fall in numbers produces decreased competition, decreased predators, and lowered environmental resistance, which produces an increase in numbers and the optimum population is sustained.[39] All organisms are within Nature's system for maintaining the balance of an optimum population, which continuously reasserts itself in a form of homeostasis.

Co-operation and Symbiosis in "The Great Chain of Being"

The self-running, self-organising, self-perpetuating system operates in a co-operative way as well as in a competitive way. All species regard each other as allies against prey when they are not hunting or being hunted. To the Universalist philosopher the self-running, self-organising, self-perpetuating system appears to advance the interests of the Whole, the One, through the massive display of co-operative behaviour which keeps the species healthy between the hunting culls which keep the populations of species stable, and therefore contributes greatly to the order in Nature. It looks as if maintaining the Whole's checks and balances, which require limits on species' populations, is the important things, and that preying on individuals within the species is unimportant, a sacrifice to achieve the overall aim of maintaining the self-running, self-organising system harmoniously. The workings of the order principle can therefore be detected in such co-operative behaviour.

The place in the ecosystem occupied by organisms feeding on each other and competing for food is called a niche. To be more exact, a niche is the smallest unit of a habitat an organism occupies, and its role in the community of organisms found in the habitat. An organism's niche is how it feeds and where it lives.[40]

Competition is at its most intense between species occupying identical or similar niches (or habitats). For example, aphids and caterpillars compete for leaves; competing caterpillars intimidate each other; and competing caterpillars and beetles intimidate each other.

All organisms also avoid direct competition and share resources so that food goes round. Species that do not inhabit identical niches may co-exist. Organisms instinctively operate within their ecosystem to utilise its resources efficiently. The organisms know at birth instinctively what they have to do and slot into a system, feeding in the right part that is "allocated" to them and guarding themselves against predators.[41]

Our four ecosystems have instances of this co-operative behaviour.

In the ecosystem of woodlands, birds share trees. Nuthatches crawl *down* trees and eat insects in the crevices at the top of segments of bark. Treecreepers crawl *up* trees and eat insects in the crevices at the bottom of segments of bark. They occupy different niches and so there is enough food to go round. The greater-spotted woodpecker's beak enables it to peck for grubs on the larger parts of trees whereas the lesser-spotted woodpecker's beak enables it to peck for grubs on the smaller parts of trees. They occupy different niches and so there is enough food for all. Jackdaws and rooks both feed in ploughed fields, but rooks' beaks enable them to feed well below the surface, whereas jackdaws' beaks enable them to feed close to the surface. They occupy different niches and so there is enough food for all.[42]

The same co-operative avoidance of competition extends to caterpillars. Two species of leaf-miner caterpillars live inside a leaf-blade. One lives in the top layer, the other in the spongy part underneath. They occupy different niches and so there is enough food for all.[43]

Plants communicate with each other in a co-operative way. Endangered clover plants undergo chemical changes when attacked and send chemical signals to other clover plants, which then protect themselves from predators by producing a horrible taste. If a clover leaf is munched by a slug, the leaf being devoured sends chemical signals to undamaged clover leaves, which then produce toxins to make them unpalatable. It is as if leaves think of others altruistically as they are eaten, and do not only display self-interest. Something similar happens in acacia bushes. Acacia bushes confronted by predator giraffes which munch their leaves give off ethylamine, a volatile gas with an odour of ammonia, to deter the giraffes. Uneaten acacia leaves have received chemical signals from leaves being munched and make themselves unpalatable.[44]

In the ecosystem of oceans some fish feed at the top (such as blue marlin, Pacific sailfish, yellowfin tuna and dolphins), while some fish feed at the bottom[45] (such as catfish, flat-fish, flounders, soles and stingray), and so there is enough food to go round. In the UK, cormorants (*Phalacrocorax canbo*) and shags (*Phalacrocorax aristotelis*) share a nesting habitat on cliffs and feed in the same water. Cormorants dive and catch fish on the sea bed, such as flatfish, whereas the smaller and slimmer shags feed on surface-swimming fish such as herrings. As they feed differently they have different niches and so there is enough food to go round.[46] (In fact, as cormorants and shags have been reclassified as belonging to the same species, the distinction between shags being top-feeders and cormorants being bottom-feeders has been largely discarded world-wide in favour of a distinction between the feeding habits of males and females.)[47] It is amazing that most cormorants and shags have an oil gland in their backs which secretes oil. They wipe their beaks on the oil and then wipe oil on their feathers, making them waterproof so they can fly when their feathers are wet after diving. This device enables bottom-feeders to feed co-operatively at the bottom, and the intricacy of this device is yet another sign of an orderly system.[48]

All these creatures are born instinctively knowing what to do (like hatching turtles), and when and how to co-operate. It seems they are born into an ecosystem that has complex order and that they instinctively co-operatively take their place in it.

Besides occupying and keeping to their niches, organisms actively assist each other, often against predators, in a form of co-operation called symbiosis, when two or more organisms of different species live in intimate association with each other.[49] Symbiosis takes three forms in ecosystems: (1) mutualism; (2) parasitism; and (3) commensalism. As symbiosis reveals the order behind the co-operative behaviour between niches more startlingly than any other aspect of Nature's ecosystems, we need to take a close look at each.

(1) Mutualism.[50] In mutualism, two organisms of different species live in an

association that provides advantages or benefits to both. This co-operation between species makes common cause within the ecosystem where there are always predators and prey within a food web. Some mutualisms are defensive (involving defending a plant or herbivore), and others are dispersive (involving the dispersal of seeds).

There are many instances of humans, birds, insects, sea fish and plants associating with another species to the benefit of both as, working down the "great chain of being" in descending hierarchical order, the following 70 examples[51] demonstrate:

(1) Humans feed pet dogs which warn of danger and are also useful for hunting. Benefits: warning/food.

(2) Humans feed horses and cows, which work in fields and feed them with meat and milk. Benefits: work/food.

(3) Human coalminers feed canaries and are warned by them of gas leaks in mines. Benefits: warning/food.

(4) Humans beat out grains from heavy sheaves of wheat for bread and help them to propagate. Benefits: food/propagation.

(5) Humans derive pleasure from cultivating plants, which would not survive without man's tending. Benefits: pleasure/survival.

(6) Humans have bacteria living in their intestines that feed on and destroy toxins harmful to humans. Benefits: health/food.

(7) Aborigines set light to forests and hunt and eat kangaroos feeding on the new shoots after the fires. Benefits: food/food.

(8) Horses, cows and buffalo eat grass and fertilise it. Benefits: food/fertilisation.

(9) Herds offer safety in numbers and protect individuals, which alert them to danger. Benefits: safety/warning.

(10) Beavers build homes from wood and fertilise flood meadows. Benefits: shelter/fertilisation.

(11) Squirrels, mice and voles eat hypogeous (underground) fungi and disperse their spores over a wide area. Benefits: food/dispersal.

(12) Squirrels bury nuts to eat and then forget about them, helping nuts to take root and sprout. Benefits: food/dispersal.

(13) Lava lizards perch on Galapagos sea lions and catch and eat flies which are attracted to sea lions' skin. Benefits: food/cleaning.

(14) Bats feed on fruit and propagate seeds over vast distance. Benefits: food/propagation.

(15) Gila woodpeckers hollow out nests in the Saguaro cactus and are one its primary pollinators. Benefits: shelter/pollination.

(16) Galapagos mockingbirds feed by picking parasites off marine iguanas, whose skin is thus kept clean. Benefits: food/cleaning.

(17) Darwin finches feed by picking parasites from Galapagos tortoises, thus cleaning them. Benefits: food/cleaning.

(18) Common noddies perch on the heads of brown pelicans after they have dived and surfaced, and pick bits of fish from their bills, thus keeping them clean. Benefits: food/cleaning.

(19) Crocodile birds (Egyptian plovers) pick parasites from Nile crocodiles for food and by their cries warn crocodiles of approaching danger. Benefits: food/cleaning/warning.

(20) Birds eat berries and cherries and spread seeds and stones which take root and sprout. Benefits: food/dispersal.

(21) Cleaning/oxpecker birds perch on rhinos, giraffes, elephants, cape buffalo, impala and oxen, pick parasites off their teeth and fly off to alert them to approaching danger. Benefits: food/cleaning, warning.

(22) Hummingbirds feed off nectar and pollinate flowers. Benefits: food/pollination.

(23) Chickens eat and what they excrete is used as an alternative pesticide in organic gardens. Benefits: food/pesticide.

(24) Daphnia or water fleas feed off duck droppings and are food for fish which ducks eat. Benefits: food/food.

(25) Woodpeckers eat parasites off and clean trees, from which they receive food and shelter. Benefits: food, shelter/cleaning.

(26) Honey guide birds lead a ratel badger (honey badger) or human to a bee's nest by chattering and flying ahead. After the badger (or human) takes honey, the birds eat the wax and bee larvae. Benefits: food/food.

(27) Butterflies gather nectar from flowers and pollinate them by transferring pollen from their stamens to their stigma. Benefits: food/pollination.

(28) Monarch butterflies feed on and pollinate milkweed whose juices make monarch larvae distasteful to predators. Benefits: food, safety/pollination.

(29) Moths feed off orchids and fertilise them. Some orchids are so specialised that only one species of moth can fertilise them. Benefits: food/fertilisation.

(30) Each yucca moth species is adapted to a particular species of yucca plant. The yucca can only be fertilised by yucca moths which gather pollen from no other plant. (They gather pollen from one yucca flower, roll it into a ball, lay 4-5 eggs on another yucca flower and insert the pollen ball for the larvae to feed. They eat half the 200 seeds produced by a yucca plant.) Benefits: food/ fertilisation.

(31) Honey bees gather nectar from flowers and turn it into honey, and the flowers are pollinated and fertilised when the stamens, stigma and hair brush the underside of bees. Benefits: food/pollination, fertilisation.

(32) Since 250-150 million years ago aphids have developed bacteriocyte cells to house Buchnera bacteria, which supply amino acids to aphids whose diets are deficient in these. Benefits: food/shelter.

(33) Ants eat the honeydew excreted by certain aphids, gently stroking their abdomens with their antennae. Benefits: food/cleansing.

(34) Red ants protect swollen-thorn acacia trees from creatures and are provided with a home with thorny protection. Benefits: shelter/protection.

(35) Worker ants are females and do the work of the nest. Soldiers defend the colony. The queen spends her life laying eggs. All work for the community. Benefits: work, defence/reproduction.

(36) A colony of 9 million leaf-cutter ants (of which there are 210 species) chew vegetation into pulp and store it underground where fungi can grow. They "farm" the fungi on a mulch of leaves which harbour a bacterium that secretes a chemical that kills moulds that might feed on the fungi. It ensures the survival of the fungi, which provides them with gingylidia, which serves as a food. Benefits: food/growth.

(37) Maggots eat dead meat/flesh, leaving living flesh, cleaning bone for humans. (Flies eat harmful bacteria and are dangerous when they spread them to humans by contaminating human food.) Benefits: food/cleaning.

(38) Earthworms fertilise and aerate soil by tunnelling in ground and eating dead organic material. Plants grow on their casts. Benefits: food/fertilisation, growth.

(39) Sharks feed, and cleaning fish, remora (a bony fish), and hagfish remove parasites from their teeth and eat them as food. Benefits: food/cleaning.

(40) Wrasse clean bass of parasites which they eat as food. Benefits: food/cleaning.

(41) Clownfish, crabs and shrimps live in the tentacles of sea anemones and

lure small fish for them to feed. They are immune to the anemone's sting and are protected from predators not immune to the anemone's sting. Benefits: shelter/food.

(42) Nomeous fish gain food and safety from man-o'-war jellyfish, and in return lure fish into their tentacles. Benefits: shelter/food.

(43) Eels protect corals by deterring fish that might harm them, and in return corals give eels a home from which to hunt. Benefits: protection/shelter.

(44) Carnivorous moray eels hide in crevices in shallow water and catch fish in their sharp teeth, from which Red Sea cleaner fish (notably a species of reef-associated grouper, the roving coralgrouper) eat parasites. Benefits: food/cleaning.

(45) Goby fish live in burrows dug in sand by shrimps, who share their food and maintain the burrow. The goby flicks the shrimp with its tail when alarmed to warn it of approaching danger, and is given a safe home to lay its eggs. Benefits: food, shelter/warning.

(46) "Cleaner gobies" remove and feed on parasites from the skin, fins, mouths and gills of large fish such as groupers and snappers, which leave healthier. Benefits: food/cleaning.

(47) Pompeii worms, deep-sea polychaetes, reside in tubes near hydrothermal vents and survive their scalding temperatures by "wearing" "fleece-like" thermophilic bacteria on their backs as protection. The bacteria, which thrive at temperatures above 45°C, gain a home and feeding opportunities from the worms' mobility. Benefits: protection/shelter, mobility.

(48) Sponges are provided with oxygen and sugars (which account for 50 to 80 per cent of sponge growth in some species) by green algae, which live close to some sponges and are protected from predators. Benefits: protection/food.

(49) Some crabs have sponges growing on their backs, which act as camouflage. In return the crabs move about, taking the sponges into the path of food. Benefits: protection/mobility, food.

(50) Hermit crabs carry sea anemones on their shells which protect them from predators with their stings. In return the crabs provide the anemones with food wafted from their meals and allow them to feed in a wider area. Benefits: food/protection.

(51) Zooxanthellae, yellow-brown algae, live in the gastrodermis or waste of reef-building corals and supply photosynthetic nutrients (oxygen and food) that make the corals grow and reproduce to create reefs. In return, the corals provide

them with protection and access to light. Benefits: protection/food.

(52) Millions of corals bind themselves together to provide a fixed environment that helps individuals survive. Benefits: mutual protection.

(53) Azollae, tiny aquatic water ferns, and a microscopic filamentous blue-green alga anabaena (cyanobacteria), grow together at the surface of quiet streams and ponds. The azolla gives protection while allowing sunlight, anabaena provides nutrients. Benefits: protection/food.

(54) Aromatic herbs' seeds are tangled in goats' coats when goats feed on them, and are dispersed. Benefits: food/dispersal.

(55) The branched tubular filaments of mycorrhizal fungi (mushrooms) obtain sugars from the roots of plants, woodland trees, young scots pines and corn and pass on excess nutrient ions to them. Benefits: food/food.

(56) Lichen algae associate with leaf-like lichen fungi on the bark of trees, soil or rocks. The fungi supply the algae with water, shade them from sunlight and provide mineral nutrients. The photosynthesizing algae supply food to the fungi. Benefits: shelter, food/food.

(57) Algae grow and photosynthesise in convoluta worms and in return provide food and camouflage. Benefits: growth, shelter/food, protection.

(58) Algae cover the hair of sloths and obtain shelter and access to sunlight. Sloths are camouflaged to look like logs and so are left alone by predators. Benefits: shelter, sunlight/protection. Benefits: growth, shelter/protection.

(59) Chloroplasts, aerobic bacteria found in plant cells and eukaryotic algae, absorb sunlight and produce sugars and energy for the cells and receive nutrients from them. Benefits: food/food.

(60) Glow-worms hold bioluminescent bacteria inside them and provide them with a safe place to live and a source of food. In return they use the light produced by the bacteria for camouflage, as an aid in hunting, to attract food and to attract mates. Benefits: protection, food/shelter.

(61) Deep-sea anglerfish and euprymna squid have specialised organs that provide bioluminescent bacteria with a safe place to live and a source of food. In return they use the light produced by the bacteria for camouflage, as an aid in hunting, to attract food and to attract mates. Benefits: shelter, food/protection, food.

(62) Enteric or gut bacteria feed off and break down the indigestible cellulose in a ruminant's (such as a cow's) gut into easily digestible carbohydrates, and help digest fats and proteins. Benefits: food/digestion.

(63) The genome of the polydnavirus, a virus, is integrated into the genome of parasitoid wasps. The virus only replicates in specific cells in the female wasp's reproductive system and it is injected along with the wasp's eggs into the body cavity of a caterpillar, whose cells it infects. The toxicity of the virus particles devastates the caterpillar's immune system and prevents it from destroying the eggs before they hatch in the caterpillars. Without the virus, the wasp's eggs would not survive. Benefits: shelter/protection, survival.

(64) Cycads, roots, provide carbon and a stable environment to cyanobacteria, which live within the roots of cycads, in exchange for nitrogen. Benefits: shelter/food.

(65) Foraminifera, plankton that live in warm, shallow seas, often near coral reefs, provide algae with nutrients and in return the photosynthesizing algae supply food. Benefits: food/food.

(66) E. coli bacteria eat harmful bacteria in the human bowel but are dangerous in the human bloodstream and cause fatality. Benefits: food/protection.

(67) Endophytic fungi inhabit tropical grasses and supply nutrients which help the grasses grow and increase water ratios. They acquire nutrients from the grasses and are protected from pathogens and insects. Benefits: food, protection/food.

(68) Wood-feeding termites digest wood with the aid of intestinal bacteria, gut microbiota. In their guts, protists contribute to the digestion of the wood, which resists the normal chemical digestive processes. Benefits: food/digestion.

(69) Rhizobium bacteria live in the root tissues of leguminous plants (such as soybeans and clover) and by nitrogen fixation convert atmospheric nitrogen into nitrates in the soil which feed the plants. Benefits: shelter/food.

(70) Bacteria eat and break down impurities in water and purify sewage. Benefits: food/cleaning.

All these examples of co-operative arrangements between species indicate the workings of a vast, ordered system in which they all play a part.

(2) Parasitism.[52] Parasitism is an association in which one organism lives on or in another organism, its host, for much of its life cycle and depends on the host for food. In other words, one organism benefits at the expense of the other, usually without killing it. Some flowering plants lack chlorophyll and depend on a host plant for water and nutrients. Some depend on a host for water alone. Some

live in the host's body, some on the outside. The host apparently receives no benefit. There may be a reason for the negative influence parasites have on their host, one that has not been understood until now. Do parasites contribute to the running of a balanced system by keeping the host's population down?

Below are 40 examples[53] of parasites and their host, at whose expense they benefit, working down the "great chain of being" in descending hierarchical order:

(1) Cuckoos invade nests of small nesting birds and lay eggs that mimic the colouring of the host bird's eggs, and cuckoo chicks push all other chicks out of the nest. Benefit: shelter, food.

(2) Woodpeckers probe for insects in the bark of trees and cut nest holes in dead wood. Benefit: food, shelter.

(3) Large blue butterflies' caterpillars mimic the smell of meadow ants' larvae and are adopted by meadow ants, whose larvae the caterpillars then eat. Sheep eat grass and keep it short for ants, whose young are preyed on by the large blues' young. Benefit: food.

(4) Tsetse flies and mosquitoes preserve wild animals by feeding on and reducing the human population. Benefit: food.

(5) Ambrosia beetles bore into trees and construct a long gallery off which are egg chambers, and in the main chamber cultivate fungi, which grow on the gallery walls, for food. Benefit: shelter, food.

(6) Bark beetles make sure their young feed by farming the bark. The larvae of bark beetles feed under bark where the food is not nutritious, so the female bark beetles inject into the tree a type of fungus which produces protein of a kind the larvae need. Benefit: food.

(7) Female Dutch elm beetles burrow into elm trees to excavate an egg-laying chamber between bark and wood. Hatched beetles emerge with sticky spores of the diseased Dutch elm fungus on their bodies and spread the disease to healthy trees, which are destroyed. Benefit: shelter.

(8) Two-lined chestnut borers' larvae (*Agrilus bilineatus*) bore into the inner bark of oaks and feed, forming meandering galleries and attacking roots, causing armillaria root rot and eventually oak decline in which foliage yellows, top leaves thin and an oak (which can live 700 years) dies top downwards, its trunk bleeding a black puss-like fluid. Benefit: food.

(9) Leaf-boring or leaf-mining beetles bore into the leaves of horse chestnut

trees, destroy leaves, damage photosynthesis and cause the death of trees after some years. (They interrupt the production of conkers and thus control the spread of horse chestnut trees.) Benefit: food.

(10) Vine weevils feed on the roots of potted plants, e.g. geraniums. Benefit: food.

(11) Human lice, *pediculus humanus*, exist in two varieties. The head louse lays eggs (nits) attached to hair on a human scalp. They hatch into larvae (nymphs), which feed by piercing the scalp's skin and sucking blood. The body louse lays eggs on clothing, which hatch into larvae that feed on the body's skin. Benefit: food.

(12) Deer ticks feed on deer, suck human blood and introduce spirochaete bacteria and other parasites that give their host fever and Lyme's disease. This can attack the brain and result in death if not treated immediately. Benefit: food.

(13) Mites live on the oil in the feathers of starlings. Benefit: food.

(14) Bot-flies lay eggs under horses' skin where the larvae turn into maggots. Lice suck the blood of horses. Benefit: food.

(15) Fleas bit rats, which acted as a reservoir of infection for Black Death, and transferred the disease by hopping onto humans. (They contributed to controlling human population growth.) Benefit: food.

(16) Bilharzia flatworms and blood flukes live off the blood of snails in fresh water. They prey on humans by attacking their blood and causing schistosomiasis (fluke) and eventually death. (One of its functions is to help control human population growth.) Benefit: food.

(17) Tapeworms living in the bowel of humans (and pigs) attack the lining of their stomachs, eat bacteria and absorb nutrients, causing humans to be thin and ill and scratch the irritation. Benefit: food.

(18) Fluke flatworms and nematodes live inside sheep, weakening them and causing disease. They infect shepherdesses and damage their reproductive organs, causing sterility. (One of their functions is to help control human population growth.) In a variant, adult flukes produce eggs inside a cow, which are passed in the cow's faeces to be eaten by snails. The flukes hatch in the snail's intestines, offspring are passed in slime balls which an ant eats. Some flukes enter the ant's brain, causing the ant to climb to the tip of a blade of grass where it is likely to be eaten by a cow, which in turn is affected by flukes. The whole process is evidence of the amazing order in Nature's system. Benefit: food.

(19) Female elephantiasis nematode worms propagate by laying eggs in a hole

pierced in the sole of a human's foot and grow in the lymphatic system, causing massive enlargement of the limbs. Benefit: food.

(20) Horned oak gall wasps and gouty oak gall wasps irritate twigs on oak trees and produce gall on which their larvae feed. Benefit: food.

(21) Flesh-eating Venus-fly traps catch insects in a sticky glue and then digest them. Benefit: food.

(22) The leaves of pitcher-plants hold liquids and trap insects when they drop down into the plant's water, drown and are then digested. Benefit: food.

(23) Ivy wraps itself round trees, gradually competes for light and chokes them to death. Benefit: support.

(24) Mistletoe imbeds in the cambium of spruce, poplar trees, apple trees, willows, lindens, hawthorns and oaks, taking nourishment from them. Benefit: food.

(25) Shelf fungus, polyporales, is parasitic on hickory-tree roots and produces large edible fruiting bodies at the base of trees. Benefit: growth.

(26) White-rot fungus grows on oak trees and destroys them. Benefit: growth.

(27) Honeydew fungus infects trees and their cambium, causing death. Benefit: growth.

(28) Amoebic dysentery protozoan can cause ulcerations in human intestines and can cause fatalities in children and adults. Benefit: food.

(29) The Aids virus destroys the human immune system. (One of its functions is to help control human population growth.) Benefit: food.

(30) Mosquitoes bite humans and transmit malaria, which keeps human population down and protects wild animals man might hunt. Benefit: food.

(31) The hepatitis virus infects the human liver and can cause death. (One of its functions is to help control human population growth.) Benefit: food.

(32) Bacterium-like micro-organisms grow in the tissue cells at the roots of human eyelashes and cause eye irritation that can lead to blindness (trachoma). Benefit: food.

(33) Plasmodium (parasitic protozoan) transmitted by female mosquito bites infects red blood cells of mammals (including humans), reptiles, birds, mice and rats, and causes human malaria. Benefit: food.

(34) Botulism bacteria in decomposing carcasses are scooped up by the beaks of seagulls, grebes and other waterfowl, eat into their systems and can kill them in flight. One gram could kill the entire population of London. Benefit: food.

(35) Wolbachia bacteria, a genus of inherited bacterium and one of the world's

most common parasitic microbes, infects a high proportion of all insects. It is the most common reproductive parasite, and more than 16 per cent of neotropical insect species carry it. Benefit: food.

(36) Agricur, an anti-bacterial clay originating in French volcanoes in the Massif Central mountain range, preys on and kills up to 99 per cent of colonies of superbugs such as MRSA and E. coli within 24 hours, and may provide an antiseptic antibiotic against superbugs until they develop resistance by natural selection. (The E. coli virus is good in human bowels but bad in human blood, and its eradication may upset the checks and balances of the Whole in which all species have population controls.) Benefit: food.

(37) Orange mosaic viruses affect leaves and fruit and wipe out orange trees in California. (One of its functions is to stop orange trees dominating.) Benefit: food.

(38) Leaf curl virus weakens peach trees, causing the leaves to fall, eventually leading to death. Benefit: food.

(39) Mosaic virus (blackspot) affects rose leaves and kills rose plants. Benefit: food.

(40) *Phytophera infestans* mould, potato blight, affects whole crops of potatoes and caused the potato famine in Ireland which killed over a third of its population. Benefit: food.

(3) Commensalism.[54] Commensalism is an association between two species in which one organism benefits, the commensal, which is usually the smaller. Its benefits can include nutrients, shelter, support, locomotion or transportation (which is called phoresy). The host receives no benefit but is not harmed. Below are 25 examples[55] of commensals and the benefit they derive from their host, working down the "great chain of being" in descending hierarchical order:

(1) Scavengers such as hyenas wait for a lion to kill and they eat part of what the lion leaves. Benefit: food.

(2) Brightly-coloured tree frogs look poisonous and so are avoided by other animals. Benefit: protection.

(3) Cattle egrets feed in fields among cattle, whose movements stir up insects which the egrets eat. The egrets benefit from the association, the cattle do not. Benefit: food.

(4) Robins wait for human gardeners to dig, then eat worms and grubs.

Benefit: food.

(5) Cowbirds feed on insects around bison. Benefit: food.

(6) Wrens nest in the protective nooks of osprey nests. Benefit: shelter.

(7) Vultures eat large or small dead animals. Benefit: food.

(8) Nesting birds gain a safe place in trees where they can build their nests. Benefit: shelter.

(9) Dung beetles feed off animal droppings, rolling them as a ball into their burrows. Benefit: food.

(10) Insects mimic the look of a wasp and so avoid being eaten by other animals. Benefit: protection.

(11) Remora (a bony fish) attaches itself to a shark or swordfish and feeds on morsels of their food. Benefit: food.

(12) Crabs open clams and molluscs for fish to feed on. Benefit: food.

(13) Hermit crabs use shells of shellfish as their homes to protect their soft bodies. Benefit: protection.

(14) Flower-inhabiting mites scuttle into the nostrils of humming-birds to relocate from a dying flower to a distant healthy flower. Benefit: mobility.

(15) Barnacles attach themselves to whales, whose movement wafts food towards them. Benefit: food.

(16) Orchids and bromophides grow in forks of tropical trees. Benefit: protection.

(17) Bee orchids mimic the shape of a female bee, deceiving male bees into copulating with them and transferring pollen, but receiving no nectar. Benefit: pollinisation.

(18) Grass-pink orchids produce no nectar but mimic nectar-producing rose pogonias and are still visited by bees, which transfer pollen. Benefit: pollinisation.

(19) The Saguaro cactus's primary pollinator is the Gila woodpecker. Benefit: pollination.

(20) Burdoch seeds lodge in the fur of a white-footed mouse and are dispersed. Benefit: dispersal.

(21) Spanish moss grows on the branches of trees, especially oak trees. Its threadlike stems can be 20-25 feet long. Benefit: growth.

(22) Moss grows against the bark of maple trees. Benefit: growth.

(23) Algae and moss grow on the bark of trees. Benefit: growth.

(24) Seaweed algae cover a manatee so that it looks like a rock covered with

algae. Benefit: growth.

(25) Algae attach themselves to turtles whose movement brings them to warm currents and to food. Benefit: food.

All these 135 co-operative relationships have been established over many generations, and they operate across the variety of species. All organisms within the various species instinctively know what to do from birth to keep the complex orderly system going.

Ailments and Co-operative Healing Plants

Another aspect of the self-running, self-organising, self-perpetuating system is the way Nature grows cures for all ailments, illnesses and diseases. Humans have always known which herbs or plants help their digestion when they are unwell. It could be that plant-eating animals also know instinctively from birth which specific plants to seek out by smell when they are ill. More research needs to be done on this. Nature's provision of growing cures is further evidence of the workings of an order principle which has created an orderly environment for all living species and has anticipated every eventuality and contingency by providing an antidote.

We have seen that the universe is bio-friendly. The medieval monks knew and homeopaths still know of the bio-friendly connection between humans and plants.[56] Nature's system which permits ailments to arise also provides their cure. Nature acts as a primitive pharmacy. It may be eye-opening to list 70 examples[57] of the main diseases and ailments humans suffered from in the Middle Ages before the opening of pharmacies and provision of pills and tablets, and which they still suffer from (by no means an exhaustive list), and the various plants which (according to traditional belief and lore) were, and still are, held to provide natural remedies within a complex order.

(1) Abdominal cramps and inflammations: apple bark, ginger, lilac, prickly ash, rhubarb, sweet cicely.

(2) Acid indigestion: belladonna, bitters, caraway, chicory, cinnamon, lilac, sweet cicely.

(3) Acne: amaranth, asparagus, birch, burdock, celandine, pine, poplar, rhubarb, roses, thyme, turmeric, walnut, wheat grass.

(4) Allergies: agrimony, Brigham tea, chamomile, eyebright, osha, thyme, uva

ursi, wood betony.

(5) Arthritis: angelica, ash, ashwagandha, bay, bean, belladonna, cayenne pepper, feverfew, pine, prickly ash, rhubarb, sassafras, skullcap, turmeric, wormwood, yarrow, yucca.

(6) Asthma: aconite, adder's tongue, agrimony, black haw, camphor, chamomile, Chinese cucumber, coffee, coltsfoot, elderberry, elecampane, eucalyptus, frankincense, ginger, ginkgo biloba, juniper berry, lobelia, mountain mahogany, mullein, onion, osha, peony, thyme, yellow dock.

(7) Bleeding: amaranth, aster, atractylis, bayberry, birthwort, bugle weed, cayenne pepper, cotton, horsetail, moss, plantain, shepherd's purse, smartweed, stinging nettle, turmeric, white oak.

(8) Blood clots: cayenne pepper, ginger, goldenseal, motherwort, pine, red clover,

(9) Bronchitis: aconite, adder's tongue, agrimony, baby's breath, bay, black haw, camphor, castor bean, Chinese cucumber, coffee, coltsfoot, dandelion, elderberry, elecampane, eucalyptus, frankincense, ginger, marjoram, mountain mahogany, osha, pau d'arco, pine, thyme, yellow dock.

(10) Bruises (abrasions): adder's tongue, arnica, birch, bluebottle, buchu, calendula, cayenne pepper, comfrey, daffodil, lily, parsley, peony, roses, St-John's-wort, thyme, witch hazel.

(11) Burns: adder's tongue, aloe, beech, birch, box elder, calendula, curry, elderberry, garlic, lavender, melaleuca, mullein, onion, poplar, St-John's-wort, thyme, willow.

(12) Cancer: ashwagandha, astragalus, beetroot, birch, bloodroot, chaparral, cinnamon, coffee, comfrey, dill, echinacea, evening primrose, flaxseed, fungus, garlic, magnolia, mountain mahogany, parsley, pau d'arco, pokeroot, poplar, red clover, rhubarb, St-John's-wort, suma, tea, turmeric, wheat grass, wild Mexican yam, yucca.

(13) Constipation: agave, aloe, barberry, bayberry, birch, bouquet garni, brier rose, cascara sagrada, citrin, elderberry, flaxseed, mountain mahogany, psyllium, rhubarb, senna, tarragon, zedoary.

(14) Coughs: agave, anise, bay, bryony, eucalyptus, eyebright, hyssop, iris, lavender, mustard, spruce, wild black cherry.

(15) Depression: alfalfa, ashwagandha, borage, corydalis, ginkgo biloba, lavender, lily, St-John's-wort, tarragon.

(16) Diabetes: alfalfa, atractylis, bean, bilberry, cactus, cayenne pepper,

dandelion, garlic, goldenseal, guar gum, pine, psyllium, turmeric.

(17) Diarrhoea: agave, amaranth, arnica, aster, bilberry, birch, Brigham tea, blackberry, butternut, cardamom, cattail, gladiola, rhubarb, sagebrush, walnut, white oak, willow.

(18) Dysentery: agave, apple bark, butternut, gladiola, hops, lobelia, sassafras.

(19) Eczema: amaranth, ash, birch, burdock, club moss, evening primrose, poplar, rhubarb, sage, sagebrush, St-John's-wort, sassafras, slippery elm, turmeric, walnut, watercress, witch hazel.

(20) Epidemics (plague): camphor, garlic, goldenseal, onion, thyme.

(21) Eye ailments: agrimony, basil, brooklime, burdock, catnip, chervil, chrysanthemum, eyebright, fennel seed, flaxseed, goldenseal, morning glory, mountain mahogany, roses, sage brush, snapdragon, squawvine, thyme, turmeric.

(22) Fatigue: astragalus, barley grass, cattail, cayenne pepper, club moss, dang-shen, evening primrose, fungus, gentian, ginseng, guarana, kava kava, kola nut, mahuang, maple, pine, prickly ash, spruce, suma, thyme.

(23) Fever: agave, anemone, apple bark, barberry, basil, bayberry belladonna, bergamot, birthwort, black pepper, boneset, borage, castor bean, catnip, cinnamon, dandelion, dogwood, eucalyptus, garlic, ginger, lilac marjoram, oak, pine, prickly ash, rhubarb, sassafras, slippery elm, thyme, wintergreen, wormwood.

(24) Gout: ash, asparagus, bean, buchu, lilac, peony, pine, rosemary, sassafras, tarragon, watercress, wood betony, yarrow.

(25) Headache: asafoetida, basil, brooklime, bupleurum, cayenne pepper, chamomile, chrysanthemum, cinchona, coca, ginger, lavender, mint, mistletoe, morning glory, sagebrush, spruce, thyme, watercress, wintergreen, wood betony.

(26) Heart attack: cayenne pepper, garlic, ginkgo biloba, mistletoe.

(27) Heartburn (intestinal ailments): anise, annatto, belladonna, bitters, bouquet garni, caraway, cinnamon, sweet cicely, turmeric, willow.

(28) Hepatitis (infections and liver ailments): barberry, Brigham tea, calendula, chamomile, dandelion, milk thistle, rhubarb, wormwood.

(29) Herpes: aloe, arnica, arum, astragalus, bearded darnel, birch, burdock, cascara sagrada, chickweed, cotton, garlic, henna, melaleuca, mint, morning glory, rhubarb, sassafras, slippery elm, turmeric, wild Oregon grape.

(30) Hypertension (high blood pressure): asparagus, black cohosh, chrysanthemum, dandelion, digitalis, evening primrose, hawthorn,

(31) Indigestion: agave, anise, annatto, apple bark, bitters, blessed thistle, bouquet garni, caraway, cardamom, ginger, licorice root, mints, nutmeg, osha,

rhubarb, tarragon, turmeric.

(32) Infection: alfalfa, birthwort, blue flag, castor bean, dandelion, echinacea, garlic, goldenseal, nasturtium, onion, osha, roses, tea, thyme, wild Orgeon grape, wormwood.

(33) Inflammation: aswagandha, beech, bird-of-paradise, borage, box elder, calendula, castor bean, Chinese cucumber, coltsfoot, evening primrose, lobelia, prickly ash, skullcap, snapdragon, tea, turmeric, wild Mexican yam, wormwood, yarrow.

(34) Influenza (common cold/fever): blue vervain, boneset, cayenne pepper, cinchona, cinnamon, cotton, elderberry, eucalyptus, mountain mahogany, pine, spruce, thyme.

(35) Insect bites and stings: adder's tongue, aloe, apple bark, asafoetida, bird-of-paradise, cattail, celery, curry, jewelweed, lavender, morning glory, onion, sage, sagebrush, shepherd's purse, turmeric, wild Orgeon grape, witch hazel.

(36) Insomnia: bearded darnel, belladonna, catnip, chrysanthemum, dill, fungus, hops, lavender, lily, passion flower, squawvine, tarragon.

(37) Kidney ailments: agave, cornsilk, digitalis, mustard, nutmeg, parsley, rhododendron, tarragon, uva ursi, wood betony.

(38) Kidney stones: apple bark, buchu, burdock, cornsilk, ginger, parsley, rhododendron, rosemary, sorrel, uva ursi.

(39) Laryngitis: aconite, arnica, Chinese cucumber, clary sage, eucalyptus, hyssop, sage.

(40) Liver ailments: apple bark, artichoke, bitters, blackberry, blessed thistle, chamomile, chicory, cumin, dandelion, eucalyptus, milk thistle, roses, tea, turmeric, wormwood, zedoary.

(41) Lung ailments (respiratory disorders): astragalus, black cohosh, bracken fern, castor bean, mountain mahogany, sassafras, shepherd's purse, yellow dock.

(42) Measles: Chinese cucumber, dandelion, lilac, marjoram, mullein, pennyroyal, pine, turmeric, wheat grass.

(43) Menstrual ailments: angelica, apple bark, aster, bayberry, black haw, blue cohosh, bugleweed, caraway, citrin, cotton, doong quai, lemon balm, marjoram, tarragon, yarrow.

(44) Migraine: basil, belladonna, bergamot, elderberry, feverfew, lavender, melaleuca, mint, pine, tea, thyme.

(45) Mucus accumulation: arnica, black alder, black haw, elderberry, eucalyptus, nasturtium, rhododendron, watercress.

(46) Mumps: Chinese cucumber, dandelion, lilac, marjoram, mullein, pennyroyal, pine, turmeric.

(47) Muscle ailments: calendula, camphor, daffodil, ginger, licorice root, marshmallow, prickly ash, roses, rosemary, thyme.

(48) Nervousness: basil, blue vervain, cattail, celery, hops, Japanese honeysuckle, licorice root, lily, sage, skullcap, tarragon.

(49) Obesity: cayenne pepper, celery, chickweed, citrin, guar gum, kelp, mahuang, psyllium, stinging nettle, wild Mexican yam, yohimbine.

(50) Pneumonia: castor bean, cayenne pepper, dandelion, elecampane, prickly ash.

(51) Psoriasis: amaranth, angelica, birch, burdock, pennyroyal, rhubarb, sage, sagebrush, St-John's-wort, sassafras, turmeric, walnut.

(52) Rash: beech, chickweed, pennyroyal, pine, poplar, sassafras, slippery elm, sumac, thyme.

(53) Respiratory disorders: dandelion, elecampane, eucalyptus, frankincense, mahuang, mountain mahogany, purslane, yellow dock.

(54) Rheumatism: agave, camphor, lemongrass, lilac, motherwort, prickly ash, roses, rosemary, sassafras, tarragon, wild Mexican yam.

(55) Sinusitis: camphor, eyebright, goldenseal, horehound, horseradish, pine, walnut, watercress, yellow dock.

(56) Skin ailments: agrimony, birch, calendula, castor bean, chamomile, club moss, coffee, elderberry, garlic, henna, horseradish, horsetail, jewelweed, marshmallow, mullein oak, pennyroyal, pokeroot, poplar, purslane, roses, rosemary, thyme, turmeric.

(57) Skin sores: agave, anemone, arnica, arum, blackberry, brooklime, burdock, henna, marshmallow, melaleuca, oak, thyme, witch hazel.

(58) Skin ulcers: adder's tongue, marshmallow, mullein, pine, poplar, shepherd's purse, slippery elm, turmeric.

(59) Sore throat: aconite, amaranth, arnica, arum, basil, birthwort, black alder, blazing star, boneset, cayenne pepper, cinchona, cinquefoil, clary sage, eyebright, fungus, garlic, goldenseal, hyssop, myrrh, oak, osha, ramps, red clover, roses, sage, sorrel, thyme, walnut, wintergreen.

(60) Sores: adder's tongue, agave, alfalfa, anemone, arnica, arum, birch, blackberry, brooklime, burdock, calendula, elephant grass, eucalyptus, oak, poplar, ramp, St-John's-wort, slippery elm, thyme, willow.

(61) Sprains: bay, calendula, cayenne pepper, daffodil, lemongrass, roses,

turmeric, wormwood, zedoary.

(62) Stomach pains: agave, asafoetida, calendula, cardamom, cumin, elderberry.

(63) Stomach ailments: eucalyptus, lily, mint, nutmeg, sweet cicely, wood betony, zedoary.

(64) Stroke: aconite, cayenne pepper, ginkgo biloba, pine.

(65) Tonsillitis: cinquefoil, echinacea, lobelia, mullein, pokeroot, sage, sorrel, walnut, willow, yellow dock.

(66) Toothache: allspice, apple bark, black pepper, catnip, cattail, cloves, coriander, garlic, ginger, melaleuca, plantain, prickly ash, ramps.

(67) Tuberculosis: castor bean, echinacea, elecampane, hops, mountain mahogany, thyme.

(68) Urinary tract infection: agave, garlic, marshmallow, mullein, purslane, uva ursi.

(69) Varicose veins: bay, bayberry, calendula, gotu kola, oak, pine, zedoary.

(70) Wounds: adder's tongue, agave, arnica, bilberry, birch, brier rose, cactus, calendula, comfrey, elephant grass, eucalyptus, lily, marshmallow, melaleuca, morning glory, mullein, poplar, sagebrush, St-John's-wort, shepherd's purse, slippery elm, smartweed, thyme, turmeric, wheat grass, wild Oregon grape, willow, yarrow.

It is as if every conceivable ailment a human "ex-ape" suffers from is taken care of by a plant in the natural world. In bygone times, these were intimately known by everyone, and in giving out plants for ailments (rosemary, pansies, fennel, columbine, rue and daisy) mad Ophelia acted out a country lore that was widely shared and a profound part of popular culture. The point here is that the complex orderly system in Nature has arranged for plants to grow which contain balms from which humans can benefit.

In conclusion, the vast ecosystem which includes all local ecosystems, the ecowhole, its food chains and food webs, of which we are part, its symbiotic relationships (mutalism, parasitism, commensalism) and the interacting between plants and humans who benefit from the healing properties of plants all provide countless instances of unrandom, seemingly bio-friendly order in Nature's ecosystems. If these instances illustrate complex order rather than randomness now, could they really have begun in a random way? It is the same question I asked (on p101) of the regular rising of the sun, moon and stars: the laws of

Nature seem so orderly, could they have come into being by random accident? And do our mathematics, which describe the running of the universe so precisely and with such unrandom complexity, describe a random structure?

Philosophical Summary

What, then, does ecology contribute to the "new scientific interpretation of the universe and Nature"? It reveals a self-running, self-organising, self-perpetuating system. Also, the workings of the order principle. Each creature is driven to fulfil its place within the system, hunting, co-operating and breeding at the right time. It knows instinctively from birth what to do and many species return to their own birthplace to breed. The complex food chains and webs are ordered and balanced to keep populations stable. Species co-operate through symbiosis. Plants provide medicinal relief for all ailments. Such a complex, ordered system does not seem to be the product of a random accident.

To sum up, Universalist philosophy can take from ecology:

a self-running, self-organising, self-perpetuating system;

the workings of an order principle;

each creature is driven by instinct to hunt, co-operate and breed at the right time;

each creature knows from birth its place and what to do in relation to the system;

an ordered system of food chains and webs keeps populations stable;

species co-operate through symbiosis;

plants give medicinal relief;

the self-organising system does not seem the product of random accident.

The "new scientific interpretation of the universe" must look at physiology, the systems that control the bodies and brains of individuals of the species in the ecological self-running, self-organising system, in particular those of humans. The Universalist philosopher must now see if science can detect self-organising systems, intelligence and the principle of order in the systems of the body, brain and consciousness.

8

PHYSIOLOGY: THE SELF-REGULATING BODY, BRAIN AND CONSCIOUSNESS

We should expect the workings of the order principle in the vast whole ecosystem of Nature to be paralleled in the physiological make-up of each individual of the many species that operate within it, but nothing prepares us for the amazing degree of self-organisation and self-regulation in each organism of the many species within the hierarchical "great chain of being". It would be instructive to pursue this idea in a selection of minerals, vegetables, insects, reptiles, birds and mammals, but space is short and, bearing all the other species in mind, the Universalist philosopher must confine his attention to the body, brain and consciousness of humans as he concludes his "new scientific interpretation of the universe" by revealing the order principle in the physical workings of the human body and mind.

The species that came out of one point contain an astonishing intricacy and complexity. Physiology is "the science of the functions of living organisms and their parts" (*Concise Oxford Dictionary*). There are a number of self-regulating "mechanisms"– a reductionist word, as I have said, whose mechanistic implications the Universalist does not accept, hence the inverted commas – "in the living organisms" (mammals, birds and reptiles as well as humans) which indicate self-organising systems. They suggest a highly complex degree of order that cannot have been arrived at by random chance.

The body's complexity can be judged from the fact that there are 10^{360} possible combinations of DNA strands;[1] there are 10^{23} bits of information in a human being;[2] and there are 10^{29} particles and (7×10^{27}) atoms in the whole human body. As we saw on p132 there are 10^{15} cells in each human body. There are 10^{12} bacteria on the surface of the human body and 10^{15} bacteria inside it.[3]

The Body's Homeostasis

The Universalist philosopher, seeking the "mechanism" that keeps organisms functioning within Nature's vast ecosytem needs to focus on the body's homeostasis which enables all organisms to take part in Nature's ecosystem, and on the organisation, self-organisation and self-regulation by which this is achieved.

Physiologists divide the organism into seven layers of organisation in which each higher layer integrates the layers below it:[4]

(1) Chemical – many chemicals, or atoms combined into molecules, combining to form organelles (parts of cells such as the nucleus, mitochondria);

(2) Organelle – many organelles combining to form cells (for example, heart-muscle cells);

(3) Cell – many cells combining to form a single type of tissue (for example, heart muscle);

(4) Tissue – one or more tissues combining to form an organ (for example, the heart);

(5) Organ – one or more organs separately or jointly forming a system (for example, the cardiovascular system).

(6) System – all the systems combining to form the organism (for example, the cardiovascular, renal, respiratory, digestive, nervous, endocrine systems);

(7) Organism (for example, a human being).

The higher layers' integration of lower layers is essential rather than incidental. Thus an organelle is identifiable in terms of its integrative capacity to combine to form cells. Many scientists state that organisms combine to form an eighth organisational layer, one involving interaction with society, which acts as a yet more complex living entity as many organisms combine to form society (or societies). Organisms interact with the population of their local community, habitat, ecosystem and biosphere.

In mammals, birds and the higher vertebrates, and in particular human organisms, body temperature, blood-sugar level, the oxygen and carbon-dioxide concentrations in blood, water potential and blood pressure are all regulated or held nearly constant within the "mechanism" known as homeostasis (Greek for "staying similar").[5] Sweating, urine formation, heart rate and breathing rate vary but are controlled and regulated to maintain the variables nearly constant.

An organism self-regulates to maintain an internal environment of constant conditions. The self-regulatory system eliminates any deviation from the normal by negative feedback.[6] In negative feedback, a stimulus or input – a deviation from the normal or a need – is detected responded to, fed back or put forward to a control centre or regulator, and then to an effector or output which responds to and eliminates the deviation from the normal or need and, having coped with it, then returns the system to the detector.

In a negative feedback loop, then, there is a challenge to homeostasis, a change to the variable being regulated, which is picked up by a sensor that alerts an integrator (often located in the brain), which compares input from the sensor to the normal condition or set point. It instructs an effector (a skeletal muscle) to produce a response that eliminates the problem and return the system to the sensor.

For example, a decrease in body temperature challenges homeostasis. The sensor, a group of temperature-sensitive nerve cells in the skin, detects the stimulus and sends a message to the integrator in the brain, which compares input (less than 37°C) to the normal condition or set point (37°C). The integrator passes the temperature drop on to the effector (the skeletal muscle), which produces a response (shivering, which increases heat production). This response compensates for the change induced by the challenge to homeostasis and the organism is returned to its normal homeostasis, under the care of its sensor.

Alternatively, if blood pressure falls due to blood loss sustained in a fight, blood-pressure arteries detect the drop in pressure and send a signal to the brain that activates pathways leading to the secretion of hormones. The effector (heart, blood vessels) produces a response (raising blood pressure back towards normal) that removes the stimulus to the sensor and returns the system to the sensor's watching care.[7]

Homeostasis is a fundamental characteristic of living organisms. It is the maintenance of the organism's internal environment within narrow, life-conducive tolerances (allowable variations). These tolerances vary only minutely between individuals, and then only rarely.

The ability of the internal environment of organisms to sustain life is affected by temperature, acidity and nutrient, ion, oxygen and carbon dioxide levels. As these directly influence the chemical processes essential for life there are numerous complex "mechanisms" to keep them at effective levels. Blood pressure, heart rate, absorption, excretion and breathing rate are all controlled, and the thyroid regulates metabolic rate and bone remodelling. A "mechanism" responsible for governing function in one organ or system may have components that work in other organs or systems: for example, the removal of excess acid from the blood (the circulatory system) by expulsion via the lungs (the respiratory system) under direction of the medulla (the central nervous system). Virtually nothing in physiology occurs in isolation, and the organism is dependent on this interaction between systems for survival.

Homeostasis is not the cause of these perpetually fluctuating changes; rather, it is an umbrella term given to the many independent but interacting processes at the chemical, organelle, cell, tissue, organ, and system levels that, in concert, keep the organism alive and healthy.

Below is a representative collection of 62 examples[8] of homeostasis covering each human organ system. The examples do not by any means provide a complete picture of human physiology, but they provide evidence for the ordered systems of the body's homeostasis. Some examples are of feedback "mechanisms" in which a stimulus, detecting and responding to a deviation from the normal or a need, is conveyed to a regulator, which then responds. Other examples are simple "features" of physiology, where a simple condition facilitates continued function and health. Both demonstrate that the "mechanisms" favour the survival of the organism and indicate a high degree of "order" rather than a chance set-up or random "accident". In fact, the feedback "mechanisms" reveal order too complex and sophisticated to be a random accident.

The narrow tolerances (variations) of these self-regulating, self-organising systems and processes are interdependent. This is demonstrated by the rapid onset of illness that results if these tolerances are transgressed and/or if there are communication errors within or between physiological layers, as we see in the numbered paragraphs beginning "failure" at the end of each system.

General homeostasis: temperature or blood loss[9]

(1) A fall in body temperature stimulates the thermoreceptors in the skin and the hypothalamus. Nerve impulses to the preoptic area of the brain and to the pituitary gland cause nerve impulses to be sent to the blood vessels in the skin and muscles and the release of thyroid-stimulating hormone (TSH). TSH causes the thyroid to release thyroid hormones which increase the metabolic rate, thus creating heat. The skin's blood vessels constrict to save heat, while muscles shiver to generate heat.

(2) Systemic shock due to blood loss stimulates the kidneys, which detect reduced blood flow. Baroreceptors in the large blood vessels detect a pressure drop. The kidneys release renin. This creates angiotensin II, which in turn causes blood vessels to contract, raising blood pressure. The kidneys also release antidiuretic hormone, which prevents the removal of fluid from blood for excretion. The baroreceptors signal to the brainstem to constrict blood vessels to

internal organs, excluding the brain and heart, so as to raise blood supply to the musculoskeletal structures.

(3) Failure of general homeostasis's self-organising system can cause hypothermia due to a lowering of body temperature to 35°C or below, for example as a result of adrenal insufficiency or an underactive thyroid gland. The effects of hypothermia are shivering, confusion, falling heart rate, falling blood pressure and build-up of acid in the bloodstream due to reduced respiration, which can ultimately lead to coma and death.

Homeostasis in the embryological system

(4) The human growth hormone stimulates the liver, which then produces more insulin-like growth to increase cell division in the embryo's growing bones.

(5) Bone-matrix (the framework for bones) stimulates the osteoblasts (bone precursor-cells) which stop producing bone-matrix for the embryo and change into osteocytes (bone-cells), showing that bone-growth is self-regulated. (Approximately 50 per cent of all conceptions fail within two days, with the host never becoming aware she was even pregnant. This illustrates the extent and strictness of the organism's self-regulatory control in policing the quality and viability of offspring.)

(6) Failure of the embryological self-organising system can cause giantism (whose victims include Abraham Lincoln and actor Richard Kiel), which is due to the over-production of human growth hormone. Its effect is excessive growth to abnormally great height and size in which cardiovascular problems are common due to increased demand on heart and circulation.

Homeostasis in the pregnancy, birth and post-natal system

(7) Release of the hormone relaxin by the placenta stimulates and softens the mother's ligaments, allowing her pelvis to expand during childbirth.

(8) Release of the corticotropin-releasing hormone by the placenta stimulates the pituitary gland in the brain, which releases more adrenocorticotropic hormone (ACTH). This causes the adrenal glands to release more cortisol, which reduces inflammation and pain during childbirth.

(9) Pressure of the baby's head in the cervix causes nerve impulses from the uterus to stimulate the hypothalamus (a control centre in the brain). This releases the hormone oxytocin, which in turn stimulates the uterus muscles to contract harder.

(10) Increase in the blood flow from the baby's lungs caused by the baby's first breath stimulates the left atrium of the baby's heart. Pressure then shuts the valve between the left and right atria and redirects the blood flow through the baby's lungs to supply it with oxygen direct so the baby is not dependent on obtaining oxygen via the umbilical cord from its mother's blood.

(11) Breast-suckling by an infant causes nerve impulses from the mother's breast-stretch-receptors to stimulate her hypothalamus. This releases more prolactin-releasing hormone, causing the pituitary gland to release more prolactin, which in turn increases milk production.

(12) Failure of the pregnancy, birth and post-natal self-organising system can cause ectopic pregnancy due to the obstruction of the fertilised egg's passage through the Fallopian tube to the uterus. As a result it lodges in the Fallopian tube and, unless removed, it will rupture the Fallopian tube, often leading to the mother's death.

Homeostasis in the immunological system

(13) A hostile reaction of one of an organism's cells stimulates an immature, developing T cell (an immune-system killer cell). This then undergoes apoptosis, which is programmed self-destruction. Thus T cells that seek to attack host cells never mature into active killer T cells.

(14) An infectious agent such as a virus (an antigen or foreign substance that causes the body to produce antibodies) stimulates plasma B-cells (immune-system cells). They are converted into memory B-cells, which recognise the antigen more quickly in future and execute a more specific and powerful attack upon it.

(15) An antigen stimulates killer T cells, which release toxins into the antigen to kill it and also release chemicals to attract macrophages, immune cells that engulf and dissolve the antigen.

(16) Failure of the immunological self-organising system can cause multiple sclerosis due to some unknown genetic susceptibility. Multiple sclerosis causes the development of immune cells that attack motor nerve sheaths, and demyelination, or destruction of motor nerve sheaths, leads to the loss of motor power. Death is usually by asphyxiation as motor nerves to the diaphragm fail.

Homeostasis in the neurological system

(17) The sudden stretching of a muscle stimulates a muscle spindle receptor, which sends a nervous impulse to the spinal cord. This sends a motor impulse to

the stretched muscle, which contracts rapidly, preventing damage by over-stretching. (This mechanism is used by clinicians to test spinal cord function, for example via knee-jerk reflex.)

(18) A sudden sound from one side stimulates the brainstem, which receives an impulse from the vestibulocochlear nerve in the ear and sends an impulse via the oculomotor and abducens nerves to the lateral rectus muscle of the nearest eye and to the medial rectus muscle of the farthest eye, and, via an accessory nerve, to the sternocleidomastoid muscle in the neck. The muscles turn the head and eyes quickly towards the source of the sound in a response that is purported to be a survival mechanism.

(19) Dynamic, "see-saw" balancing of activity between sympathetic and parasympathetic branches of the autonomic nervous system (i.e. the involuntary nervous system responsible for controlling bodily functions not consciously directed, e.g. the heartbeat) nervous system achieves appropriate levels of activity in the vegetative processes (digestion, tissue repair, relaxation of heart and diaphragm etc) and the excitatory processes (energy release, fight/flight response, perspiration etc).

(20) Nerves reporting muscle movement and joint position take precedence over pain nerves for the same area so as to suppress pain following injury during physical activity when there are frequent muscle movements and changes in joint position. (This was important in *Homo*'s early existence when injuries that occurred while he was being pursued could not be allowed to prevent him from fleeing.)

(21) Pain from surface structures such as the skin travels along much faster nerves (taking 0.1 seconds) than pain from deeper structures such as organs, which takes a second or more, so that the nervous system can react quickly to breaches of the skin that can lead to rapid blood loss and initiate vascular changes. Injured organs tend to experience constriction or blockage, which are less immediately fatal than breaches of the skin.

(22) The nerves in the nose (olfactory nerves) are the only nerves in the human body that regenerate after damage from an infection from a cold virus, so that the nose can continue to detect predators, food and/or toxins, a response purported to protect survival.

(23) Failure of the neurological self-organising system can cause Raynaud's disease due to the involuntary nervous system's excessive stimulation of small arteries in the fingers and toes, which are constricted and feel very cold. Some

people suffer tissue death of the fingers and toes.

Homeostasis in the haematological system

(24) The release of thromboplastin in response to mechanical cell damage (such as a breach of the skin) stimulates clotting in the blood. Platelet aggregation at the site of the injury creates a platelet plug, or clot, to stem the loss of blood.

(25) The kidneys detect reduced oxygen and release more erythropoetin into the bloodstream, and stimulate immature cells called proerythroblasts in bone marrow. These therefore mature more quickly than they normally would and enter the bloodstream as red blood cells, which transport oxygen. (Normally, approximately 2 million red blood cells are destroyed and replaced every second.)

(26) Red blood cells' lack of nuclei allow the cells maximum flexibility, permitting them to flex through the smallest capillaries and thus deliver oxygen to all tissues.

(27) A spasm in the muscular wall of damaged blood vessels constricts blood vessels, reducing blood loss.

(28) Failure of the haematological self-organising system can cause anaemia due to the reduced ability of red blood cells to carry oxygen, for example as a result of lack of dietary iron or loss of blood through wounds. Anaemia results in fatigue and intolerance of cold.

Homeostasis in the cardiovascular system

(29) Baroreceptors in the carotid artery in the neck detect reduced blood pressure and stimulate the cardiovascular centre in the brainstem. This sends nerve impulses to increase the heart rate and the force of heart contractions, and to constrict arteries, thereby increasing blood pressure.

(30) Chemoreceptors in the brainstem and arteries detect increased carbon dioxide and reduced oxygen in the bloodstream due to a sudden increase in physical activity. They stimulate the cardiovascular centre in the medulla of the brainstem. This sends nerve impulses to the heart to increase the rate and force of heart contractions so as to pass blood through the lungs more quickly for carbon dioxide to be exchanged for oxygen so as to supply oxygenated blood to the muscles more quickly.

(31) The kidneys detect reduced rate of blood flow, release renin into the bloodstream and stimulate an enzyme called angiotensin-converting enzyme. The

renin converts this into angiotensin II, which causes blood vessels to constrict, thus increasing blood pressure, and the kidneys to filter out less sodium and water, thus increasing blood volume.

(32) Failure of the cardiovascular self-organising system can cause congestive heart failure due to long-term high blood pressure. As a result more blood is left in the heart after each contraction due to weakened muscle and the heart is stretched by the increased blood volume and tries to contract harder to empty itself, thus tiring itself further.

Homeostasis in the respiratory system

(33) Chemoreceptors in the brainstem and arteries detect increased carbon dioxide and/or reduced oxygen in the bloodstream due to sudden increase in physical activity. They stimulate the inspiratory centre in the brainstem, which as a result sends nerve impulses to the diaphragm to contract more frequently and more forcibly, thus increasing the breathing rate and the exchange of carbon dioxide for oxygen.

(34) Failure of the respiratory self-organising system can cause respiratory acidosis due to reduced respiration as a result of emphysema. As a result the kidneys attempt to excrete more acidic hydrogen ions into the urine and kidney damage can be caused if acidosis is untreated.

Homeostasis in the gastro-intestinal system

(35) Chemoreceptors in the stomach wall detect an altered Ph (Potential of Hydrogen, a measure of acidity or alkalinity in a solution) of stomach contents due to food entering the stomach. They stimulate the submucosal nerve plexus in the muscle of the stomach wall, which as a result sends nerve impulses to the parietal cells in the stomach lining to secrete hydrochloric acid to break down the stomach contents for easier absorption.

(36) Stretch receptors in the stomach wall detect food entering the stomach and stimulate the submucosal nerve plexus in the muscle of the stomach wall. This sends nerve impulses to the muscle of the stomach wall to contract more forcibly and churn the food, eventually emptying the stomach.

(37) The pyloric sphincter (which separates the stomach from the small intestine) remains closed during the churning of stomach contents to extract the maximum possible nutrition from food and maximise the breakdown of ingested food for later absorption of waste by the intestines. "Design" by random

accident rather than by a principle of order would not achieve "mechanisms" as subtle as these.

(38) Toxin-filtering by the liver destroys and removes from the bloodstream ingested substances that would otherwise be deleterious to health. "Design" by random accident rather than by a principle of order would not create a liver that removes only toxins; it would "randomly" remove some toxins and some nutrients.

(39) The presence of *Escherichia coli* (E. coli) bacteria in the large intestine assists the breakdown of waste, the production of vitamin K, and the absorption of nutrients. E. coli causes potentially fatal infection if it breaks out into the bloodstream Its confinement to the gut, where it serves a vital function, is therefore conducive to life.

(40) Failure of the gastro-intestinal self-organising system can cause lactose intolerance due to cells lining the small intestine producing insufficient enzyme called lactase, which digests lactose in dairy products. Undigested lactose holds fluid in the faeces in the bowel and bacterial fermentation of undigested faecal lactose causes gas, bloating and potentially dangerous abdominal distention.

Homeostasis in the renal system

(41) Stretching of the blood vessels in the kidneys due to increased blood pressure as a result of stress or exertion stimulates the muscle fibres in the kidney blood-vessel walls. As a result they contract, reducing the blood flow through kidneys to "normal", thus maintaining the normal blood filtration rate.

(42) Increased delivery of sodium ions to the kidneys due to increased blood pressure stimulates juxtaglomerular apparatus in the kidneys. This increases the secretion of the vasoconstrictors, which as a result constrict arterioles delivering blood to the kidneys, thus maintaining the normal blood filtration rate.

(43) Falls in the water content of blood and intercellular fluid stimulate osmoreceptors in the hypothalamus. This instructs the pituitary gland to release more anti-diuretic hormone (ADH), which in turn causes the kidneys to remove less water from the bloodstream in the form of urine.

(44) Failure of the renal self-organising system can cause polycystic kidney disease due to filter-tubes in the kidneys becoming riddled with cysts. As a result, the amount of functional kidney tissue is eventually reduced to the point where kidney failure occurs and waste substances accumulate in the bloodstream. Without dialysis or a kidney transplant death occurs.

Homeostasis in the endocrine system

(45) Low blood levels of thyroid hormones (which maintain the metabolic rate) stimulate the hypothalamus which as a result releases thyroid-releasing hormone (TRH). This causes the pituitary gland to release thyroid-stimulating hormone (TSH).This in turn causes the thyroid gland to produce more thyroid hormones, which in turn increase the metabolic rate.

(46) Low blood levels of glucocorticoids (hormones that deal with stress) stimulate the hypothalamus, which releases corticotrophin-releasing hormone (CRH). This causes the pituitary gland to release more adrenocorticotrophic hormone (ACTH), which in turn causes the adrenal glands to release more glucocorticoids.

(47) High blood levels of glucose (sugar from food) stimulate the islet cells in the pancreas, which release insulin hormone into the bloodstream. This causes all cells to absorb and store more glucose, reducing the blood level of glucose to "normal".

(48) Low blood levels of glucose (sugar from food) stimulate the alpha cells in the pancreas, which release glucagon hormone into the bloodstream. This causes the liver to convert protein in the bloodstream into glucose, elevating the blood level of glucose.

(49) Sudden stress or fright stimulates the hypothalamus, which sends a nerve impulse to the adrenal glands on kidneys, which release stored adrenalin into the bloodstream. This increases the force and rate of the heart contractions, releases glucose into muscle cells, raises alertness, diverts blood from the digestive system to the skeletal muscles, and dilates the airways, all of which prepare the body for sudden and extreme exertion such as "fight or flight".

(50) Failure of the endocrine self-organising system can cause diabetes mellitus (Type 1) due to the failure of islet cells in the pancreas to produce insulin. As a result, glucose cannot be absorbed by cells, which use fat for energy instead. This produces waste in the form of ketones, which are acidic. Raised blood acidity leads to coma and death.

Homeostasis in the reproductive system

(51) The release of oocytes (which produce oestrogen) by ovaries causes a drop in blood-oestrogen, which stimulates the hypothalamus. This releases more gonadotrophin-releasing hormone (GnRH). GnRH causes the pituitary gland to release more follicle-stimulating hormone. This causes follicles in ovaries to

develop into oocytes, which in turn produce oestrogen, the level of which then returns to "normal".

(52) DNA replication in developing sperm cells in the testes stimulates the meiotic spindle (a "guy-rope" in the cell) which attaches to a random* (either paternally- or maternally-derived) chromosome, ensuring unique genetic material for the offspring of each sperm.

(53) Elevated blood testosterone due to a change in diet, exercise or aggression stimulates the hypothalamus (a control centre in the brain), which produces less GnRH, a fall in which causes the pituitary gland to produce less luteinising hormone (LH). A drop in LH causes the testes to produce less testosterone, levels of which return to "normal".

(54) A drop in external environmental temperature stimulates the skin of the scrotum, as a result of which contraction of the cremaster muscle pulls the testes closer to the body to maintain the correct temperature for sperm production.

(55) Prostaglandins (agitating chemicals) in semen stimulate the uterus, whose muscular contraction aids the propulsion of semen.

(56) ZP3 (a glycoprotein in the female egg, or oocyte) stimulates sperm heads, which release enzymes that dissolve the oocyte (egg) wall and allow a sperm to penetrate.

(57) The fusion of a sperm head with the oocyte stimulates the oocyte membrane, which is depolarised, making it impermeable to further sperm.

(58) Calcium ions released by oocyte membrane depolarisation stimulate secretory cells in the oocyte. They release enzymes that inactivate ZP3, making the oocyte chemically unresponsive to further sperm, ensuring that only the fittest (the first to arrive) fertilises the oocyte.

(59) The alkalinity of seminal fluid protects sperm from the acid environment of the vagina.

(60) Sperm contain an antibiotic, seminalplasmin, whose purpose is to protect sperm against bacteria in the male urethra and the vagina.

(61) The temporary coagulation of semen 5 minutes after ejaculation is followed by reliquefication 10-20 minutes later. Its purported purpose is to form a protective "plug" that prevents other males' sperm from fertilising the egg.

(62) Failure of the self-organising reproductive system can cause cryptorchidism due to failure of the testes to descend into the scrotum during foetal life. This can cause the destruction of sperm precursor cells due to the temperature inside the body, causing permanent sterility.

I question the use of the word "random" (asterisked in (52) above). Conventional wisdom holds that the splitting of the centromeres during DNA replication results in a random distribution of chromosomes to dividing oocytes. (This wisdom is reflected in standard works such as Langman's *Medical Embryology* and Tortora and Grabowski's *Principles of Anatomy and Physiology*, both of which use the word "random".)

The processes involved are highly complex, and revealing them in itself indicates an awesome degree of order in the microcosmic world. We should remind ourselves that each human body contains 6 billion strands of DNA. We have both our mother's and father's DNA in our sperm cells/egg cells which are produced within the gonads, and each chromosome within each cell/egg has come either from our mother or from our father. We have 23 pairs of chromosomes, only one of which determines the cell's gender.[10] Each chromosome pair (one from our mother, one from our father, both the same "type" of chromosome, for example chromosome 17) in the developing oocyte (sperm or egg) is attached at the midpoint of each chromosome's tubular length by a small spherical structure called a centromere, which is connected by two microtubules to opposite sides of the cell wall. The centromere splits, allowing each microtubule to pull "its" chromosome to its side of the cell. When all the chromosome pairs have been split and are now all occupying opposite sides of the cell, the cell "pinches" across the centre and divides into two oocytes, each having a different combination of chromosomes. For example, oocyte A has chromosome 2 from our father and chromosome 14 from our mother, while oocyte B has chromosome 2 from our mother but chromosome 14 from our father.

There is more variability than just male/female combinations of each of the 23 chromosomes. Once they pair up and are connected by the centromere in the developing oocyte (before being pulled to either side) they swap thousands of segments of their DNA, to form two completely original chromosomes, but still of the same "type". This happens for all 46 chromosomes in each sperm or egg cell. The oocyte now has two completely original chromosomes of the same type (for example, chromosome 17), formed from very complex mixing of our father's and mother's DNA from their versions of the same chromosome. So oocyte A actually has maybe 38.27 per cent of chromosome 2 from our father, 65.03 per cent of chromosome 14 from our mother, and so on. There are over 100,000 genes divided across 46 chromosomes, now mixed-up in varying combinations. Each newly created chromosome is unique.

As one of my main themes is the interplay between order and randomness, I am entitled to ask whether the complex interchange of DNA between the male and female versions of each chromosome that produce a unique new chromosome is in fact "random". (See pp277-278.) Or is "randomness" an assumption that may one day be replaced by "orderly" when the complex process is better understood? There may, for example, be a lower-level "mechanism" at work, hitherto undetected, that in accordance with the universal principle of order decides which genes are swapped and therefore when the centromere splits – that controls DNA exchange and chromosome division. Both these processes would need to be proved to be controlled to remove the idea of randomness from the standard textbooks. Perhaps the science of the 21st century will locate this lower-level mechanism. Until such evidence is produced we should suspend judgement as to what controls the process of DNA exchange and chromosome division.

The narrow tolerances involved in these thousands of bodily processes, and their crucial interdependence indicate that the human organism is a highly complex, ordered "system of systems" that has evolved as a self-improving, self-bettering response to changing environmental stimuli over thousands of millennia rather than an almost exponentially complex "accident" of incalculably improbable, interdependent coincidences. The workings of these systems are too complex, interdependent and sophisticated to ascribe to "accident".

To a Universalist philosopher, all these highly complex self-regulating, self-organising systems and "mechanisms" of the body are indications of the workings of the universal principle of order and its drive for self-improvement within each organism.

The Brain's Homeostasis

Further evidence for an order principle controlling the self-regulation of bodily homeostasis can be found in the homeostasis of the brain. There are different levels of brain function, some of them autonomic or automatic, others more "conscious", more associated with thought. The Universalist philosopher needs to approach the higher level of activity from the lower, more autonomous brain functions which involve the brain's self-regulation of homeostasis.

All species came out of one point, and there are self-regulating "mechanisms" in the brains of all mammals, birds, reptiles and humans which suggest order. The complexity of brains' "mechanisms" can be judged from the fact that the human brain has 10^{11} (100 billion) neurons.[11] There may be 10^{12} neurons and as many non-excitable neuroglial cells.[12]

These "mechanisms" are connected with: the organisation of the central nervous system (i.e. the brain and spinal cord); the brain's self-regulation of its own health (for example, the production and reabsorption of cerebro-spinal fluid which nourishes the brain, and the protection offered by the blood-brain barrier; the brain's plasticity which makes possible learning, experience and habit and its ability to transfer function (for example, sensation, movement or memory) from an injured area of brain tissue to another area of the brain – after a stroke or head injury; the survival value of the link between memory and emotion; and the brain's control over the immune system, and the effects of cognition and mood on health and disease.[13]

The brain has a physical, structural organisation that is described by the way in which its physical constituents are arranged to form a whole. It also has a functional organisation that determines how specific areas of the brain provide particular functions such as sight, memory, sensation and emotion etc.

The brain's structural organisation has seven levels which correlate to seven levels of organisation within the body (see p219):

(1) Chemical – many chemicals combining to form organelles;

(2) Organelle – many organelles combining to form neurons (nerves, particular types of cell);

(3) Neuron – many neurons combining to form a neuron circuit or tissue (such as a reflex, for example the knee jerk reflex);

(4) Circuit – many circuits combining to form a structure such as the hypothalamus (hormone control) or the medulla oblongata (control of vital functions such as breathing, heart rate);

(5) Structure – many interconnected structures combining to form a function (for example, memory, emotion, cognition);

(6) Functions – all the functions combining to form the brain, a system;

(7) Brain – not a separate organism from the body, but head of its own system.

As in the case of the body, there is an eighth organisational layer, one involving interaction with other brains to form a yet more complex living entity as many brains combine to form a society (or societies). The brain interacts with the population of the local community, country, region and humankind.

The brain's functional organisation is now recognised to be more complex than was first thought. In the past the brain was regarded as a single "organ". It is now more accurate to view it as a collection of many systems and structures, all of which carry out many different specialised functions.[14] At the same time all are interconnected with, and regulated by, each other to work together towards the optimal functioning of the brain's organism: a specific mammal, bird, reptile or human being. Thus, an area of the brain may have a role pertaining to memory when working with another area, but when working with yet another area may also have a reasoning function or perhaps a sensory one such as hearing.

Our current understanding of the brain's functional organisation is based on this model of the brain as a network of multi-functional areas. We no longer see it as having specific functional areas, each performing a specific task in isolation. This multi-functional approach[15] has important ramifications for the protection of brain function when the brain suffers injury, and for the organism's survival. The complex order of the brain's functioning can best be seen in action, in examples of how the brain regulates its own environment and thereby its health; stores information (facts, memories); transfers function from damaged areas to other areas; codes emotion (particularly fear) into some memories; and influences the immune system and thereby our health. These "mechanisms" enable the brain and the organism to survive, survival being necessary for the perpetuation and evolution of the species.

The brains of mammals, birds, reptiles and humans seem to work in a broadly similar way, but I am now focusing primarily on the human brain. The human brain regulates its own homeostasis and environment via the feedback "mechanism" of the "blood/brain barrier".[16]

In a strict sense, there is no blood in the brain as blood does not come into direct contact with the nerve tissue comprising the brain. The nerves of the brain require the same nutrients as nerves throughout the body, primarily glucose and oxygen. However, whereas body cells receive their nutrients directly from the blood, the brain receives its sustenance from cerebro-spinal fluid,[17] or CSF, which is a filtrate of blood and prevents potentially harmful substances from entering the brain.

Blood circulating around the lining of the brain passes through systems of capillaries known as choroid plexi (the plural of *plexus*). The walls of these capillaries are tighter than those of capillaries elsewhere in the body and only permit certain substances such as glucose, oxygen, minerals and blood plasma to "leak" through the capillary walls into the space surrounding the brain. Red blood cells, proteins and other large molecules cannot pass into the space directly surrounding the brain. The choroid plexi filter the blood to produce CSF, the straw-coloured liquid which only contains the nutrients required by the nerve tissue of the brain.

As the pressure of CSF in the cranial vault increases it is reabsorbed by venous capillaries which form structures called arachnoid villi, and is returned, along with waste products such as carbon dioxide, to the circulating blood. There it mixes again with red blood cells, proteins, and the other substances that were originally filtered-out.

We have seen that feedback "mechanisms" control the body. A feedback "mechanism" also maintains adequate blood-flow to the brain. Although the brain only comprises around 2 per cent of the body's weight, it uses around 20 per cent of the glucose used by the body as energy.[18] Given the sensitivity of the brain to drops in glucose levels and the fact that oxygen deprivation causes brain death within four minutes, a stringently controlled blood supply to the brain is vital. Nutrient and oxygen requirements of the brain are relatively constant, even when the body is exercising strenuously, and so the brain must also be shielded from the rise in systemic blood-flow and pressure that occurs during exercise.

A feedback "mechanism" regulates the brain's temperature as well as the body's. Brain function is very sensitive to changes, especially increases in temperature. The temperature-control "mechanism" that applies to temperature drops in the body applies to raised temperature in the brain. It is significant that the body's thermostat, the hypothalamus, is sited centrally at the base of the brain, where it can respond very quickly to a change of brain temperature.

Neural circuits become formed in the brain in response to experience. This is due to the brain's "plasticity" (in its biological meaning of "exhibiting an adaptability to environmental changes"). The experience might be visual, auditory, or via any of the senses. The more often a circuit is activated, the stronger it becomes. Hence the more frequently we revise a particular fact, the more firmly implanted it becomes in our memory, the easier recall becomes, and forgetting that fact becomes less likely.

An experience can also be internal, in the form of an opinion or belief. Beliefs that are frequently reaffirmed or externally validated become more firmly entrenched. There is a natural bias towards only seeking external information that supports a belief, and so beliefs often become self-perpetuating even though the original impetus may have been weak or plainly incorrect. We can also be manipulated in this way.

Plasticity is vital for our survival. According to the Darwinian view, if our ancient ancestors had not learned from one or two occurrences of being burned when they put a hand into a camp fire, or from seeing other tribe members attacked by animals, they would probably not have fled from forest wildfires or when they heard growling from the bushes, and the species may not have survived.

The brain's plasticity is a great aid to learning through experience and habit. It might help to think of neural circuits as 'grooves', which become deeper the more often they are used. This is the basis for all learning and experience. Repetition strengthens these neural circuits, and it is this principle that athletes and other performers employ when they practise the same action many times over. They are strengthening the neural circuits that permit a perfect cover-drive, free-kick, golf-swing, ball-pitch or arabesque. Circuits that are not used often tend to weaken and can be subsumed by other circuits. In this way it is possible for us to alter our knowledge once we realise a previous belief was incorrect, and performance of skills deteriorates through lack of practice.

Similarly, circuits can form that become associations between other circuits. For example, we might have a "circuit" that recognises a familiar smell even though we have never closely encountered the smell's source. According to Darwinians, one day we may turn a corner in a jungle and come face to face with a black bear that we have previously only seen at a great distance, and experience the smell we recognise at the same time. Thus we now associate the smell with the black bear. We also probably now associate the smell with danger and fear. We have now created an "association circuit" between the circuit that allows us to recognise the smell, and the circuit that allows us to recognise something as being a black bear. Thus every time we subsequently encounter the smell we will fire the association circuit and thus the circuit that tells us a black bear is nearby. Such emotional imprinting of experiences is conducive to survival.

Habits are formed in the same way. Most people tie their shoelaces the same way every day. They do not think about how they are going to tie them, they simply fire the appropriate circuit and the hands perform the movements quickly and smoothly and without any further cognitive effort on the part of the shoe-wearer, permitting a detailed conversation with another person at the same time without detracting from either activity. This circuit has become so strong, the groove so deeply engrained, that trying to tie the laces another way requires one's full attention, feels awkward, and is likely to take measurably longer.

Our new understanding that most brain areas take a partial role in a number of functions, rather than each being responsible for a specific function, and that various combinations of functional areas collude to provide one particular function, has had beneficial medical effects.

It is now understood that the brain's plasticity helps it adapt to physical damage or injury, for example when damage to a particular area of the brain, possibly as a result of a stroke or physical trauma, leads to impairment or loss of a function provided by the damaged region of the brain.

The area in the brain responsible for conscious awareness of sensation in the body and for conscious movement is called the homunculus (meaning "little man"). The amount of neural tissue in the homunculus that is dedicated to a particular body structure directly corresponds to the sensitivity of, and/or the level of fine motor control over, that body structure, not to the size of the structure. Thus the hands have the largest amount of tissue dedicated to them in the homunculus. The eyes similarly have a large area, while the skin of the back has a much smaller area. The brain has specific plasticity that adapts to physical damage and takes over a damaged area.

The brain has considerable control over the body's health and well-being. A relatively new branch of medical science, known as psychoneuroimmunology (PNI), has revealed extensive direct links between the brain and the immune system. That the brain has some control over the body's response to disease is perhaps not surprising. What is interesting however is that the conscious mind has been shown to have a measurably significant influence over this control.

Stress can be positive or negative, in broad terms. "Eustress" is the term traditionally given to challenges that are within our ability to cope and which cause us to improve our responses and performance – it is "healthy competition". The negative, deleterious form of stress, the form to which we refer when we use the word "stress", is "distress". This is the type of stress that exceeds our ability to cope.

Our stress response "mechanism" is located deep within our primitive brain and an identical stress response "mechanism" is found in most mammals, and therefore probably predates our evolution into modern man and consciousness around 130,000 years ago. This feedback "mechanism", which is under the control of the brain, is designed to help us cope with immediate short-term threats, which it does well. It is our, and the brain's, most basic and valuable survival "mechanism".

Modern causes of stress however tend to be of a more protracted nature, such as debt worries and relationship problems. These cause stress hormones such as cortisol to be raised chronically (over long periods). Cortisol is useful in reducing inflammation and pain (its essential function) during short-lived crises, but over longer periods it depresses immune responses and mood. Mental anxiety, which originates in the neo-cortex (the "modern" part of the brain with which we think), has a profound effect on our immune function and therefore health.

Studies have been carried out on the brain's use of depression. Common cold viruses have been sprayed into the nostrils of volunteers previously screened for low-grade depression or suppressed mood. There is an overwhelmingly significant correlation between lowered mood and the development of symptoms subsequent to exposure to the virus. It is clear that the brain makes self-protective use of cognitive or emotional depression. It is now thought that the brain uses sleep to file or store memories, to make contact with the bigger picture that was not seen when awake and to regulate the body and itself so it is ready to solve the problems of the next day. It is as if the brain needs impressions to stop for a few hours so it can have some administrative time to catch up with what has been seen, heard, sensed and physically exerted; and so it can digest all the new information and bodily movements of the previous day and relate them all to the bigger picture. As every parent knows, sleep loss makes one less able to concentrate, more reckless, more emotionally fragile and more vulnerable to infection. Perhaps this is because "mechanisms" which control concentration, judgement, emotional stability and a degree of immunity from infection have not been updated and renewed, and tiredness is a signal that the brain is due for administrative time.

The higher levels of mental activity – thinking, choice, willpower – have been associated with the cerebral cortex but have not been located in one area. Do different parts of the brain also combine to give rise to these more sophisticated brain functions? We will return to this shortly when we consider consciousness.

The following 15 examples[19] of homeostasis provide evidence for and illustrate how – through feedback "mechanisms" involving the blood/brain barrier, blood flow, temperature and neural circuits – the human brain controls learning, habit, memory and its response to damage and displays highly complex order in regulating itself:

Homeostasis of the blood/brain barrier and CSF

(1) The filtration of the blood to form CSF prevents large molecules including infectious organisms such as bacteria and viruses from entering the brain. They are too large to pass through the choroid plexi. As brain infections are very often quickly fatal this "mechanism" is clearly conducive to the survival of the organism. This is fortunate as antibiotics are also too large to pass into the brain through the blood/brain barrier.

(2) The maintenance of constant pressure of CSF around the brain is to ensure an even supply of nutrients to the brain's nerve tissue. As the brain's nerve tissue is very sensitive to falls in oxygen supply and dies within four minutes of being deprived of oxygen, CSF regulation is optimally balanced to ensure survival.

(3) Failure of the brain's CSF regulation/self-organising system can cause hydrocephalus due to impaired reabsorption of CSF by the arachnoid villi, which increase CSF pressure around the brain. This damages the delicate nerve tissue, leading to brain damage and death if untreated. Treatment is by insertion of a tube or "shunt" into the cranial cavity and by emptying CSF into the main blood vessel (*vena cava*) that returns blood to the heart.

Homeostasis of blood-flow

(4) The baroreceptors in the carotid arteries in the neck detect reduced blood pressure, which means there is reduced blood-flow to the brain, and stimulate the cardiovascular centre in the brainstem. This sends nerve impulses to increase the heart rate and the force of contractions, thereby increasing blood pressure and thus the blood-flow to the brain.

(5) Failure of the brain's regulation of blood-flow can cause a stroke due to either obstruction of the blood supply to the brain by an embolism (blood clot or air bubble), or the haemorrhage of a blood vessel within the brain. The consequence of either is a critical loss of blood, and therefore of oxygen, to the brain, the starved area of which quickly dies. Depending on the extent and location of the stroke, the effects of a stroke range from loss of sensation and of movement in certain areas of the body, or impaired speech, cognition or memory, or (when the brain is starved of oxygen) death. Some effects can be temporary, often due to the plasticity of the brain.

Homeostasis of the brain's temperature

(6) A rise in blood temperature, possibly due to a fever from infection, stimulates thermoreceptors in the hypothalamus. As a result the sympathetic nervous system (SNS) causes surface blood vessels to expand, thus filling with blood and transferring heat to the skin from where it is lost.

(7) Failure of the brain's regulation of temperature can cause pyrexia (fever), usually due to infection, but sometimes due to a pituitary tumour's disrupting hypothalamic function. The effects of pyrexia are the denaturing of amines and peptides which comprise the neurotransmitters required for neuronal signalling "brain function", and also direct, possibly permanent, disruption of neural (and therefore brain) function due to temperature. Brain death occurs if one's temperature exceeds 44-46 degrees Celsius (112-114 degrees Fahrenheit).

Homeostasis of neural circuits: learning through experience and habit

(8) The frequent reactivation of a neural circuit leads to its becoming strengthened and thus more easily "fired" in future so that we learn that things that happen frequently are "normal", and thus become able to recognise normal events. This is the basis of learning and experience.

(9) A neural circuit that is not fired frequently tends to weaken and is not easily fired so that an event is kept "unusual" and if it occurs it is therefore likely to get our attention. This is how we forget, and also partly why we pay closer attention to events outside what we consider normal – behaviour more likely than if we ignore such events.

Homeostasis through memory

(10) A memory of an emotionally-charged event is connected through an area of the brain called the amygdala. This creates particularly strong neural pathways, or memories, that cause the original emotion to be experienced when the memory is "fired" so that an individual feels the emotion associated with the event that imprinted the original memory, every time the memory is recalled. This is a survival "mechanism", as the associations are strongest when fear is the "imprinting" emotion. (Counsellors claim this is why, for example, revisiting a location associated with a past trauma can often cause such strong emotional responses, as the emotions imprinted into the memory are so vivid.)

Homeostasis and damage

(11) If an area of the brain that has an important role in a particular function (for example memory) is damaged (for example as a result of a stroke or head trauma) an unrelated area takes over some of the lost function, so that the function is not totally lost. Loss of mobility could lead to death, through inability to avoid danger or eat. This adaptation does not occur for all functions, and not necessarily to a full extent. (This taking-over by an unrelated area has been observed via functional Magnetic Resonance Imaging scans in stroke patients whose speech centres have been damaged by stroke.)

(12) The loss of a body part such as a hand results in a reduction of the amount of neural tissue dedicated to that part in the homunculus. The homunculus can now reallocate the neural tissue to another body part – in the case of the loss of a hand, probably to the other hand. Some of the homunculus remains "tied" to the lost part, which is possibly the cause of phantom limb pain in amputees.

(13) If an area of the homunculus is damaged (for example, as a result of a stroke) the body structure controlled by that area of the homunculus is taken over by another area so that control over the body structure concerned is not completely or permanently impaired. This "mechanism" has considerable importance for survival.

Homeostasis and stress response

(14) Sudden stress or fright stimulates the hypothalamus, which sends a nerve impulse to the adrenal glands on the kidneys. They release stored adrenalin into the bloodstream, which increases the force and rate of heart contractions, releases glucose into muscle cells, raises alertness, diverts blood from the digestive system to the skeletal muscles and dilates the airways, all of which prepare the body for sudden and extreme exertion such as "fight or flight".

(15) Depression causes people (and other mammals) to refrain from moving around in their environment as normal, to reduce exposure to environmental hazards such as disease or predators until the depression has abated. This is an obvious survival "mechanism".

To the Universalist philosopher all these highly complex self-regulating, self-organising systems and feedback "mechanisms" of the body and brain are indications of the workings of the self-improving order principle in body and brain.

How Photons Control Body and Brain

The self-regulating homeostasis within body and brain has a surprising input from outside, which the Universalist philosopher needs to probe. Light energy is continuously being converted into chemical energy within the brain and body, and I can go so far as to say that this light energy makes homeostasis possible.

So far we have examined evidence for the workings of homeostasis in body and brain without considering how its feedback "mechanisms" could be controlled by the order principle. Seeing the order principle at work in so many instances of homeostasis in body and brain requires some consideration of the entirely new science of nanostructures; the information-bearing potential of photons; and the way light is received in the eyes.

Because the new technology is being applied by scientists we must approach the workings of photons in body and brain via the technological and industrial application of photons. It has only relatively recently been discovered that photons can bear information. Leaving the question of what information existing photons bear to one side for the time being, we must grasp that a new science of photonics is being developed. This is the optical equivalent of electronics.

Instead of using electrons to transmit and process information as in telephone and computer technology, photonics use photons or living units of light. Advanced optics such as laser beams converging inside crystals the size of sugar cubes can form holographic images that work with quantum theory to process huge amounts of information.[20] Advances in fibreoptics have made it possible to transmit data on several wavelengths of light simultaneously along the same fibre through a technology called wavelength division multiplexing. Information can be sent a thousand times faster than current internet technologies can handle. Existing telecommunications networks use classical light to transmit information through optical fibres, and have to boost signals by repeater stations. Rapid progress is being made towards quantum networking.[21]

Fibreoptic lines can transmit a thousand colours at the same time alongside masses of photonic devices such as crystal switches and processors in which light controls light inside transparent crystals. All this is happening at present in laboratories, but as yet not on the internet. Photonics will not replace electronics but will work alongside them to transmit data over long distances, via network lines. Photons have thus made possible a future technology in quantum computing. Using quantum theory in which atoms or electrons can be in two places at the same time, computers based on quantum physics would have quantum bits ("qubits") that exist in both on and off states simultaneously, making it possible to process information much faster than conventional computers. A future internet of quantum computers may self-organise itself into an intelligent network of modes, and resemble a quantum brain.[22]

Optical microchips that can store light for short periods of time before sending it onwards have been constructed by IDM researchers in the US. Light is good at transmitting data at high rates as its signal can be switched on and off quickly, but it is difficult to store as photons interact weakly with each other. Light connections require optical buffers that can prevent packets of data from collecting at switching points. Already, light can be delayed by sending it through a length of optical fibre. A light pulse must pass through 21 centimetres of fibre to be delayed by one nanosecond.[23]

Using photons, electrons and quantum particles to carry information works as follows. The sender encodes the data as the quantum state of a particle, and the recipient measures the particle to infer the original quantum state. The laws of quantum mechanics specify the maximum amount of information that can be extracted. This theoretical maximum for transmitting information via photons and other quantum particles can be approached by choosing how information is encoded, transmitted and decoded.[24]

The amount of information that photons can carry can be increased by distinguishing between photons that have incrementally different amounts of momentum. A single photon with 32 possible orbital angular momentum values could transmit data five times as fast as binary-digit computers. Combining techniques including the polarization of light can multiply the information-carrying capacity of each photon.[25]

Physicists at the Georgia Institute of Technology have achieved the transmission, storage and retrieval of single photons between remote memories. They created a photon by exciting a cloud of ultra-cold rubidium atoms which generated one photon every five seconds. As it is in resonance with the atoms from which it was created the photon carries specific quantum information about the excitation state of these atoms. The photon was sent down about 100 metres of optical fibre to another very cold cloud of trapped rubidium atoms. The control beam was switched off, and the photon came to a halt inside the dense atomic cloud, thus storing it.[26] Although it is widely believed that a photon or single-particle quantum state cannot carry more than one bit of information, it is simple to transmit more than one bit of information through a photon or single-particle quantum state.[27]

Photonic crystals are periodic optical nanostructures designed to affect the motion of photons in a way similar to the way periodic semiconductor crystals affect the motion of electrons. MIT physicists have found that photonic interactions, which are too weak to have computer applications at present, can be strengthened to control the flow of light – which has implications for the computers of the future.[28]

The unbreakable relationship between twin photons, known as "entanglement", requires that the state of one photon is strengthened by the state of the other. This has made quantum teleportation possible. It has already been achieved by the California Institute of Technology, using twin photons.[29]

The properties of photons that are currently being applied by scientists to machines can also be applied to the body and brain through the eye. Photons bombard the human retina, which is a filmy bit of tissue barely half a millimetre thick that lines the inside of the eyeball and develops from a pouch of the embryonic forebrain. The retina is therefore considered to be part of the brain.[30]

Each retina possesses about 200 million neurons and contains 120 million rod and 6 million cone photoreceptors.[31] The photoreceptors' rods are for low-light vision, the cones for daylight or bright-coloured vision. Nocturnal animals such as owls are dominated by rods. These sensory neurons respond to light and lead to intricate neural circuits that perform the first stage of image processing.[32] Photons collapse into the retina and are subsequently processed as information at the level of neural membranes. The information reaches the sensory cortical regions where it is converted into patterns of microtubule subunits and specific quantum states.[33] (See Appendix 5.)

The photoreceptors lie at the back of the retina against the back of the eyeball. They are in layers. In the second of three cell layers, called the inner nuclear layer, lie 1-4 types of horizontal cells, 11 types of bipolar cells and 22-30 types of amacrine cells. The numbers vary in each species.

The surface layer of the retina contains about 20 types of ganglion cells. Impulses from these ganglion cells travel to the brain via over 1 million optic nerve fibres. Synapses link the photoreceptors with bipolar and horizontal cell dendrites. This region is known as the outer plexiform layer. Where the bipolar and amacrine cells connect to the ganglion cells is known as the inner plexiform layer.[34]

Light rays – photons – must pass through the retina before reaching pigment molecules. This is because the pigment-bearing membranes of the photoreceptors must be in contact with the eye's pigment epithelial layer, which provides the vital molecule retinal, or vitamin A. Retinal becomes attached to the photoreceptors' opsin proteins, where it changes its form in response to photons, or packets of light. Retinal molecules are then recycled back into the pigment epithelium. This tissue behind the retina is very dark because its cells are full of melanin granules. The pigment granules absorb stray photons to prevent them from being reflected back into the photoreceptors, and causing images to blur.[35]

The retina is thus amazingly complex and plays an active part in perception. We do not understand the neural code that the ganglion-cell axons send as trains of spikes into the brain, but we are beginning to understand how ensembles of ganglion cells respond differently to aspects of the visual scene the retina sees. Ganglion cells construct aspects of what is seen, but the greater part of the construction of visual images occurs in the retina while the final perception of sight takes place in the brain.[36]

Much of the information transfer from photons to the retina depends on electrical connections among cells rather than on standard chemical synapses. Thus, the major neural pathway from the rods depends on electrical connections. Some fast-acting signals pass from amacrine cells into ganglion cells at gap junctions. Neuromodulators change the environment of the neuron circuits but act at a distance by diffusion rather than at synapses. A previously unknown ganglion cell type acts as a photoreceptor without input from rods or cones. The cell membrane of this ganglion contains light-reactive molecules known as melanopsins.[37]

All photoreceptors in humans are found in the outer nuclear layer in the retina at the back of each eye, and the bipolar and ganglion cells that transmit information from the photoreceptors to the brain are *in front of* them. Light must travel through the axons and cell bodies of other neurons before reaching the photoreceptors. A region in the centre of the retina, the fovea, which contains only photoreceptors, supplies high visual acuity or sharpness to compensate for the slow reception of photons. Each retina contains a blind spot, where axons from the ganglion cells go back through the retina to the brain. This also compensates for the way photons are received and information is dispersed.[38]

In vertebrates, including humans, the photoreceptor cell (or specialised type of neuron found in the retina of the eye) is capable of phototransduction. The photoreceptor cells absorb photons from the visual field and send this information along a specific biochemical pathway through a change in its membrane potential.

This information will eventually be used by the visual system to form a complete image or representation of the visual world.[39]

The fact that a photon created from rubidium carries within it information about the excitation of the atoms from which it was created suggests that all photons that arrive from intergalactic space contain within them equivalent information. We have seen on p241 that a photon can carry a thousand colours and many bits of information, and so it is reasonable to suppose that one of its thousands – perhaps millions – of bits of information may contain practical instructions for the order principle to photosynthesise plants and to influence the bodies and brains of all creatures via their eyes and visual system: their retinas on which photons collapse and the neural circuits along which they are fed as information to form an image of the world and, in all probability, to stimulate and direct the various feedback "mechanisms" of the body and brain.

The bodily absorption of photons has been used in the medical measuring of blood flow, in a technique focusing on bio-photon emission. Either a laser Doppler blood flowmeter or an ultrasonic blood flowmeter can be used. A glass fibre is inserted into a blood vessel and irradiated by a laser beam. Blood flow is measured in terms of a variation of the wavelength of a reflected light or of an ultrasonic wave. The point is that blood vessels emit bio-photons, and these are measured in terms of units of time by converting them into an amplified electric signal. This method is used in combating cardiovascular diseases, cancers, strokes, hardening of the arteries and transient ischemic attacks (TIAs, dysfunctional blood circulation in the brain).[40]

As this medical technique suggests, human blood is full of bio-photons which have been received as photons in the retinas of the eyes. I submit that all the many instances of the body's and brain's homeostasis and negative feedback we have considered in this chapter – each of the 62 examples of the body's homeostasis and the 15 examples of the homeostasis of the brain – can be seen as being serviced by information-bearing photons which continuously bombard the retinas, collapse and are dispersed throughout the brain's neural circuits and the body's blood as bio-photons which bear information containing instructions that direct and maintain the functioning of body and brain.

Much work remains to be done in this new area, in what will surely one day be a heavily researched, data-based and documented new science of photonology and biophotonology, but I submit that there is enough evidence within current optical and medical knowledge to substantiate the claim that photons trigger the directing and nourishment of the physical body and brain. I submit that Newton was right (see pp82-84) and that the photons of light control our bodies. I further submit that, like DNA, photons contain coded instructions through which the order principle controls our bodies and brains. I submit that the 62 examples of the homeostasis of body and the 15 examples of the homeostasis of brain are evidence for the workings of the self-improving order principle which controls all creatures' bodies and brains – and our consciousness, the high point of the evolution of body and brain.

Interaction of Brain and Levels of Consciousness

Having established a link between photons and brains, the Universalist philosopher can now turn to consciousness, which seems self-regulatory in both its lower, autonomic and its higher, more thoughtful aspects and activities. The Universalist philosopher looks to science to determine whether consciousness is an effect of the brain like homeostasis, or whether it controls the brain as homeostasis controls both body and lower brain – and whether consciousness is implicated in controlling homeostasis.

Consciousness is a psychological condition, defined as "a state of perceptual awareness" (*Concise Oxford Dictionary*). It is also an awareness of self, and awareness of being aware.

The history of the study of consciousness has led to no satisfactory outcome. In the early 19th century some scientists, perpetuating Descartes' dualism of mind and body, saw it as a substance, "mental stuff" that differed from material substance. Others saw it as an attribute characterised by sensation and voluntary movement that separated animals and man from the lower forms of life and also separated their normal waking state from sleep, coma or anaesthesia ("unconsciousness"). It was also seen as a relationship or act of the mind towards Nature or objects, which provided a stream of mental sense data. (I once asked Sir John Eccles, a physiologist and Nobel prizewinner for his discovery of the chemical behaviour of nerve cells and a Cartesian dualist, where the mind is when the eye looks at a flower, and he replied, "Between the eye and the flower," a view of perception similar to that in Donne's 'The Exstasie': "Our soules …hung 'twixt her, and mee.")

For a long while, the main method of examining consciousness was introspection, "the perception of what passes in a man's own mind" (John Locke), looking within one's own mind to discover the laws of its operation. However, different trained observers disagreed on their observations and failed to reveal consistent laws. As a result, in the early 20th century mental states – and indeed, the concept of consciousness – were rejected by scientists, who switched to behaviourist psychology as described by John B. Watson. They approached consciousness from the outside, through behaviour.

Studies of animal behaviour concluded that animals live for food and survival. Do animals have consciousness? A colony of 80,000-100,000 breeding pairs of king penguins on South Georgia, near Antarctica stand along Salisbury Bay and spread back up the snow-clad mountain. They stand all day, go out to sea and return with food in their stomachs which they regurgitate for their young. Do they only think about food, mating, feeding their young and survival? Reductionist behaviourists would say they are solely driven by such considerations. Is there no social interaction as they stand within their colony? I have already said that there are crèches for the young, policed by adult king penguins, which suggests a social structure. We have seen (on pp169-170) that pied babblers display altruistic social behaviour by acting as sentries. Do wandering albatrosses fly from the Antarctic to Australia and back or orbit the South Pole, a six-month round trip in a migratory pattern that takes them away from their chicks, solely to obtain nutrients they need for the winter which they cannot find elsewhere and to distribute the demand on food? Or is there social interaction on the way? Are animals solely their brain functions, which (as we have seen) include instincts and powers of co-ordination to fly and dive in formation without colliding? Is the belittling of animals' consciousness anthropocentric and patronising? There is much that we still do not understand.

Similarly, brain physiologists thought that human consciousness depended on the function of the brain, and that states of consciousness are correlated to brain functions. Levels of alertness and responsiveness were found to be correlated to patterns of electrical activity in the brain (brainwaves) which could be recorded on an EEG or electroencephalograph. In wide-awake consciousness, the brainwaves are irregular, of low amplitude or voltage. During sleep, the brainwaves are slower and of greater amplitude, coming in bursts. Beta and alpha rhythms are between 30 to 8 cycles per second. Theta and delta are from 8 to 0.5 cycles per second. The gateway to higher-frequency waves seems to be the meditative state of 4 cycles per second.[41]

All this electrical and behavioural activity comes from part of the brainstem known as the reticular formation. This comprises the ascending reticular system, connections between the networks of each of more than 100 billion brain cells or neurons (10^{11}). There may even be a trillion (10^{12}) neurons, but if each of 10^{11} connects with an average of 3,000 other brain cells there are over 300 trillion connections (3 times 10^{14}). If each connection is capable of 10 levels of activity, there are 3 quadrillion potential brain states per second (1 quadrillion being 10^{15}).[42] Molecules within the brain cells may act as miniature computers within each cell and take the brain's computing power to 10^{27} operations per second.[43] Also see p318. Stimulating the ascending reticular systems with an electrical impulse rouses a sleeping cat to consciousness and activates the waking pattern of its brainwaves.

It used to be thought that the neurophysical "mechanisms" that might produce consciousness and the higher mental processes – thinking and free choice – were in the cerebral cortex. It is now thought that the cortex has specialised functions of integrating sensory experience and organising motor patterns. The neural structures that might produce consciousness are now thought to be in the ascending reticular system. The brainstem reticular formation is not the seat of consciousness but acts as an integrator, and functions through its interconnections with the cortex and other regions of the brain. As we have seen in the brain's use of feedback "mechanisms", human brain function is not located in any one area of the brain but rather in a connection of areas. The same is true of the functioning of human consciousness.

The search to locate the seat of human consciousness in the brain has been a long one. Gerald Edelman, a theoretician on the brain and consciousness, located the unity of consciousness in the neural networks throughout the cortex and the thalamus. Francis Crick helped crack the double-helix genetic code and discover how DNA makes RNA and how RNA makes amino acid and therefore protein, discoveries that in the opinion of some put him alongside Copernicus, Newton, Darwin and Einstein. (It has to be said that much was known about DNA before 1953, and that the workings of protein may be more significant than the structure of DNA.) He spent nearly the last twenty-eight years of his life, from 1976 when he was sixty, researching into and seeking the seat of consciousness.

Near the end of his life Crick located the neural correlates in the claustrum, a sheet of brain tissue below the cerebral tissue cortex which connects to, and exchanges information with, nearly all the sensory and motor regions of the cortex and with the amygdala which plays an important role in emotion. (See Appendix 4.) Crick was correcting a paper which he produced with Christof Koch on this theory on his way to hospital a few hours before he died on July 28, 2004. The paper compares the claustrum to the conductor of an orchestra. (Compare Plato's charioteer and two horses.) Perhaps the claustrum is the seat of a system of light which may comprise consciousness, which Descartes located in the pineal gland. Crick's friend Vilayanur Ramachandran visited him three weeks before his death and reported:

"As I was leaving he said: 'Rama, I think the secret of consciousness lies in the claustrum – don't you? Why else would this tiny structure be connected to so many areas in the brain?' – And gave me a sly, conspiratorial wink. It was the last time I saw him."[44]

Crick's view of the claustrum must be seen in connection with the work of Hans Kornhuber, a German neuroscientist, who wired volunteers to an EEG with an electrode to their skull and asked them to move their right index finger. He found that each movement was preceded by a spark in the electrical record of the pain one second before the voluntary movement, denoting the moment of free will. In 1983 in the US Benjamin Libet developed this work. He found that the "readiness potential" happened 200 milliseconds before a volunteer felt the urge to move his or her index finger.[45] According to Crick free will is located in or near the anterior cingulate sulcus.[46] There is also a link between complex memory and the neuro circuits and pyramidal cells of the hippocampus, which is associated with the spatial awareness of rats and rabbits.[47]

Just as the brain controls and regulates the body and interacts with it so that at times the body controls and regulates the brain, is there a similar relationship between the brain and consciousness? Does the brain control and regulate consciousness with the help of photons, and at times does consciousness control and regulate the brain with the help of photons? That there is a link between the higher functions of consciousness and the brain is indisputable. But the Materialist, reductionist view that consciousness and mind are simply brain function and that consciousness can be reduced to physics and chemistry may be premature. Consciousness is still something of a mystery, and despite decades of research on EEGs there is still no obvious causal connection between experience and the molecules, neurons and neural circuits of the brain.

Just as we saw on p232 that the brain has seven levels of organisation like the body – eight if society is included – so there are as many levels of consciousness.

There are, in fact, at least twelve levels if we include the two unconscious ones at the beginning and admit two higher consciousnesses, one for rational philosophical and scientific thinking and the other for intuition and imagination, all of which between them cope with every known experience:

(1) sleep – unconsciousness;

(2) dream – dreaming while asleep;

(3) drowsiness – the state of being half-awake when ideas can be received from the unconscious immediately after sleep, readiness for sleep, drunken or drugged consciousness (though some drugs temporarily induce higher levels of consciousness);

(4) waking sleep – the robotic ordinary state of everyday automatic working consciousness, everyday passive social consciousness, reacting to external events, writing cheques, performing, travelling automatically back from work, tired and drifting in a daydream, reactive choices, negative emotions (envy, jealously, hate), attachment to objects, mindlessness while exercising or gardening, relaxing, listening to undemanding music, passive watching of films on TV (a primarily physical state controlled by the sense experience of the rational, social ego or temporal self);

(5) awakening consciousness – more alert everyday consciousness, self-assertion, more considered choices, awareness of the world, noticing every detail of what one observes, alertness when active shopping, business deals, more positive emotions, sexual activity, the poet's observational view of the world (a more psychological state of awareness and observation);

(6) self-consciousness – awareness of being aware, awareness of the rational, social ego that is doing the seeing, looking out from a centre that is not the ego, a mixture of positive and negative feelings, conscience, self-restraint and self-reproach, feeling anxious and discontented (a meditative, reflective state drawing on memory in a "wise passiveness");

(7) objective consciousness – selflessness, losing oneself in concentration, thinking, active working on papers or computer, everyday active social consciousness, active choices, dictating, active reading of books and newspapers, doing research, listening to classical or spiritual music, being absorbed in art, calligraphy or pottery, lecturing, performing, writing, concentrating on the speed of the ball while playing sport, loving, deeply involved sexual activity, being completely absorbed, responding to religious

experience, positive feelings, feeling contented and happy, seeing the universe with clarity through a centre that is below one's ego and known to poets as "the soul", and is often accompanied by a feeling of serenity or contentment (an intense soul state);

(8) higher (or deeper) rational consciousness – detachment from senses, thinking very deeply, reflecting, discussing issues, debating, profound choices, being sceptical, a more intense state of consciousness than (7), writing at a philosophical or scientific, rational level (a contemplative state);

(9) higher (or deeper) intuitional and imaginative consciousness – detachment from senses, feeling very deeply, being immersed in a creative artistic project, imagining oneself into a character in a play or a novel, being transported into a different world and living a character's detailed experience (an inspirational-imaginative, visionary state);

(10) superconscious or subconscious monitoring – an ordering, organising functioning that overnight finds and then identifies things left undone (a superconscious or subconscious editing state);

(11) transpersonal consciousness – psychic and paranormal consciousness (telepathy, precognition, prophecy, healing), awareness of memories from the womb, past lives or another world seen in the near-death experience, mediumistic contact with departed spirits, rare glimpses seen through the centre known as the spirit which traditions claim has lived before (a spiritual state);

(12) universal or cosmic (or unity) consciousness – awareness of oneness with the universe and the infinite when time and space cease to exist, the mystic oneness that is perceived through the centre known as universal being and is often accompanied by a feeling of ecstasy, contemplative experience of the metaphysical Light (a mystical, contemplative state approaching the One Reality, traditionally known as the divine, the highest spiritual state and the highest-known state of consciousness).

All these hierarchical levels (in ascending order from numbers (1) to (12)) are attested in accounts in literature, philosophy and mysticism. They may not all have been corroborated by science but these states have been reported on by those who have experienced them. They include all the experiences that a "system of general ideas" needs to interpret, including the experiences Whitehead listed (see p50).

Levels (8) and (9) could arguably be the other way round, elevating philosophy and science above intuition and imagination. Literary champions of the imagination are sometimes scathing about science's view of consciousness. Blake wrote of Newton (who as we have seen was actually very perceptive regarding the expanding force of light) that he had one level of consciousness: "May God us keep/From Single vision and Newton's sleep!"[48] Coleridge wrote in a letter "Newton was a 'mere Materialist' – Mind in his system is always passive – a lazy looker-on on an external world."[49] Blake and Coleridge were being more than a little unfair to Newton for his thinking about the universe was done in level (8). I could just as easily have designated Newton's level number (9). This would have challenged their belief in the superiority of the Romantic imagination over science. In fact, the consciousness of Blake and Coleridge was different in kind from Newton's consciousness rather than superior, and my levels (8) and (9) could both have been designated (8)=.

The monitor of level (10) may be an extension of levels (8) and (9). It combs through recent work and throws up omissions. Only this morning I woke early and, when the images of a dream had receded, my consciousness pushed forward – or my brain threw up – three amendments to the text of a particular chapter I neglected to make last night: a need to change the order of some words, to insert a phrase I thought of yesterday and a detail involving a scientific fact. I duly wrote these into my text. What "I" do is being monitored by a part of my consciousness or brain which seeks out and finds my omissions and imperfections, and literally acts as a monitor. The monitor is an ordering function, and is perhaps the universal principle of order in operation.

There could be a thirteenth level, group consciousness, one involving interaction with society at different levels (State, country, region, globe), losing oneself in crowd emotion at a sporting occasion, but it could be argued that this is not hierarchically higher than level (12) as the social reality of a group is arguably less real than the mystical experience of level (12).

It could be argued that just as the brain controls and regulates the body with the help of photons and interacts with it so that at times the body also controls and regulates the brain, so the brain controls consciousness with the help of photons and interacts with it so that at times consciousness controls and regulates the brain with the help of photons. In this interaction, at times consciousness controls and regulates the brain and organises itself between these 12 levels. We are the quality of our consciousness. Those who spend the greater part of each day and evening in level (4) are completely different people from those who spend the greater part of each day and evening in levels (5)-(12). Those who spend their lives in level (4) will drift passively, whereas those who spend their lives in levels (5)-(12) are more likely to have drive and purpose. To a very large extent the quality of our lives is measured by the quality of our consciousness.

So – to return to the point on p249 – does science tell the Universalist philosopher whether consciousness is an effect of the brain like homeostasis, or whether it controls the brain with the aid of photons as homeostasis controls both body and lower brain   and whether it controls homeostasis? No, it does not. Science is surprisingly reticent on this matter. Crick pushed the science of consciousness as far forward as it is possible to reach at present, and after twenty-eight years his findings were inconclusive. He did not even prove "the astonishing hypothesis" (the title of his book), that mind is merely brain function. That mind is an effect of the brain is still only a hypothesis.

The Universalist philosopher has consulted science on consciousness and has been left little the wiser, having established the limits and limitations of science in this field. The problem has therefore now been transferred from science to philosophy. The category has changed, and from now on Universalist philosophy can scrutinise the scientific links between photons and the brain and draw conclusions that operate within a philosophical category. The proper place for this to happen is in Part Three.

All bodies, brains and consciousnesses began in one point, the singularity that began the universe with the Big Bang, and later on in one cell, the first cell. The seed for all bodies, brains and consciousness emerged, like bio-photons and DNA, biologically along with the self-organising principle of order. In the course of the 3.8 billion years since its appearance, the first cell has evolved via the ape and self-improvement towards higher (or deeper) consciousness, which has only manifested itself during the last 2 million/200,000 years, and more significantly during the last 50,000 years.

Philosophical Summary

What does physiology contribute to the "new scientific interpretation of the universe and Nature"? Most importantly, the self-regulation of homeostasis in both body and brain. Homeostasis is a constant – compare the cosmological constant. (The universe may operate through constants.) There are many self-organising systems within the body and the brain, which have narrow tolerances. Photons enter the brain through photoreceptor cells, and light energy is dispersed through the brain's neural circuits the body. Photons can convey information. It is possible that light photons carry coded information like DNA. Crick could not prove his theory that the claustrum is the seat of consciousness. There are 12 levels of consciousness. The limitations of science as regards consciousness. That mind is an effect of the brain is only a hypothesis. It is for philosophy to offer an explanation for the scientific data regarding consciousness. Science has not explained the differences between the 12 levels of consciousness and how the transitions between them take place.

To sum up, Universalist philosophy can take from physiology:

the self-regulating of body and brain:

homeostasis is a constant (compare the cosmological constant);

many self-organising systems within the body;

photons enter the brain through photoreceptor cells;

light energy is dispersed through the brain's neural circuits to the body;

photons may carry coded information like DNA;

there is no proof that the claustrum is the seat of consciousness;

there is no proof that mind is an effect of the brain;

there are 12 levels of consciousness;

the different levels of consciousness have not been explained by science.

To return to the title of Part Two, 'The Scientific View of the Universe and the Order Principle', having examined the scientific evidence, the philosopher cannot but conclude that there is a universal principle of order in the universe which – through the atoms, cells, photons, bio-photons and "mechanisms" revealed in cosmology, physics, astrophysics, astronomy, geology, palaeontology, biology, ecology, biochemistry and physiology – has delivered with awesome precision the conditions and habitat that are ideal for the creation of galaxies and stars and the birth and maintenance of human existence; and that without the order principle in the cosmological and biological ecosystem human life could not have existed.

The universe is quite simply a stunning system. Operating as a philosopher, I have tried to find ways of summarising the enormous multiplicity and self-organising power of Nature, and have opted for succinct lists in the hope that the sheer variety and exuberance of life shines through the welter of minor detail. Reflecting on Part Two, I am stunned by the juxtaposed order the universe reveals.

It would be possible to provide hundreds of further instances of the workings of an orderly system from each of the various disciplines, going into ever greater and greater detail. I hope that others will make such a comprehensive study. Here in this introductory *Prolegomena* I indicate the principle of order in the universe that is bio-friendly to human life and provide enough evidence to establish it.

We have seen that our universe began as a point or singularity within an infinite sea-like Void containing latent, potential energy, and that its orderly structure controlled the first few seconds of the Big Bang when all four forces were united before they flew apart. We have seen that the universe's rate of expansion and the balance of gravity and electromagnetic forces provided an expanding, accelerating, nearly flat universe that may expand forever, and that the making of galaxies, stars, heavy elements, the atmosphere, climate, particles and forces are all conducive to life.

We have seen that the universe does not have a purely finite, physical origin that can be explained by a Theory of Everything expressible in a formula. I have not said that an infinite singularity is a finite, physical singularity as Materialist reductionists do, having it both ways – having the benefits of the infinite alongside a view of physics as exclusively finite. I have argued that the universe began from the infinity before the Big Bang.

There is still considerable ignorance and uncertainty about the basic principles of cosmology. More discoveries are waiting to be made, for example in dark energy and dark matter and on whether the universe will expand forever or become perfectly flat and slow to a halt. And the order principle operates in many disciplines, as we have seen. It is quite possible that this order principle is a fifth force borne in photons and minute particles that has not been identified or recognised until now. CERN's experiment in its Large Hadron Collider, which cost at least £4.4 billion,[50] from September 2008 sought an invisible field of bosons no more substantial than the order principle (see pp91-92). I shall have more to say on bosons in ch. 10. Because the order principle has not been discovered does not mean that it does not exist or that it may not be about to be discovered and evidenced. It is no less probable than some other hypotheses put forward by theoretical physicists. The first task is to set it out as a hypothesis and predict it in the hope that it can be detected, which is what I have attempted to do in this Part.

Just as the theory of atoms existed only as a theoretical proposition for 2,500 years, until the existence of atoms was objectively established in the 20th century when atoms ceased to be a theory and became the cornerstone of modern science, so the invisible energy manifesting from Reality in the microworld – the order principle – may be transmitted and conveyed in photons that control, nourish and direct the growth of plants and, I submit, of all creatures' bodies and brains. The order principle may even be carried on photon-like particles too small to have been found – smaller than neutrinos, perhaps even smaller than photinos (if they exist), so small that they have not even been given a name, and forming a kind of invisible *aither* (or ether) as the Presocratic Greeks called it   and may be waiting to be discovered, tested, measured and proved.

We shall consider how the order principle controls consciousness in chapters 9 and 10 as the universe's impact on scientifically nebulous consciousness falls under a philosophical rather than a physiological, scientific category.

Now philosophy has detected a self-organising system, life principle and order principle – indeed, Law of Order – in the operations of Nature and behind the physical world of science, philosophy can renew itself by reflecting the universe known to science and by combining it with the hidden Reality behind and within the first singularity.

PART THREE

THE METAPHYSICAL VIEW OF THE UNIVERSE IN THE NEW PHILOSOPHY

"Our minds are finite, and yet even in the circumstances of finitude we are surrounded by possibilities that are infinite, and the purpose of life is to grasp as much as we can of that infinitude."

Alfred North Whitehead[1]

"The only real valuable thing is intuition."

Albert Einstein

"Operations of thought are like cavalry charges in a battle – they are strictly limited in number, they require fresh horses, and must only be made at decisive moments."

Alfred North Whitehead, *Introduction to Mathematics* (1911)

9

THE RATIONAL APPROACH TO REALITY
AND THE SYSTEM OF THE UNIVERSE

My strategy has been to consult the scientific view on the universe and import its essential findings into philosophy (see the end of Part One, pp49-53). The metaphysical view of the universe will then include the essential findings of science. Having laid out all the facts objectively assembled in Part Two I can now piece them together into a picture, react to them and devise a philosophical explanation – a "system of general ideas" – that explains them all without imposing a preconceived idea on them. I can then approach the objective truth about the universe. For convenience I reproduce the eight Philosophical Summaries which include our findings – the facts – so far:

1.

the eternal, moving "boundless", the "infinite";

the Void as a sea of moving energy;

moving "ever-living Fire";

the "germ" within the "boundless" that became the universe;

the One;

the ordered, unified universe or *kosmos* ("ordered whole"), concept of order;

philosophers focused on the universe and Nature;

aither and process flux;

Atomism.

2.

truth about hidden Reality;

"science which studies Being";

the One;

Intelligible or Uncreated Light;

limitations of Empiricism, Rationalism and Idealism;

focus on consciousness and being;

four subdivisions of metaphysics today.

3.

the infinite;

the first singularity;

no multiverse;

the Big Bang;

inflation and shaping of the universe;

increased acceleration;

the surfer.

4.

the timeless Void;

the quantum vacuum as a reservoir of order;

a sea of energy;

order within the vacuum;

order beneath apparent randomness and uncertainty principle;

photons and virtual photons;

the expanding force of light that counteracts gravity;

the cosmological constant;

undetected ether-like, dark-energy-like network of tiny particles;

manifesting metaphysical Light;

light stimulating chemical composition and biological growth.

5.

a bio-friendly universe;

the order principle which may work with four forces as a fifth force;

the order principle as a life principle;

constants and ratios in relation to each other reveal order and are right for
    life;

physical laws in 40 conditions (gravity, dimensions, expansion rate of the
    universe, density, speed of light, anti-gravity/dark energy, flatness, size
    and emptiness of the universe) reveal order;

the four forces and many particles and their mass ratios and charges, and all
    forms of matter reveal order;

the atmosphere, glacials, climate, temperature reveal order;

the Big Bang reveals order.

6.

no evidence for the "soup theory";

no evidence that the first cell arose by accident;

no evidence that man's place in the universe is a pointless accident;

a "mechanism" for self-organising ordering has driven the origin of life and
evolution;

land mass and climatic conditions have been right for sustaining life;

evolution by natural selection is the process of development, but little is
known about the ordering thrust of evolution;

evolution may be a matter of inner will, a drive for self-improvement, a
persistent desire for self-betterment in relation to the environment, to
realise the next stage of development;

no evidence on man's descent from apes but it is likely;

DNA has revealed the oneness of all living things;

Darwinism is imperfect but unbettered;

Expansionist, ordering thrust within evolution via DNA and perhaps sunlight.

7.

a self-running, self-organising, self-perpetuating system;

the workings of an order principle;

each creature is driven by instinct to hunt, co-operate and breed at the right
time;

each creature knows from birth its place and what to do in relation to the
system;

an ordered system of food chains and webs keeps populations stable;

species co-operate through symbiosis;

plants give medicinal relief;

the self-organising system does not seem the product of random accident.

8.

the self-regulating of body and brain:

homeostasis is a constant (compare the cosmological constant);

many self-organising systems within the body;

photons enter the brain through photoreceptor cells;

light energy is dispersed through the brain's neural circuits to the body;

photons may carry coded information like DNA;

there is no proof that the claustrum is the seat of consciousness;

there is no proof that mind is an effect of the brain;

there are 12 levels of consciousness;

the different levels of consciousness have not been explained by science.

These points can be included in the metaphysical view of the universe and should be fed into the new philosophy of Universalism. From now on in Part Three, we are in a philosophical category and are scrutinising scientific views from a philosophical point of view as Universalist philosophy incorporates scientific views into the new metaphysical philosophy.

So what does science have to tell the philosopher? We have looked at cosmology and astrophysics and have found that there is a hidden infinite Reality behind the apparent finite universe. I have associated this Reality with the quantum vacuum and the Void the surfer breasts on the tip of the expanding, accelerating universe. This order predates the Big Bang for it is within the singularity and there is evidence that it has determined the structure of the universe from galactic to atomic scales, photons, DNA, quantum energy and feedback systems. We have found that the universe has proved to be bio-friendly and that the convergence of bio-friendly conditions is too improbable to be caused by random accident. We have looked at biology and biochemistry and have deduced from the evidence that a self-organising principle of order has driven evolution from the origin of life to DNA and via apes to higher consciousness. We have found that Nature is a system in which all creatures instinctively know their place from birth. We have seen that their operations are too systematised and orderly to be a random accident. We have looked at physiology and have seen that complex order underlies the feedback systems of the body and brain, order too complex and sophisticated to have arisen by random accident. We have seen that higher consciousness reflects the underlying order and harmony of the universe.

Having based Universalism on the scientific view of the universe and evidence for an order principle outlined in Part Two, I am in a position to make a coherent statement in a "system of general ideas" on the universe's system and on man's position in it. I have laid out all the relevant facts and can now arrive at an explanation which fits all the facts without making the facts fit a preconceived idea. My statement will be more accurate than the statements of those philosophers in the past who have theorised in abstract without grounding their views in Nature, cosmology, astronomy and the scientific evidence regarding the universe. My summary of what all the sciences say about the structure of the universe and "what is" – the infinite Reality and the phenomena of Nature – covers both the metaphysical and the scientific.

The points in the eight Philosophical Summaries can be distilled by combining all the lists and eliminating repetition, as follows:

1. Before the Big Bang there was an eternal, ever-moving boundless or timeless infinite Reality, the Void or One, an "ever-living Fire".

2. The universe manifested from this infinite Reality as a point or germ, the first singularity, and, soon after the Big Bang, expanded by inflation, and the expansion has accelerated. A constant may save the universe from slowing down and perishing in a Big Crunch. The universe is finite and shaped like a shuttlecock or cone, open at the top. Space-time is within it.

3. Outside the universe, all round it and within it is the boundless, infinite Reality, a sea of energy which the surfer breasts. It consists of Light (Uncreated Light), and is "*aither*" (ether) or dark energy, a moving process or flux. This contains the latent possibilities of Being and Existence from which our universe was thrown up. It manifests into a quantum vacuum, an emptiness filled with order.

4. There is some evidence that there was a predisposition to order in the very early universe. This includes land mass and climatic conditions which favour life. There are 40 bio-friendly laws and many ratios and constants that were exactly right for life and allowed human life to happen. The order principle can be detected at all stages in the universe's life and exists alongside the apparent randomness and uncertainty.

5. The order principle, or Law of Order, has been associated with and may be contained within the photons of physical light, which may travel within metaphysical Light as Newton thought (his "spirit" and "ether"). This multi-layered light may manifest into an expanding force of light that counteracts gravity, stimulates chemical composition and stirs plants and the bodies of organisms to growth. It also controls the four forces, particles and all forms of matter.

6. It is not clear how life began but the drive of evolution, an order and life principle, from one cell via apes and Ice Ages to modern *Homo sapiens'* body, brain and consciousness has perfected many self-regulating, self-organising physiological systems. It has developed higher (or deeper) consciousness, part of which can know Reality. The ordering thrust of evolution may be borne to cells by the information in photons, a tenet of Expansionism.

7. All living things share oneness through DNA, and take part in a self-running, self-organising system in which all creatures know their place and what to do. The order principle can be seen to operate through competition and co-operation: in the food webs, niches, symbiosis and healing plants of Nature's vast ecosystem. The focus of science is on the whole planet. The earth is now in an interglacial within a continuing Ice Age.

8. The constant of homeostasis is self-regulated in the self-organising systems of both body and brain. Photons carrying information enter the brain through photoreceptor cells and are converted into chemical energy along neural circuits. There are 12 levels of consciousness. Mind may be partly an effect of brain and partly a system of light.

The scientific view of "what is" can be expressed more succinctly:

The finite universe manifested from an infinite, timeless, boundless Reality or Being which surrounds and permeates it and is both transcendent and immanent. Within the infinite the photons of physical light stimulate all growth, creating bio-friendly conditions for life through a Law of Order which works in Nature's local ecosystems, keeping all creatures in their place and driving to higher consciousness.

Philosophy, then, hears a message from the scientific view of the universe and reshapes and restructures itself accordingly. If there is a universal order principle or Law of Order that develops brains through evolution and self-organisation into higher and higher (or deeper and deeper) consciousness, then philosophy needs to reflect this principle. We have seen (in ch. 2) that the previous models of philosophy are at variance with this model of the universe. Rationalism, Idealism and Intuitionism, and Empiricism, Realism and Logical Analysis by themselves do not reflect the findings of the scientific evidence and my focus on the universal order principle. The reason, sense impressions, intuition and logic are not sufficient by themselves for the task of describing the universe of the 21st century. The operation of the universal order principle or Law of Order means that only manifesting metaphysical philosophy reflects this model of the universe.

Rational Ontology: The Concept and Structure of "What Is"

Our consideration of the metaphysical (which is behind the scientific and precedes it and generates it) can now rise naturally out of our consideration of the scientific (as foreshadowed on pp49-53). We saw on pp33 and 42-43 metaphysical philosophy has both rational and intuitional strands. The Intuitionists lost faith in the powers of reason and rejected the systems of Rationalists, and brought an Intuitionist metaphysical emphasis to philosophy. In this chapter I shall follow the Rationalist tradition of metaphysical philosophy. In the next chapter I shall follow the Intuitionist tradition of metaphysical philosophy. If read in conjunction with each other, these two chapters will bring together and reunify the two opposing strands of metaphysical philosophy.

I can now apply the category of rational metaphysical philosophy to the scientific view of the universe to reveal the structure of "what is". "What is" includes the infinite timeless Reality *and* the finite universe.

The sum total of everything, both infinite and finite, can be referred to as the All or the One.[1] In the Prologue I referred to algebraic thinking as a dialectical method that reconciles all contradictions, and I set out the simple formula $+A + -A = O$. If O is the One, then the infinite + the finite = O, the All or the One. The All or One includes all things concrete and abstract, natural and supernatural, known and unknown, probable and improbable, orderly and chaotic, temporal and eternal, material and immaterial, post-Big-Bang and pre-Big-Bang.

The concept of the All or One holds all possibilities within itself. It includes every possible concept including the unmanifest and the manifested. Within the One are all the potentialities of Non-Being and Being, and so the One has been described as Supreme Being but is more accurately to be described as Nothingness.

Within the All or One, the infinite has two aspects: the Void that preceded the Big Bang and which is outside the universe, or Non-Being; and the void within the finite universe after the Big Bang, or Being. Non-Being is the Nothingness within infinity that the surfer breasts and is outside the conical universe. It includes all the potentialities of Being but in an unmanifested form. Being, on the other hand, is the manifestation of Non-Being which contains all potentialities of everything manifested and all possibilities of Existence. It is the quantum vacuum, which appears to be empty but in fact seethes within the activity of virtual particles like an invisible sea of energy. If we were to accept Plato's Ideas as being virtual, they would be found in Being along with the blueprints, templates and germs of existing things in their potential form. In our algebraic, dialectical thinking we can say that just as $+A + -A = O$, so Non-Being + Being = the infinite All or One.

Within the All or One, the finite has two aspects: Being, which as we have just seen contains the invisible potentialities of Existence and is an intersection between the infinite and finite, between the timeless and time; and Existence, the manifestation of Being, the multiplicity which is contained within the unity of Being, all the phenomenal forms of Nature, all atoms and cells, all matter and organisms, all existing things we can see or touch in the universe of space-time. In our algebraic, dialectical thinking, we can say that just as $+A + -A = O$, so Being + Existence = the finite.

Within a philosophical category, the process of manifestation from the metaphysical to the scientific proceeds from the infinite to the finite, from the hidden Reality to the universe, from the boundless the surfer breasts to the phenomena of Nature and its ecosystems. Manifestation originates in Nothingness and becomes Non-Being and then, like a nothingness becoming a virtual particle, becomes an idea in Being, whence, like a virtual particle becoming a real particle, it becomes a form in Existence, the world of evolution and organisms. Existence pours out of Being like the multitudinous plenty pouring from a *cornucopia* (or horn of plenty). The totality – the infinite *and* the finite – is the All or One which permeates everything, both infinite and finite, as a Nothingness or Emptiness that is also a Fullness, a *pleroma*, a latent Non-Being and Being – an idea that is found in Eastern thinking and in Gnosticism – and that includes their manifestation into Existence.

For sake of clarity I can sum up the process of manifestation as four tiers:

Nothingness/Fullness, or the All or One – the potentialities of infinite Non-Being, infinite/finite Being and finite Existence;

Non-Being – the infinite Void outside the universe that preceded the Big Bang, the potentialities of Being and manifestation of Nothingness;

Being – the infinite/finite void which succeeded the Big Bang within the finite universe, the potentialities of Existence and manifestation of Non-Being; and

Existence – the finite manifestation of Being.

At the philosophical level, within a philosophical category, the rational approach to Reality therefore recognises that the infinite the surfer breasts which lies outside our universe is Nothingness and Non-Being, and that within the finite universe it manifests into Being. At the philosophical level, the rational approach to the system of the universe focuses on finite Existence and space-time which began after the Big Bang. At the philosophical level, we are looking at a manifestational process from Nothingness or the One, through the Void of Non-Being into the quantum vacuum or unity of Being and then to the diversity and multiplicity of Existence.

I must now return to the rationalist-reductionist position on holism (see p173), that the whole of the universe cannot be abstracted from its parts and reified to suggest it has an existence over and above that of its parts. If the whole is something more than its interconnected parts, reductionists ask, what is this "something more" and where does it come from? The rational Universalist answers that the whole is a manifestation from the All and is the infinite Reality behind the finite universe that manifests into it and permeates it with its ordering principle. In a word, it is Being.

We have seen on p262 that Being is a cosmic sea of energy, the potentialities of Existence. The study of Being in philosophy is ontology, which is not Materialistic or phenomenal (i.e. the study of visible existence). Many philosophers have seen Being as fundamental to the universe. In the 13th century, Robert Grosseteste wrote of Being as the Uncreated Light[2] which is the potentiality of form, physical light, and which, he claimed, is God. In this he was following a tradition that since Plato has seen Being as metaphysical Light. Heracleitus wrote of the world order as an "ever-living Fire", an idea close to the Uncreated Light. Plato wrote of "Universal Light",[3] which he imaged as the fire in the cave, meaning the Light behind the phenomenal world which appears as shadows. Augustine wrote of the "Light Unchangeable"[4] ("God is the Intelligible Light"). Mechthild wrote of the "Flowing Fire and Light of God".[5] Dante wrote of "Eternal Light",[6] "pure intellectual Light"[7] and the "sempiternal rose".[8] The Light of Being is known in enlightenment in the world's religions. We have seen that Newton thought that metaphysical Light was associated with physical light, and sought an expanding force of light that would counteract gravity.

Physical light was extremely important to the early universe. For the first 10^{-37} seconds there was no matter or radiation, just a force that was perhaps dark energy hurtling the universe into expansion. A short time later this force or dark energy in the vacuum transformed itself into radiation and empty space was filled with physical light and matter, mostly physical light.[9] During inflation the universe suddenly increased in size by a factor of ten trillion quadrillion (10^{28}), a factor a hundred thousand times larger than the number of galaxies in our observable universe (10^{23}).[10] The initial radiation cooled into protons, neutrinos and photons, which coalesced from the early physical light that filled the vacuum. Protons decay into photons, positrons and neutrinos. Before a second had elapsed, protons were generated and when nuclear activity processed a quarter of these into helium there were a billion times more photons than baryons. The microwave background radiation is composed of photons. Physical light, radiation and photons were fundamental while matter formed. When matter coalesced it comprised a tiny amount of what filled the empty space caused by inflation. Today, as we have seen, matter comprises 4 per cent of the universe, with dark energy (an anti-gravity force) and dark matter (matter that remains undetected because it does not emit electromagnetic radiation) thought to fill the rest.

In today's universe, light has wave-particle duality, sometimes behaving in waves and other times as particles. Matter also has wave-particle duality. De Broglie first saw the wave properties of matter in terms of photons in view of the wave-particle nature of electromagnetic radiation.[11] Both light and matter emerged from the radiation of the early universe, and both have wave-particle duality. Atoms have electrons and exist among charges and magnetic fields, and it is still not fully understood how the atoms of matter and the cells of organisms emerged alongside the photons of light.

The Universalist philosopher proposes that the essential potential substance of the finite universe – the dark matter, if it exists, the glue which holds galaxies together and which exerted the force that may have been dark energy of the first 10^{-37} seconds – is *aither*-like metaphysical Light, from which, he proposes, manifests physical light. Newton saw light as an expanding force which counteracts gravity, and since dark energy (if it exists), thought to comprise 73 per cent of the matter of the universe, is just such an antigravity force it is reasonable to associate dark matter/energy and light and to see dark matter/energy as perhaps being a latent potential of light. Dark matter (if it exists), thought to comprise 23 per cent of the matter in the universe, does not emit electromagnetic radiation and may consist of neutrino weakly interacting massive particles rather than electrons and protons. Neutrinos, detected experientially in a nuclear reactor in 1953, have no charge and have a mass close to zero (perhaps about a millionth of the mass of an electron), always travel at the speed of light and can pass through the earth and our bodies.[12] Photinos also have a mass close to zero.

Both photons and neutrinos are created from the decay of protons. Neutrinos may be associated with photons in being emitted by the radiation following the

Big Bang and there are the same number of neutrinos as photons (10^{87}).[13] It looks as if as both travel at the speed of light and exist in equal numbers, and as both are created in decay there is a symmetry between the equal amounts of neutrinos and photons, neutrinos representing the dark and counterbalancing photons, light. Dark matter and dark energy may be the manifestation of the metaphysical Light, the potential of photons, and may be composed of particles smaller than neutrinos such as photinos if they exist. (According to supersymmetry, photinos are ½ spin particles that partner photons.)

In our algebraic, dialectical thinking, just as $+A + -A = O$, so photons + neutrinos = a balance between light and dark, a physical *yang* (light)-*yin* (dark).

It will be helpful to list the four kinds of light that operate during the manifestation process:

Nothingness/Fullness – infinite metaphysical Light;

Non-Being – the infinite Void outside the universe: the potentialities of infinite/finite Being, infinite metaphysical Light, virtual tiny photons;

Being – the potentialities of finite Existence, perhaps dark matter/energy, virtual neutrinos or photinos manifesting into actual neutrinos or photinos beyond gamma rays on the electromagnetic spectrum which convey the order principle into consciousness, seen as metaphysical Light, Uncreated Light or Intelligible Light;

Existence – the finite manifestation of Being: photons which photosynthesise and convey the order principle into bodies and brains via physical eyes/retinas.

In terms of $+A + -A = O$, virtual neutrinos/photinos of Being + actual photons of Existence = all manifested metaphysical-physical light.

It will also be helpful to list the parts of the self that receive the four kinds of Light:

Nothingness/Fullness – purest infinite Light of finest density, too pure to be seen clearly in universal being, the level of consciousness that has an infinite component;

Non-Being – a denser metaphysical Light but still too fine to be seen in the universal being;

Being – neutrinos or photinos or tiny photons smaller than neutrinos received in the universal being (level (12)) as illumination in the Western tradition and in the base-of-spine *chakra* within the subtle body in the Eastern tradition;

Existence – photons received by retinas and transmitted to brains and bodies via objective consciousness/the soul (levels (7), (8) and (9)).

On this view, the metaphysical Light manifests from the very fine and subtle to tiny, denser "consciousness" neutrinos or photinos and to much denser physical photons.

Form from Movement Theory: The Origin and Creation of the Universe

The view that subtle and physical light comprise the essential stuff of the universe has arisen from and is supported by the Universalist proposal regarding the origin of the universe: a Form from Movement Theory which looks back to Heracleitus's eternally moving Fire c.500BC, the Fire of Reality which has always existed: "This world-order did none of gods or men make, but it always was and is and shall be: an ever-living fire, kindling in measures and going out in measures."[14]

The Form from Movement Theory differs from most theories and ideas of creation, which begin with stillness and simplicity that later become movement and complexity. Xenophanes' motionless unity is an early Presocratic example of this stillness. The infinite which threw up the universe with the Big Bang is a moving timeless "boundless" that has always existed, and I begin with an eternal and infinite movement and complexity as an ontological basic condition. This infinite movement is like Anaximander's eternally moving "boundless" *apeiron*[15] and Heracleitus' eternally moving "ever-living Fire" which may also have been *aither*, Newton's ether.[16] Aristotle was half-way to this position in seeing an eternal and immutable "unmoved mover" (my Nothingness) that is infinite and excited a continuous motion in the infinite (my Non-Being), a motion that Bohm called a "holomovement".[17] The eternal movement was how it always has been, and rational objections to a moving beginning by Xenophanes and Parmenides ignore the basic ontological condition.

The view of Anaximander, Heracleitus and Aristotle that the infinite is eternally moving is echoed at the beginning of *Genesis*: "In the beginning God created the heaven, and the earth. And the earth was without form, and void; and darkness was upon the face of the deep. And the Spirit of God moved upon the face of the waters. And God said, Let there be light: and there was light" (1, 1-3). The formless Void and dark deep were moving like waters, and God acted as an unmoved mover. This text "echoes" Anaximander and Heracleitus as *Genesis*, like the rest of the Pentateuch, has Yahwist, Elohist and Priestly strains. Scholars are united in attributing *Genesis* 1 to the Priestly strain, and whereas the Yahwist strain can be dated as early as 950BC and the Elohist strain to 900-700BC, the Priestly strain is usually dated to the 5th century BC, *after* Anaximander and Heracleitus.[18]

Manifestation is a reduction or limitation of this infinite movement rather than an additional quality, and is passed through four distinct stages: from Nothingness to Non-Being, to Being and Existence.

Nothingness had always been a completely self-entangled energy of virtual, infinite, irregular movement in all directions (although it preceded space-time and therefore direction), a real nothing or Oneness, a latent "ever-living Fire" or metaphysical Light. I postulate that this Nothingness was infinitely self-aware and a non-locality. I can formally denote this infinite movement as M,[19] which is absolutely the most subtle substance.

All limitations of M are potentialities within M, and in a region of M which I can denote as S a limitation of movement occurs so that there is a regular movement, possibly in the form of a multidimensional (in the sense that it had more than three dimensions) spiral (Non-Being). This regular movement arises as all irregular movement dies away. S, or Non-Being can be said to be a subset of M, where both M and S are infinite. Everything in M not in S is denoted M−S (M minus S). There is therefore now in place of Oneness a duality, that of M−S versus S (or pre-manifestation versus first manifestation). Whereas M−S entails all sorts of infinitely complex movement, S is a fairly simple and symmetrical form of movement. As in set theory both M and S are infinite, yet M is larger than S and contains the whole of S. In the same way there is a more definite yet still infinite self-entanglement (or non-locality) in all of S. There has to be a creative tension or pressure between M−S and S, between irregularity and regularity, and as a result of this interaction new forms of order arise between M−S and S, one such possible order being symmetric points or pre-particles. I postulate that a pair of pre-particles arises in S as a result of the pressure of M−S on S. These pre-particles, which can be termed +p and −p, are symmetrically entangled (i.e. in an unbroken relationship) and are both opposite and complementary. I can state this entanglement succinctly by drawing on algebraic, dialectical thinking $(+A + -A = O)$. In sum, the two pre-particles equal zero. Thus, $S: +p + -p = O$.

I now postulate a reduction of symmetry because of the disturbing influence of the infinitely complex and irregular movement of M–S upon S. This manifests an annihilation of one of the pre-particles, $-p$. In other words, the regular movement constituting a pre-particle is absorbed back into the vast movement of M and ceases to have any presence in S. The symmetry is broken, for the other pre-particle, $+p$, is not absorbed into M but derives energy from the movement of M–S upon S. This pre-particle is an empty point or vacuum or singularity. This point potentially contains Non-Being. I can say that the point is "a defined Non-Being" in contrast to the undefined Non-Being which is S, and that both aspects of Non-Being were always potentials within the unmanifest M. This point, $+p$, begins to spread like a multidimensional (or more-than-three-dimensional) wave in all directions in the field of S, and expands. The point, $+p$, is gradually becoming Being, the first subtle structure in the vacuum field of S: $+p \rightarrow B$. I can thus say that from Non-Being happens Being.

Due to the interaction between B and M–S in the field of S, B evolves more and more structure. M–S, Nothingness, is the source of infinitely complex impulses, a self-aware Non-Being, primitively intelligent, apparently random – but perhaps fundamentally purposive – generator. S, in contrast, provides a background of silence and a field in which the reverberations of the structure can manifest into form, and be displayed, within M, without disturbance. In the language of physics I can say that the emergence from Nothingness (M) of Non-Being (the vacuum field of S) and then Being (B within S) is activity within the quantum vacuum, and that B, which is implicit (in the sense that it is not plainly expressed but implied as it is hidden), evolves more and more explicit (in the sense that it is expressly stated or manifested) orders within its implicitness. I can denote this series of more and more explicit implicit orders which B contains by $I_1, I_2, I_3,...I_n$; where I_n corresponds to the most explicit order that is still implicit to us.

The process of Being (formerly $+p$) becoming Existence (E) is one in which potentialities become actualities, pre-particles become particles, pre-matter becomes matter, pre-organisms become organisms, pre-consciousness becomes consciousness and possibility becomes factuality, completing the process of manifestation. I can denote Existence or the explicit order by I_{n+1}. I can say that $E = I_{n+1}$, which is an event in M. I can postulate the possibility that there are one

271

or more loops between these implicit orders, such as between I_{n+1} (the explicit order) and I_{n-2} (an earlier implicit order). These loops provide the rudiments of primitive consciousness entangled in and manifesting from within matter. There is a possibility that there is "horizontal" division into two manifesting orders, one eventually leading up to explicit matter and the other leading up to aware consciousness and creative Intelligence, with one or more "vertical" bridges between these two orders.

As a result of +p deriving energy from M–S, the infinite and implicit +p or Being, which has spread in all directions in the field of S and expanded, has become pre-matter (and perhaps as we have just seen also pre-consciousness) in place of a first pre-proton (+p), and form is ready to arise from the infinite movement. Virtual particles emerge from the quantum vacuum (B in the field of S) in pairs, and one particle in each pair has the potentiality to become a real particle in explicit Existence if it draws energy from the pressure of M–S on S. Through quantum processes, the emergence of virtual particles happens in many regions of S, all over the expanded field of +p or B, and, deriving energy from the pressure of M–S on S, many ephemeral virtual particles become enduring real particles.

As a result of the proliferation of simultaneously emerging real particles, the process of the Big Bang or hot beginning – the actual creation of the universe as we know it – takes place.

Space-time emerged as part of the physical universe as Einstein has shown in his general theory of relativity which holds that gravitation is an effect of the curvature of the space-time continuum. The anti-Newtonian Leibniz, in his second letter to Clarke, saw space and time in terms of events. He held that space is an order of co-existence of events, and that time is an order of succession of events. However, Einstein's work supersedes this event-based view which preceded, and in itself cannot account for, the curvature of space causing gravitation.

To the Universalist philosopher, then, the process of the origin and creation of the universe begins with a first principle or first cause, an eternal and infinite moving Nothingness or latent "ever-living Fire" or metaphysical Light. From this formed a spiral (Non-Being), from which emerged or manifested a pre-particle that became a point or empty vacuum or singularity (Being). Out of this, Existence – Form – manifests as energy becoming matter through the Big Bang, inflation and the expanding universe Hubble discovered. Matter and consciousness have thus both manifested from ever-living Fire or subtle Light.

This Form from Movement Theory can be set down mathematically: $M \rightarrow M-S + S$. In S, $+p \rightarrow B$, which evolves $I_1, I_2, I_3, \ldots I_n$. $B \rightarrow E$, which is I_{n+1}. (M=movement; S=spiral; p=pre-particle; B=Being; I=implicit order; E-Existence.)

Beginning with an infinite, self-aware, self-entangled movement like Anaximander's boundless raises the question as to whether there is an arbitrariness about the process I have described, which could in theory be one of infinitely many possible processes unless the self-aware movement is purposive – in which case this origin and creation was the only possible process that could have happened, and life too had to happen. The universal order principle may indicate that the process may not have been arbitrary and random.

This account of creation and causality – which draws on the tradition of scientific cosmology and ontology – postulates that the invisible substance behind the universe manifests into radiation and physical light, and into the *yang-yin*-like symmetry of neutrinos and photons and the ordered system of Nature.

The Rational Universe as a System of Light

From the rational viewpoint, the Universalist philosopher holds that the universe is a system in which invisible substance has manifested into visible Nature and matter. Energy from the invisible Reality of Being manifests into Existence, whose phenomenal forms and systemic workings bear witness to an orderly origin. The intricacy of Nature's ecosystem, whose photosynthesising growth is spurred on by photons, suggests organising behind it as shadows suggest a sun. The Universalist philosopher holds that the metaphysical Reality is hidden behind the world of appearances and phenomena which conceals it just as the Chinese characters *Chin Hua*, "Golden Flower", conceal a central character *Kuang*, "Light". (In the 9th century AD it was heretical to speak of metaphysical Light in China, and so the character for Light was hidden between two other characters.)[20]

Universalist philosophers hold that the manifesting energy comes into the universe along with coded instructions for ordering phenomena, which may be borne by light. These instructions may include a drive for self-improvement or self-betterment in relation to the environment, so that each organism realises the next stage in its development. The electromagnetic radiation of different frequencies interacts with matter in different ways, and has done so since the very early universe when both electromagnetic radiation and matter emerged from the infinite via the Big Bang. The electromagnetic spectrum (see Appendix 3) reveals an ordered system. The nearest rays are long radio waves (with a frequency of $10-10^5$ cycles per second). Normal radio waves have a frequency of 10^6 cycles per second. Short radio waves have a frequency of 10^7-10^{12} cycles per second. The cosmic microwave background radiation has a wavelength of about 1.1 millimetres (about 10^{-3}) and a frequency of 217 million cycles per second – over 10^8 cycles per second, which places it in the radio region.[21] Radar and TV come next, and then infra-red (with a frequency of $10^{12}-10^{14}$ cycles per second). Beyond is visible light with a frequency of 10^{14} cycles per second. Each colour within the spectrum of light has a different wavelength. Beyond that is ultraviolet (with a frequency of $10^{14}-10^{16}$ cycles per second) and beyond that X-rays (with a frequency of $10^{15}-10^{20}$ cycles per second) and gamma rays (with a frequency of $10^{18}-10^{23}$ and beyond cycles per second), all of which are absorbed by molecular oxygen, ozone and molecular nitrogen. Universalists hold that the invisible substance that manifests from the infinite and is present throughout the universe is beyond gamma rays on the spectrum.

Within this spectrum, and working alongside the four forces (the strong and weak forces, the electromagnetic force and gravity), Universalists hold, is an order principle, possibly in light as Newton thought, which via sunlight is responsible for all growth in our solar system. Not enough is known about the transmission of light over the vast distances of intergalactic space, and the theory of light overlaps with the science of cosmology. Today scientists are searching for one logical theory that includes all known terrestrial phenomena of light – and the intergalactic sources of light and their origin.

But it is known that light is converted to chemical energy. We have seen on p115 that in plants light is absorbed by chlorophyll, which is in the elongated chloroplasts of cells, and produces energy. In harvesting sunlight energy through the process of photosynthesis, plants also harvest order.[22] See Appendix 3. (In plants there is one exception to the rule that all photosynthesising cells contain chlorophyll: the bacterium *Halobacterium halobium*, which photosynthesises by using a coloured protein resembling the visual pigment rhodopsin).[23] Nature's vast plant ecosystem is controlled by light, whose information-bearing photons are perpetually converted into chemical energy.

It is also known that in animals and humans light is converted to chemical energy by being received in light-sensitive cells and organs, notably the retinas of the eyes, which are designed to receive light from the lenses.[24] We have seen on pp243-246 that eyes are not just for looking out and for receiving images, they are crucially also for receiving light. The light-sensitive cells transform the light signals into a photochemical reaction which leads to electrical charges in cells and eventually nerve signals.[25] Sunlight is crucial to the maintenance of the feedback systems of the body and brain.

The reception of light in the light-sensitive cells of animals seems to be universal. Iguanas and lizards warm up by basking in the early morning sun. Sunlight penetrates through their eyes and radiates their internal circulatory system until their inner thermostat prevents the warming process from making them too hot. These animals are motionless until they have absorbed the light they require, and only then do they go about their daily business. For an iguana, the most important and essential event of their day is basking in the sun in complete stillness to absorb the light they need to function.[26]

Humans acquire the light they need without having to sunbathe. Even in blind humans, light penetrates to the light-sensitive or photoreceptor cells (see pp243-246).[27] The light-sensitive cells in the vicinity of the retina still catch light even though the eyes of blind humans have lost their ability to see visual images. What happens if a man has his eyes gouged out like Gloucester in *King Lear*? Does light still penetrate to light-sensitive cells in the eye sockets? It apparently does if the removal of the eyeball has left sufficient photoreceptors near the back of the eyeball intact. What happens if a man has to live in the dark for a long period of time, in a windowless underground dungeon? Does he become like a plant wilting in a dark room? A plant needs light to survive, and is this also true of humans and animals? Humans and animals can be expected to live longer than plants under conditions of light deprivation. What happens to citizens of the Arctic region when there are not many hours of sunlight for parts of the year? All these questions need further investigation in relation to the light-sensitive photoreceptor cells.

This system of living things converting light into chemical energy operates universally throughout Nature. Light-sensitive photoreceptor cells and organs specialised to received light are found in every creature, even in unicellar forms. Primitive eyes are found at the edge of jellyfish, in the armpits of starfish and the gills of tube worms. Earthworms have light-sensitive cells within their skin that are connected to their nervous system, many near their front end, a few at their back and not many in the middle. From worms developed molluscs (for example, octopi), anthropods (spiders and insects) and vertebrates (birds and man).

Molluscs have some sort of eye. The eye of the octopus and the human eye represent independent lines of evolution as do the wings of insects and of birds.[28]

The vertebrate eye in animals and humans has a lens which increases the eye's ability to collect light (besides making the image sharp). The iris is pigmented and opaque, the pupil is a transparent hole in its centre. The size of the pupil depends on light intensity (and also on psychological factors such as excitement or boredom). The eye is protected by the cornea and eyelids (or in birds, winking membranes). The lens focuses a sharp image of the world on the retina. Humans can focus on objects at different distances by changing the shape of the lens, which is achieved by changing the tension in the fibres supporting the lens. Octopi, fish and jumping spiders do likewise. The lens refracts light: blue light (short wavelengths) is deviated more than red (long wavelengths). Sensitivity to short-wave light and visual acuity are held in balance. An eagle's visual acuity is six times that of humans, a bee's one hundredth. The eagle sacrifices short-wave vision and the bee long-wave vision.[29]

But while such adjustments are being made, light is turned into chemical energy via the eye. The absorption of photons in the eye initiates a photochemical reaction which leads to other chemical reactions and to an electrical change in cells. Photons cause chemical reactions in the eye's fumaric acid and turn it into maleic acid, and vice versa. Light-sensitive cells have to be combined with pigment-rich cells which extinguish light from one direction, assisting directional progress.[30]

Light also stimulates physical and chemical processes to have a profound effect on the direction of movement in animals and humans, and (in ways not yet known) stimulates an inner clock which enables organisms to behave predictably as regards the time of the day and the time of the year.[31] There is also a photobiology of the skin, and light can counteract rickets and nourish the skin with Vitamin D, which develops the skeleton. Light deprivation results in seasonal affective disorder ("SAD"). Sunlight can, however, have a deleterious effect as it can cause skin cancer and DNA can be damaged by exposure to ultraviolet light. In short, sunlight is central to the vast biological ecosystem which includes local ecosystems, for it controls growth and an orderly way of life.[32]

Looked at in philosophical terms from the perspective of the reason, evolution has taken place within bio-friendly, orderly conditions amid a continuous process of manifestation from the boundless infinite into Being and Existence of invisible metaphysical Light becoming physical light via the cosmological procedures which created suns, and our sun. The system provides for the development of bodies and brains with an order that, if not purposive, attends to every need and deficiency.

The Law of Order and the Law of Randomness

In philosophical terms, within a philosophical category, I have frequently mentioned the universal principle of order or the order principle, to which Universalism draws attention. A principle in physics, such as the uncertainty principle, is "a fundamental truth" or "a highly general or inclusive theorem or 'law', exemplified in a magnitude of cases" (*Shorter Oxford English Dictionary*). Part Two contains evidence for a magnitude of cases in which a principle of order is at work. A law, as in "the laws of Nature" or "the law of gravity" is "a theoretical principle deduced from particular facts, expressible by the statement that a particular phenomenon always occurs if certain conditions be present" and "the order and regularity in Nature expressed by laws" (*Shorter Oxford English Dictionary*). ("Regular" means "conforming to some accepted rule or standard".)

Part Two has shown that order is always present in an ecosystem's conditions, and I have already defined order as "the condition in which every part or unit is in its right place, tidiness" (p4) – which covers how fish, birds, reptiles and animals regularly share feeding and breeding sites to make food go round. The instances of order in Part Two confirm a regularity of natural occurrences of order in particular instances ranging from 40 bio-friendly conditions to symbiosis, medicinal plants and physiological systems, all of which conform to a rule. I can therefore now speak of "the *Law* of Order" in place of "the principle of order". I can say that Universalism draws attention to the Law of Order.

In philosophical terms, within a philosophical category, approaching the universe objectively through the reason, I can in fact identify two conflicting laws:

(1) The Law of Order, which manifests from the infinite into the universe, perhaps as a fifth force, and probably operates through light via the eyes, is conveyed to all creatures. It controls the growth of all creatures, regulates their systems and pervades and controls their DNA. The Law states that order creates organised, methodical, highly structural, complex systems which require a lot of information to describe them and which are maintained by the active co-operation of the participants in the whole ecosystem which includes all sub-systems (as outlined in Part Two).

(2) Alongside it is the Law of Randomness which states that the universe is also permeated by an unmethodical tendency for the energy of organised complex systems to deviate from their self-organised directions. It states that the systems deviate either through adaptation or through faulty DNA replication into accidental mutations or into chaotic breakdown due to chance mishaps which may bring starvation and early sudden death. "Random" can be defined as "made, done without method or conscious choice".

The Law of Randomness is different from the Law of Entropy, another name for the second law of thermodynamics, which states that organised systems become less and less organised as time advances. The Law of Entropy[33] is about the running-down and decay of systems as a result of disorder at the molecular and atomic levels. It is not about their being subject to chance mishaps as when a chick is a random victim of a predator or a careless driver.

The Law of Order can be seen at work in a range of natural phenomena from bio-friendly conditions to contours of landscape, ecosystems, physiological systems, mammals, reptiles, insects, vegetation and plants. All living organisms show patterns of complex order that shape and ease their lives and their struggle against the conflicting Law of Randomness, which brings chaos, starvation and sudden accidental death.

As in my algebraic, dialectical thinking, just as $+A + -A = O$ the two laws, though in apparent conflict, are two sides of a fundamental unity and they are held in balance like *yin* and *yang* by a constant. This can be expressed as +Order + −Randomness = unity. One law does not triumph over the other. Order contains within itself a propensity for the disorder of randomness – and, indeed, entropy – but due to the constant, neither randomness nor entropy succeed in destroying the universe at the end. Rather, randomness is a deviant from order's energy, which enables an organised system to adapt and mutate into a new form that is beneficial to the whole ecosystem, or hastens its end in the interest of maintaining a balance in the population of species, which is also beneficial to the whole ecosystem.

Thus the two apparently conflicting forces and laws, the Law of Order (which suggests an energy methodically organising all creatures from birth into their rightful place within a system) and the counter-balancing Law of Randomness (that chance mishaps affect the same range of creatures within the whole food-web without method) are complementary like *yang* and *yin*, and myths about

Chaos and Order are found in many ancient cultures. The point is that both laws exist, and a view of the universe in terms of the Law of Randomness alone (the view of extremist, anti-Creationist Darwinists) does not truly and objectively reflect the universe.

Order may one day be established as a fifth force, an expanding force of initially dark, subtle light that is invisible to the outer eye like electricity but manifests from the infinite into the universe at the level of Being in so far undetected particles even smaller than neutrinos such as photinos. When explicit – manifested in Existence – it counteracts gravity, permeating all Nature, is passed on through generations by DNA and shapes evolution to an end that is self-organising and self-regulating. It may one day be as measurable and testable as the other four forces and be regarded as the most fundamental of all the forces as it balances gravity and is responsible for the growth of organisms via eyes and chloroplasts in chlorophyll. CERN has sought the field of Higgs bosons, which may in fact be the field of the order principle. It may have been discovered, measured and tested in 2008/9, although the results may not be known until a year or two later. (See Postscript.)

Is there a teleological purpose behind the Law of Order in the universe? Aristotle in *Metaphysica* declared that a full explanation of anything must consider not only the material, formal and efficient causes, but also the final cause, the purpose for which the thing exists or was produced.[34] There was a move away from Aristotle's teleological explanations in the mechanistic 16th and 17th centuries, which did not echo Aristotle's argument that things developed towards ends internal to their own natures but held that biological organisms are machines devised by an intelligent being. In *Kritik der Urtheilskraft* (*Critique of Judgement*), 1790, Kant held that teleology can only guide inquiry rather than reveal the nature of reality.

Teleology is "the doctrine or study of ends or final causes, especially as related to the evidence's design or purpose in nature" and "such design as exhibited in natural objects or phenomena" (*Shorter Oxford English Dictionary*). To proceed rationally, I have set to one side the idea of intelligent design to focus on "what is" rather than speculate as to why it is as it is. However, while "order" can be defined and described in terms of regularity and the functioning of a system in the present ("every part or unit in its right place") it has other meanings which include "a state of peaceful harmony under a constitutional authority", and the question of the purpose of the present harmony cannot be far away. It is not known why Nature's system exists. We can only speculate as to its end and final cause, and human evolution is not yet complete. If we are not careful, thinking about the purpose of the universe can take us readily from scientific description, philosophical exegesis and metaphysical rationalism – whose language is neutral – into theology, religion and faith, different categories of thought.

I can however say that the many instances of order in Nature's system I have cited in Part Two rationally suggest that in the case of order rather than in the case of randomness there must be a teleological principle at work for if the ordering was not towards any end it would not have been successfully concluded in terms of the purpose for which the ordering was taking place. In the case of randomness, of course, there is no teleological principle at work for randomness just happens without necessarily leading to a goal. Furthermore, I can say that Nature manifested from the infinite and that humankind has evolved in a partly ordered, not totally random way, and that there may well be a purpose linked to the infinite substance, the "ever-living Fire" or metaphysical Light, the subjective experience of which we shall consider in the next chapter.

The rational answer to whether the universe has a purpose, in contrast to the intuitional answer, must be that in view of the Law of Order the infinite must have manifested into Nature for a reason, and that to the extent that evolution is controlled by order this reason must be associated with the evolution of humankind. It is therefore not impossible that each human being, linked by physical light to his or her source of growth and by metaphysical Light to the infinite, is fulfilling an infinite plan in his or her everyday finite living. In this chapter I am being rationally objective and the final answer must await a consideration of subjective intuition via the universal being in the next chapter, and is necessarily beyond the competence of the reason and the rational, social ego to fathom. Logically, however, if the universe is ordered it is purposive.

Rational Approach to Metaphysics

The leading 20th-century metaphysician Whitehead declared, as we saw on p50, that "speculative philosophy" (i.e. metaphysics) "is the endeavour to frame a coherent, logical, necessary system of general ideas in terms of which every element of our experience can be interpreted". Every element of our experience must include all our experience of the finite world and all our experience (such as it is) of the infinite, all the elements of experience I listed under the 12 levels of consciousness (see pp250-251). My system of ideas – manifestation from the infinite into Nature and bio-friendly conditions which helped life evolve – does include these.

A "system of general ideas" must be dominated by one of four metaphysical perspectives: Materialistic monism (that matter gives rise to mind, or scientific Materialism); dualism (that matter and mind are separate substances, the Cartesian position); transcendentalist monism (that mind or consciousness give rise to matter so that the ultimate stuff of the universe is consciousness, an Idealistic view);[35] and metaphysical non-duality (that a metaphysical Reality gives rise to both matter and mind, that a unity gives rise to a diversity). Whereas monism implies no diversity, non-duality implies both a unity at the transcendent metaphysical level of the infinite and diversity at the immanent physical level.

My "system of general ideas" is a non-duality. There is unity at the metaphysical, infinite level and diversity at the finite level. My system of coherent general ideas explains how unity became diversity by manifestation into matter, light, consciousness and the vegetative world of plants, how the One manifested into many.

Metaphysical philosophy includes both the infinite and the finite, as we have seen. (See the four-tiered process of manifestation on p265.) Just as in algebraic, dialectical thinking $+A + -A = O$, so the infinite (which is untestable and unmeasurable) + the finite (which is testable and measurable) = the All or One. Metaphysics is the science of the All; a "system of general ideas" that covers the universal Whole, which includes all known experiences and concepts. Universal science was sketched by Plato and followed by Aristotle,[36] but is now the science of the infinite and the finite – the science I have outlined in Part Two. "Every known concept" includes the paranormal in relation to a sea of Reality, and the after-life, whether or not there is proof. The very fact that it is a concept makes it worthy of consideration.

But here, as I am following the rational strand of metaphysical philosophy, a cautionary note. The reason concocts a system of general ideas which explains the reason's place in the rational universe. The reason's approach to the experiences of the 12 levels of consciousness on pp250-251 does not embrace *all* the experiences encountered in those 12 levels. It does not include levels (9), (11) and (12) and may not include (10). We must wait until the next chapter to include the remainder of the experiences of all levels of consciousness. But the reason does see the individual in relation to a rational universe.

I said on p53 that today metaphysics has four subdivisions, which cover different approaches to Being. They are in descending hierarchical order:

ontology (the study of Being), which includes traditional metaphysics;

psychology (what can be experienced of Being) which includes some traditional moral philosophy;

epistemology (what can be known about Being), which includes some ethics; and

cosmology (the structure of the universe), which includes traditional natural philosophy.

The reason has its own approach to this metaphysical scheme, as we shall now see.

Rational metaphysical Universalism offers a universe which is permeated by the infinite but ruled by reason as Reality is a sea of Being which follows mathematical laws. The rational study of ontology is therefore a study of Being manifested into the universe whose structure is rational. To the rational philosopher, the study of cosmology is of its compliance with rational, mathematical laws. The rational metaphysical philosopher uses a rational psychology and sees the rational, social ego, which uses the reason or higher mind, as *thinking* about infinite Being. His scheme allows the inclusion of the soul but his medium is the reason. He is more concerned with Existence and, as our moral judgements are psychological, with practical ethics and conduct (the branch of philosophy known as moral philosophy). Ethical problems, what is right and what is wrong, have social contexts and therefore involve the rational, social ego. He sees language, and therefore linguistic analysis, in terms of communicating socially, the world of the rational, social ego. The rational metaphysical philosopher uses a rational epistemology by which to seek certainty and probability that the infinite exists. It is rational to consider the position of the spacesuited surfer who "knows" the infinite by crouching on the "wave" of the advancing universe. The rational philosopher, within the psychology of his reason, thinks that infinite Being of ontology must be in front of the surfer, and knowing it epistemologically he sees ontology (Being) as giving rise to Becoming (the universe of cosmology). His universe is up-to-date but still has a strong affinity with the rational Enlightenment.

It is important to grasp that the four-tier manifestation process, the four subdivisions of modern metaphysics, the 12 levels of consciousness and non-duality are all integrated in the reason's system. The last two stages of the manifestation process correspond to all the subdivisions of metaphysics and to non-duality of matter and consciousness – whose 12 levels correspond to the same last two stages. The process can be summed up as follows:

Manifestation process (see p265)	Subdivisions of metaphysics (see p282)	Experiences of 12 levels of consciousness	Non-duality
Nothingness/ Fullness			
Non-Being			
Being { Existence {	ontology	levels (11)-(12)	unity
	psychology } epistemology }	levels (5)-(10)	
	cosmology }	levels (1)-(4)	diversity (consciousness/ matter)

Rational Universalism looks back to Classicism or neo-Classicism, which emphasised order, clarity, balance, restraint and sense of beauty. Classicism trusted in the powers of the mind, especially in the reason. Rational metaphysics has Classical attitudes in emphasising the order in the universe, and with it a clarity about "what is" and the system of the universe. It balances the infinite and the finite (on which it focuses) on the one hand, and on the other hand the contradictions in Nature: $+A + -A = O$. Classicism is restrained in being objective about Nature, and perceives the beauty of bio-friendliness and the symmetry I remarked on in Part Two. Rational metaphysics, trusting in the reason, sees philosophy as "thought-based", as did Classicism. Classicism has a social view of humans in relation to evolution.

The rational approach to metaphysics looks for laws within the system, which are determined by the framework of the system. The reason analyses and fragments the All or One into bits, whereas intuition perceives the Whole and reunifies the fragments. The All is a whole and has laws. In rational metaphysical psychology (what can be experienced of Being according to the knowledge of the rational, social ego) there are 14 basic laws which are part of the invisible fabric of the universe and applications of the metaphysical perspective:[37]

the Law of Unity – all within infinity and eternity co-exist;

the Law of Interpenetration – an infinite scale cannot be divided into finite sections, all of which must contain infinity;

the Law of Symmetry – the infinite is symmetrical in all dimensions;

the Law of Cause and Effect – if all things are One, every cause has an effect on everything else and most things that happen to human beings have a cause within human thought;

the Law of Existence and Non-Existence – as everything that exists has a cause, if a cause is created then its desired effect will follow;

the Law of Equilibrium – if two systems are created out of the One, their energy will return to unity like a pendulum coming to rest;

the Law of Degrees – in an infinite series of levels, all "facts" or parts of the same dimensionality are on the same level;

the Law of Pyramidic Construction – the individual parts of the All or One are joined together in a pyramidic structure;

the Law of Parallelism – all know everything obeys the same laws;

the Law of Enclosure – the higher encloses the lower as the Absolute encloses the universe, but the lower cannot enclose the higher;

the Law of Convergency (or "narrowing") – progress proceeds from chaos to perfection;

the Law of Finite Perception – the infinite and eternal can only be perceived from a finite degree, in apparently isolated parts;

the Law of Reciprocal Action – harmony depends on both sides giving and receiving;

the Law of Factual Area – the fewer facts, the higher the degree as in a pyramid.

All these 14 laws assume that the All or One – the infinite and the finite – is a unity in which everything corresponds and everything has its allotted place. Our perceptions of the infinite are blinkered as we are finite, but as we shall see in chapter 10 we can know the infinite through universal being.

Possibility of Survival after Death

The rational Universalist philosopher has to consider all questions which may affect his view of man's role in Nature and the universe. That means laying out all that is known – "every known concept" – whether it has been verified or not, and then arriving at an explanation in terms of a "system of general ideas" that satisfies every conceivable eventuality, including possible survival after death. There have been scientific experiments to monitor and test the phenomena of parapsychology. These tests have provided some scientific support for a multi-levelled view of consciousness and in some cases have confirmed the experiences of the transpersonal or universal/cosmic consciousness (levels (11) and (12)) reported by those who have had experiences of telepathy, precognition, communication from the dead, near-death experiences and immersion in the mysterious Light.

Mediums report a world in which the departed are still alive but in a different state of "existence". They speak of their hair colour, or living in bungalows, of being able to smoke, share precise memories and have detailed up-to-the-minute awareness of what is happening in our lives. They are present in our rooms, in this universe. If what they convey is true, the departed must be within this universe to be present with us and not in an inaccessible multiverse. So where is "the other side"? It can only be within the level of Being.

It is rationally and theoretically not impossible that on death spirits return from Existence to a "land", or an "island", within the ocean of Being, where they can be contacted by the living and that they can communicate by a kind of radio wave. It may be that for the initial time after their departure many are present, unseen, in our midst, inches away from us in the invisible air, within Being rather than Existence but able to witness what we do and say. Many of the features of the paranormal may be found in this ocean of Being, for example, telepathy (or the power of minds to communicate at a distance). Also precognition, for past, present and future belong to time and are surrounded by a timelessness to which the future has already happened.

The quantum vacuum is not another dimension. It is present all round us now, though invisible. We have seen that the same is true of the infinite. Nothing has to "unfold" from the infinite for Universalism sees the universe as being governed by three dimensions of space and one of time – and by the infinite, which is hidden and all about us like the quantum vacuum. The infinite we encounter may *be* the quantum vacuum in which Being becomes Existence on the border of the infinite and the finite, and which has manifested from the infinite the surfer breasts on the tip of our expanding, accelerating universe, perched on the crest of a shuttlecock-shaped wave.

Without invoking other dimensions, I should point to the possibility that, just as all atoms are interconnected in the quantum vacuum (as was pointed out by Bohm, who held that consciousness is inherent in matter – I would say that the unitary infinite manifested into a diversity of finite consciousness and matter), so may the consciousnesses of humans act like atoms and be similarly interconnected. Waves between them may account for telepathy, where one atom-like human connects with another atom-like human on an identical frequency and a thought is transmitted from one brain via a radio-like wave and is received in another brain that is tuned to receive it far away, non-locally over a great distance, and simultaneously.

Just as particles such as bosons surface from the quantum vacuum, live a short time and then disappear back into the quantum vacuum where they continue their "virtual" existence in a potential rather than a material state, so the consciousnesses of humans may emerge from the quantum vacuum on conception, live and then return to it on death – and continue a virtual, potential existence there. And perhaps the disembodied, disembrained consciousnesses of the physically, materially dead can send telepathic thoughts like radio waves which are transmitted and received in a living medium's mind when it is tuned to the transpersonal level of consciousness (level (11)).

The assumption of reductionist, Materialist science is that all consciousness and mind is brain function and that when the brain dies there is no consciousness to survive death and transmit messages in the quantum vacuum. To the rational Universalist there is a universal being and a spirit, as all the religious traditions (see pp17-21) have taught, which return to the infinite and timelessness – eternity – below or behind space-time on death and perhaps live in the quantum vacuum of Being within the infinite timelessness. Perhaps we do not "pass away" from the earth on leaving our body but remain here unseen, tied or magnetically drawn to the place which dominated our interest in life and witnessing succeeding events until we cultivate wider interests which allow us to move on elsewhere. Being may be the invisible side of Existence, and the infinite the invisible side of the finite. The spirit (*pneuma*) St Paul wrote of[38] was supposed to have only one life. Hindus and some other traditions believe it reincarnates, though not immediately, and therefore lives many times. And that we have far memories, sometimes in the form of traumas, of our past lives. According to the teaching of traditions that affirm reincarnation, when the body it occupies becomes old and, its skin tattered, wears out, the spirit transmigrates and is reborn into a fresh body of a baby to make a fresh start. Perhaps eventually our core is reborn to learn new things on a planet which is a "classroom of the soul". It may be that the spirit re-lodges in a new human consciousness and anchors itself in the brain, somewhere below the claustrum.

Shamanism, the religion of Central Asia at the dawn of civilisation c.50,000 years ago and still to be found in certain societies today such as the Eskimos, has always held that animals have spirits.[39] If so, the same principle may work in biology. Considering "every known concept", the rational Universalist holds that it is not inconceivable that the mammals, reptiles, birds, insects and plants we have considered from a Darwinist point of view may similarly die back into the quantum vacuum when their bodies are old and worn out, to transmigrate to fresh young bodies on conception and make a fresh start. To the rational Universalist the true reality of evolution may be found in the quantum vacuum rather than in the vegetation and forests of continents emerging from Ice Ages.

To the rational Universalist it is not impossible that a part of our consciousness survives death and that we have spirits that are immortal. There is a consensus in religious literature that the spirit is the enduring, timeless part of our consciousness which may return to timelessness, an ocean of Being, on death to live in the quantum vacuum like a fish. Tradition, on the other hand, sees the soul (levels (7), (8) and (9)) as a human's individual centre that connects feeling and thought. According to the Christian tradition it survives death, but according to the tradition of the Jewish Kabbalah, which is as old as Abraham, at least c.1700BC and probably older,[40] the soul perishes with the body whereas the spirit survives death. It is not impossible that the *genus Homo*, who evolved from the apes, developed consciousness that has spiritual awareness and has an enduring spirit (level (11)) that returns to the infinite and timelessness – the quantum void, or Being, or collective unconscious to which all consciousness returns from space-time on death.

Because metaphysics includes "every known concept" the rational metaphysical philosopher admits such possibilities even if he has no direct evidence of them from his own experience. To the reason, these are valid possibilities and at the rational, philosophical level may have validity as postulates even though they cannot be verified.

Rational Solutions to Philosophical Problems

The rational approach to the scientific view of the universe and the universal order principle has cleared up a number of problems that philosophers have pondered since the time of the Presocratic Greeks, the topics that fall under the definition of "metaphysics" on p2. By way of summary I can now list 20 problems on the left and solutions given by the rational approach of metaphysics after the dash:

the "first principles of things", i.e. the source, first element – the "ever-living Fire" or metaphysical Light, the Void/sea of Being which manifested into light, atoms and consciousness;

the "first cause" of the universe – the Big Bang out of a pre-particle in the infinite/particle, rapidly followed by inflation;

non-duality – unity in transcendence/the infinite, diversity in immanence/the finite without the difficulty of monism (that everything is either matter or mind, i.e. Materialism or Idealism);

dualism – false dichotomy of mind and body, light and darkness, false as all opposites are reconciled in non-duality, emerging by manifestation from the One into diversity;

the origin of the universe – in the infinite pre-manifestational "ever-living Fire" or Light;

knowing – epistemological knowing of the infinite through the concept of the surfer;

substance, "the essential material", the first material – the latent pre-manifestational, subtle "ever-living Fire" or metaphysical Light in the real unity of Nothingness, Non-Being (the Void, the vacuum field within S) and Being (B within S) which manifests into the diversity and multiplicity of physical light, matter and consciousness;

ultimate Reality – Nothingness which manifested into Non-Being and Being, which then manifested along with the universal order principle into Existence;

ultimate material – the "ever-living" Fire or metaphysical Light which manifested into physical light, matter and consciousness;

essence (to Aristotle that which a thing really is when it exists and which if altered would destroy the unity of the thing) – the derivation of all phenomenal forms of Nature from the latent "ever-living Fire" or Light of pre-manifestational Being;

Being – the potentialities of Existence in the infinite "ever-living Fire" or Light (B within S);

Existence – manifestation from the infinite into the finite;

the reason for Existence – within Order rather than Randomness, the goal of the universal order principle within evolution and consciousness;

the purpose of the cosmos – within Order rather than Randomness, to create bio-friendly conditions in which self-regulating, self-organising organisms can develop and humans can develop higher consciousness;

personal identity – to the rational philosopher, the rational, social ego which is the centre of personality and identity, with regard given to the soul (but to the intuitional philosopher, the whole of the multi-levelled self whose deepest seat is universal being);

change – the flux of Becoming which has manifested from the moving sea of Being;

time – part of a unified space-time fabric curved by gravity, the flux or succession of spatial events which manifested from the timeless infinite;

space – relational to a unified space-time fabric curved by gravity, also co-existence of events or qualifications, not as primary as Newton held;

free will – the power to act in the finite world without the constraint of necessity or fate, the choice of independent consciousness;

mind – the seat of the rational, social ego that thinks and of all the 12 levels of consciousness.

The Big Bang settles the debate in Plato (*Laws* X) as to whether the motion in the universe is imparted or self-orginated and the debate in Aristotle (in *Physics* and *Metaphysics*) as to whether the first cause was a Prime Mover or Unmoved Mover. The infinite was an Unmoved Mover, and the Big Bang was a Prime Mover which set the world of creation in flux, and it was imparted from the infinite Unmoved Mover rather than self-originated accidentally. The Unmoved Mover is the infinite One in which exists Non-Being, from which Being emerged into Existence in an emanational manifestation. The universe could under different circumstances conceivably not exist, and so to that extent it is contingent. Its existence has a cause which can be associated with the 40 conditions, suggesting it needed many right conditions to apply before life could be created – and that they *did* apply. As the right conditions applied the universe does not appear contingent. The universe's causation was *in esse*, within Being, but also *in fieri*, within Becoming, for as soon as Being became Existence with the Big Bang it became a process, a Becoming. Universalism supports Aquinas in seeing the universe's causation as being both *in fieri* and *in esse* – and as I have just said, in contingency – rather than Aristotle who saw it as solely *in esse*. I leave aside Plato's argument (in *Timaeus*) that there was a demiurge or creator, for this requires us to leave the philosophical category and stray into theology or faith.

In a celebrated image, as we have seen on p35 Descartes saw philosophy in terms of a tree. In one of his last writings, the Preface to the French edition of the *Principles*, Descartes wrote that "all philosophy is like a tree, whose roots are metaphysics, whose trunk is physics, and whose branches, which grow from this trunk, are all of the other sciences, which reduce to three principal sciences, namely medicine, mechanics and morals". Descartes was re-establishing the Aristotelian–Christian synthesis of metaphysics, physics and the sciences.

Universalism, to adapt Descartes' metaphor, is a tree whose roots are in the infinite and metaphysical, whose trunk is the finite universe and whose branches are the sciences which express the infinite and the finite. In this emphasis on the infinite context of the finite, Universalists differ from Aristotle and Whitehead.

Philosophical Summary

So what is rational metaphysical philosophy or rational Universalism? Having studied the scientific universe with its infinite singularity, the reason sees "what is" as a continuous process of manifestation from the infinite into the finite universe and Nature through four tiers of manifestation, each of which has its own form of light. This ontology is revealed by the reason's Form from Movement Theory, which accounts for the origin and creation of the universe. Manifestation happens through a system of light via the electromagnetic spectrum as photons enter plants and retinas, obeying the Law of Order which counterbalances the Law of Randomness. Matter and consciousness manifest from unity into diversity as a non-duality. Organisms have drives towards self-improvement, an order principle in evolution. In a rational universe in which the reason acknowledges 14 metaphysical laws, the reason sees that it is possible for part of consciousness to survive death by returning to the Void, the infinite aspect of the quantum vacuum.

The rational, objective approach to Reality and to the system of the universe has produced a "system of general ideas" in terms of which the lower end of the four tiers of manifestation correspond to the four subdivisions of metaphysics and many of the experiences of the 12 levels of consciousness, and in terms of which the 20 perennial concepts and problems of metaphysical philosophy can find solutions. However, the rational approach to the universe is only half the story.

10

THE INTUITIONAL APPROACH TO REALITY
AND THE SYSTEM OF THE UNIVERSE

Crucially, metaphysical Universalism is also intuitional and subjective. Besides looking at the universe through sense data and the reason, Universalism also relates to it intuitionally for we are part of the oneness of the universe, not separate from it. The proposals of the reason are confirmed by the direct experience of individual intuition. The intuitional Universalist accepts the rational view but introduces "add-on" or "overlay", intuitional experience which strengthens it. I shall now follow the "add-on" or "overlay" Intuitionist tradition of metaphysical philosophy.

Whitehead spoke of speculative philosophy (i.e. metaphysics) as "the endeavour to frame a coherent, logical, necessary system of general ideas in terms of which every element of our *experience* can be interpreted". We are now looking to see how the system of the universe puts our *experience* in touch with Reality, and how our experience of Reality gives us a framework in which all our experience can be fitted.

It will help to remind ourselves of the rational view of the integrated manifestation process, metaphysical tiers, levels of consciousness and non-duality:

Manifestation process (see p265)	Subdivisions of metaphysics (see p282)	Experiences of 12 levels of consciousness	Non-duality
Nothingness/ Fullness			
Non-Being			
Being { Existence {	ontology	levels (11)-(12)	unity
	psychology	levels (5)-(10)	
	epistemology		diversity
	cosmology	levels (1)-(4)	(consciousness/ matter)

Intuitional Ontology and the Experience of What Is

Intuitional Universalism holds that, quite simply, it is possible for humans to experience and *know* Reality, the "ever-living Fire" or metaphysical Light which I established rationally in the previous chapter. This hidden "ever-living Fire" or Light can be experienced individually and existentially in a manifesting form. Ontology (see the four subdivisions of modern metaphysics on p53), such an important aspect of the structure of the universe, can be known within our consciousness when our consciousness is at level (12): universal (or cosmic, or unity) consciousness.

This is a most important consideration. Communion between consciousness and the metaphysical "ever-living Fire" or Light cannot happen unless the individual enters the highest, or rather deepest, level of consciousness, which is below the rational, social ego. The rational metaphysical philosopher of the last chapter can only *think* about Reality because he remains in his rational, social ego in level (8). He describes Reality mathematically. The intuitional metaphysical philosopher goes deeper, or transcends the rational and reaches a transpersonal level, and can open existentially to the "ever-living Fire" or Light in inner depths.

All the early religions – and the more recent ones – know and report on the experience of the Fire or Light, which they interpret as Being and often as "God". We considered the early religions' contribution to philosophy on pp17-21. The established religions formalised the process of knowing the Light as enlightenment. Christ is "the Light of the World", the experience of the Light is known in Christian and Orthodox transfiguration. The Buddha is "the Enlightened One". Mahayana Buddhists call enlightenment *sunyata*; Zen Buddhists call it *satori*. Hindus call it *samadhi*, and Taoists "the Void". All cultures and civilisations report on the experience of Being as Fire or Light.[1]

Being is unitary and is perceived as One from the diversity of different cultures and disciplines – science, philosophy, history and literature – just as the one sun is perceived from different countries and has different names in different languages. Returning to the Whole that reductionists object to (see p173), the intuitional Universalist supports the rational Universalist (see p264) in asserting that the Real, the One or Being in the finite universe and its universal order principle pervade the parts of the finite universe and organise it into a self-regulatory equilibrium, so that the vast ecosystem which includes all local ecosystems can be a self-regulating whole.

Mystics speak of illumination – experiencing the Light as a sun known within

– as a stage on the Mystic Way. The traditional stages are: awakening, purgation, illumination, the Dark Night of the soul and the unitive life. On p266 we considered Plato's "Universal Light", Augustine's "Light Unchangeable" or "Intelligible Light", Mechthild's "Flowing Fire" and Dante's "Eternal Light", which were all mystical descriptions of the mysterious Fire or Light that can be contacted deep within ourselves.[a]

The intuitional metaphysical philosopher does not approach "the sun" of hidden Reality through religion but existentially (or Universalistically) via his own contemplating universal (or cosmic, or unity) consciousness. This is an essentially mystical vision. Physicists are cautious of seeing quantum theory in terms of the mystical as Fritjof Capra did in *The Tao of Physics*. However, there is a tradition far broader than physics that sees Reality and the quantum void as a sea of energy with a force that *can* be known mystically, as has been attested by mystics in the West and East during every generation during the last five thousand years. The mystical vision has always been of profound interest to philosophers. Hence Whitehead wrote: "The purpose of philosophy is to rationalize mysticism" (*Modes of Thought*, 1938, from an address to Harvard in 1935).[3] Whitehead, the philosopher of organism, in effect held that the hidden Reality within Nature is what the mystics have known, that Nature has no "bifurcation" and that the philosopher must reveal "the totality", the One, by using a poet's intuition, as Heidegger asserted, and then rationalize it so that the One, already grasped by intuition, can be understood by the reason. To the intuitional philosopher, philosophy rationalises an intuitional perception of Being or Reality.

Intuitional Epistemology and the Intellect

Intuitional epistemology (the study of what can be known about Being) differs from rational epistemology, which is based on a concept of the rational, social ego's: the concept of the surfer breasting the infinite. Intuitional epistemology is based on direct experience of the "ever-living Fire" or Light from a different centre, the universal being in which it is possible to see with universal consciousness. It is therefore appropriate to deal with intuitional epistemology before intuitional psychology.

Sitting quietly with closed eyes, the philosopher waits patiently for the inner darkness to open, and, as he goes deeper in contemplation and his breathing becomes slower, sitting as still as a stone, Light breaks within his inner dark, first shafts from below the horizon and then the full white intensity of bright light like a sun. All the cultures and civilisations record the powers this mysterious energy pours into the universal being: powers of knowledge, healing and vision. In universal or cosmic consciousness (level (12)) *Homo sapiens* is able to open to the Reality behind the universe in this life, as first shamans and later mystics have always taught. This Reality is stated in many traditions as a hidden Light (see pp17-21). It seems that the universal being within consciousness – a deeper brain centre than the cortex – is able to admit an energy from the quantum vacuum which fills it with health, a sense of meaning and purpose as a harbour admits the sea.

A crucial point is that the universal being is in fact "the intellect". This originally had nothing to do with the reason. "*Intellectus*" in Latin means "perception" (and therefore "understanding", "comprehension"), and it is an intuitional and intuitive faculty which perceives universals and meaning, and therefore understands, whereas the reason is a logical faculty that analyses particulars and often sees meaninglessness. The "intellect" is a perceptive faculty that lies outside the five senses. The claim is that when it perceives the "ever-living Fire" or metaphysical Light (which is Shelley's "Intellectual Beauty")[4] it receives wisdom in an area of the brain which is separate from the sensory combinations of neurons, outside the five senses, and this infused, revealed, intuitional knowledge then influences the reason. Today the word "intellect" has been corrupted. It has almost lost its original meaning, and is used as a synonym for "reason" to mean "the faculty of reasoning, knowing and thinking, as distinct from feeling". An "intellectual" is regarded as having a highly developed reason. The movements of Logical Positivism and Linguistic Analysis marked a further turning from the intellect to empirical reason.

It seems that the experience of illumination is linked to energy in the quantum vacuum, which flows into the intellect. It seems that this Light is on the electromagnetic spectrum (see Appendix 3) but so far towards the short-wave end, beyond ultraviolet and gamma rays, and in particles or bosons tinier than neutrinos or photinos, that science has not so far been able to measure it. It seems that universal or cosmic consciousness (level (12)) opens to an energy that has not yet been identified. Three "eye-witness" accounts ascribe its origin to "eternity" and "heaven", but might just as well have ascribed it to the infinite:

"I entered within myself. I saw with the eye of my soul, about (or beyond) my mind, the Light Unchangeable. It was not the common light of day.... What

I saw was something quite, quite different from any light we know on earth. It shone above my mind.... It was above me (or higher), because it was itself the Light that made me, and I was below (or lower) because I was made by it. All who know the truth know this Light and all who know this Light know eternity."
(St Augustine, c.400.)[5]

"When I was forty-two years and seven months old, a fiery light (or burning light) of tremendous brightness coming from heaven poured into my entire mind. Like a flame that does not burn but enkindles, it influenced my entire heart and my entire breast, just like the sun that warms an object with its rays. These visions which I saw, I beheld neither in sleep nor in dream, nor in madness, nor with the eyes of the body...I perceived them in open view.... From my infancy up to the present time, I am now being more than seventy years of age, I have always seen this light, in my spirit (or soul, Jung's translation) and not with external eyes, nor with any thoughts of my heart, nor with the help from my senses.... The light which I see is not located but is yet more brilliant than the sun,...and I name it 'the cloud of the living light'.... But sometimes I behold within this light another light which I name 'the living light itself'. And when I look upon it every sadness and pain vanishes from my memory."
(St Hildegard of Bingen, c.1140.)[6]

"I saw a point that sent forth so acute a light, that anyone who faced the force with which it blazed would have to shut his eyes, and any star that, seen from earth, would seem to be the smallest, set beside that point.... Around that point a ring of fire wheeled.... O Highest Light (or Light Supreme).... I presumed to set my eyes on the Eternal Light so long that I spent all my sight on it!.... Whoever sees that Light is soon made such that it would be impossible for him to set that Light aside for other sight; because the good, the object of the will, is fully gathered in that Light; outside that Light, what there is perfect is defective.... The Living Light at which I gazed – for It is always what It was before.... Eternal Light, You only dwell within Yourself, and only You know You; Self-knowing, Self-known, You love and smile upon Yourself!"
(Dante, *Paradiso*, cantos XXVIII, XXXIII, 1318-1321.)[7]

Just as photons bombard the retinas of the eyes and trigger the nourishment of the physical body and brain (see pp242-246), so, I submit, the tinier-than-neutrino "metaphysical photinos" of the Light bombard the intellect of consciousness and trigger the nourishment of the soul and spirit, organs which can see beyond the physical.

It seems that there a cosmic order as all the religions have affirmed. An order which harmoniously manifests in the universe and Nature reflects itself in universal harmony through human truth, a process reflected in religious rituals – as in the Hindu Vedic concept of *rta* (a Sanskrit word meaning "order" or "law"). It seems that this order influences the earth through the spectrum of light – sunlight in the case of plants and humans' bodies, and Light energy in the case of human minds. Mixed with the mysterious Light, Universalists hold, are rays from the universal order principle which cleanse and transform the pathways to the universal being, deepening consciousness so the philosopher lives (at different times) in levels (6)-(12).

In terms of the physiology of the brain, during illumination it is as if there is a shift from the controller of the sensory, motor areas of the superficial cortex to a deeper level, to what may be the claustrum's control of other regions of the brain, including the amygdala which is significant in evolution.

To be aware of the infinite involves a shift from one network of interconnected brain cells in different parts of the brain to another level and another network of interconnected brain cells in different parts of the brain, so different areas are involved. It is like the digestive and excretory functions associated with the gastro-intestinal system. During digestion the excretory functions are sealed off by feedback "mechanism", as we have seen. It is as if during our everyday life we are sealed off from below, but when the time comes the seal is removed and we are able to open to the infinite, as when harbour gates open and a harbour opens to the sea. The point that needs to be stressed is that the level of consciousness one is in at any particular time is reflected in – reductionists would say determined by – the centre within the brain where consciousness is operating, controlling and regulating. And the highest levels of consciousness – those at the end of the list on pp250-251 – are deepest in the brain as the brain seems to connect with the infinite *beneath* the cortex.

The highest or deepest levels of consciousness, particularly universal consciousness, reflect the order and harmony in the universe. Our consciousness is a spectrum that receives light energy at differing frequencies. This light energy may include natural light and infinite metaphysical Light, which the non-finite high-frequency part of ourselves can receive along with the healing energy and, perhaps, various "paranormal" energies. Our consciousness, which is anchored and lodges among electrical brain rhythms, may well turn out to be a system of electrically-connected photons of light, a spectrum that has a dense, low-frequency part and a subtle, high-frequency part. It may be that when we shift the level of our consciousness we shift the frequency of our light and open to high-frequency currents of energy. As a result we find an increase in our energy.

As we saw on p247 our brainwaves vary from 30 cycles per second to 0.5. Beta and alpha rhythms are between 30 to 8 cycles per second, theta and delta from 8 to 0.5. It seems that the gateway to higher frequency is opened when we shut down our own beta and alpha interference (the rational, social ego) and go into contemplative higher, or deeper, consciousness in the intellect at 4 cycles per second. We are then filled with high-frequency invisible rays which energise us. In terms of the electromagnetic spectrum (see Appendix 3) the gateway to universal consciousness is just off the bottom.

Our brains are surrounded by gamma rays, radio waves and invisible rays of natural light. Among them are waves of manifesting metaphysical Light. At the scientific level, though they can be known epistemologically by direct experience these waves are at present unmeasurable. This may be because when they manifest into existence they are off the top of the electromagnetic spectrum.

It may be that we receive the full frequency range of the electromagnetic spectrum, including the healing and paranormal energies, when we still ourselves to around 4 cycles per second. This gateway opens up our own receiving station to our own consciousness. This is able to receive the manifesting metaphysical Light which we receive along with gamma rays, and which we can see with our "eye of contemplation".

The shorter the wave length the higher the frequency and the greater is the energy of every photon. Receiving the metaphysical Light may be like receiving healing energy. When a patient receives healing, molecules with low energy absorb photons of high energy which have been channelled by the healer, and they become more energetic and excited. When a healer *gives* healing, the process is reversed: photons of high energy (in the healer) are transformed into lower energy, and chemical energy is transformed into radiant energy.

Mystics assert that the infinite – which in the concept of the rational philosopher the surfer breasts, the moving sea of energy from which the point or singularity that became the Big Bang emerged, which underlies space-time – can be known by consciousness in the here-and-now. It cannot be known by the rational, social ego which controls the consciousness of waking sleep and awakening consciousness but through objective or cosmic consciousness when the centre of consciousness is deeper in levels (7), (8) or (12).

Intuitional Psychology and the Multi-Levelled Self

In intuitional psychology (what can be experienced of Being and in what part of the self we experience it) the intuition of the intellect or universal being *experiences* Being, the "ever-living Fire" or Light, in contrast to rational psychology in which the rational, social ego *thinks* about Being.

The human multi-layered consciousness and personality are anchored in the brain. The human, rational, social ego is above ape-old instinct, while somewhere beneath the cortex are the hierarchical levels of consciousness (1)-(12). To see where the intellect or universal being is located within the structure of the self of identity and personality we need to consider the seats of consciousness, which correspond to differing permutations of brain activity and which I arrange in descending hierarchical order:

universal being or intellect (*intellectus*, see p295) – in universal consciousness, into which illumination shines from the One (level (12));

spirit (*pneuma*) – in transpersonal consciousness, possibly relating to Non-Being (level (11)), more probably relating to Being (level (11));

reason (*nous*) – higher mind, (levels (7), (8));

soul (*psyche*) – higher feeling (levels (7), (8), (9));

ego – rational, social ego (levels (4), (5), (6), (7));

sense – controlled sense-impressions (level (5));

body (*soma*) – body consciousness of body-brain system, robotic automaton (level (4));

instinct – (level (1)-(3)).

This arrangement solves several philosophical problems regarding the mind which have traditionally been posed as questions. Is mind brain-function? Or does mind use the brain? Is it true that neuroscience cannot explain I-hood? Is the "I" or self a construct of memory? Is the self lacking in substance, is it a mass of impressions? How do we achieve mental and bodily activity, and a stable self?

The answer to all these questions is surely that different levels of consciousness use different parts of the brain and that different permutations and combinations of the neurons operate from different centres which connect to differing layers or strata of Reality at different times. At different times, all these questions may be answered "Yes". Mind seems to be brain-function in levels (1)-(4). Mind seems to use the brain in levels (5) (12) "I"-hood seems to defy neuroscience in levels (1)-(4). The "I" seems to be a construct of memory in level (5). The self is a mass of impressions in level (5). The self is stable in levels (6)-(9).

So it is time to ask again the question on pp249 and 253, whether consciousness is an effect of the brain like homeostasis. The answer is "No", not in levels (5)-(12). Does consciousness control the brain with the aid of photons as homeostasis controls both body and lower brain? Possibly, but this has not been proved. There is philosophical case for seeing this as happening, for consciousness should be within the system of the One which includes Nature's visible ecosystem and the more invisible system of air like *aither* (or the Higgs field of bosons) full of massless photons entering brains and controlling bodies, and entering plants and controlling photosynthesis. Does consciousness control homeostasis? Possibly, but again a case can be made out in philosophical terms as the science on this matter is in its infancy – even though photonology is being applied successfully to machines.

I submit that consciousness controls the brain and homeostasis in levels (5)-(12) within the philosophical category of the rational metaphysical system, and the system I offer includes this partial aspect of consciousness's behaviour. For the autonomic parts of consciousness, from levels (1) to (4) are an aspect of the autonomic functions of the brain. Lower or shallow consciousness is enmeshed in the body and physical brain's electrical activity, while higher or deep consciousness contains within itself the possibility of controlling the brain and homeostasis, and through them the body. The saying "mind over matter" captures the serene consciousness imposing the calm of homeostasis on the brain – rather than being a mere physiological effect of the physiological calm of homeostasis.

Having made clear the two-tier nature of consciousness (which is partly autonomic and partly controlling) I can list the seats of consciousness with the levels of manifestation to which they connect, referring to pp268-269:

Nothingness, the All or One – infinite metaphysical Light of finest density, too
    pure to be seen in universal being or intellect (level (12)), mystical/"divine";

Non-Being – infinite, denser metaphysical Light but still too fine to be seen in the universal being (level (12)) or spirit (level (11)), spiritual;

Being – infinite/finite, sea of neutrino-like or photino-like particles of metaphysical Light received in universal being or intellect (level (12)), and psychic/paranormal energies received in spirit (level (11)) as discarnate entities can be expected to be in the sea of Being, spiritual; soul (levels (7), (8), (9)), archetypes (including Ideas if they exist), psychological;

Existence – natural phenomena and natural light seen by the physical, rational, social ego (lower mind, levels (5), (6), (7)) and bodily instinct known to levels (1)-(4), physical.

Functions of course overlap. Soul blends with reason and also ego, which are next to it. The relationship between soul, reason and ego is graphically illustrated in an exercise that is taught in Kabbalistic schools,[8] where ten chairs are arranged in the shape of the *sefirot* in the Kabbalistic Tree of Life. Chair 5 represents *Teferet* (Beauty), the psychological soul, and this controls chair 2 (the physical ego) and also the spiritual and divine pathways of the *sefirot* 6-11.

Plato was aware of the overlapping functions of the soul, and in *Phaedrus* (253d) he shows the soul as two horses, one white, one black, and a charioteer who represents intellect, the part of the soul that must guide the rest of the soul to truth and enlightenment. The white horse represents rational, moral impulse and the positive passions of the higher levels such as righteous indignation. The black horse represents the appetites, concupiscence and irrational passions and instincts of the lower level. The charioteer directs the chariot or soul to prevent the two horses from galloping in different directions and to proceed in harmony towards enlightenment.

The early philosophers made the Greek names of the levels of the self into principles. Anaxagoras of Clazomenae elevated *Nous* to the first intelligence which set particles whirling, a rational principle that foresaw and intended the production of the cosmos and of living and intelligent beings – a view that would endear him to Intelligent Design. In Plotinus's thinking, Soul forms, orders and maintains the material universe, and is, I suppose, an early antecedent of my universal principle of order. Gnosticism and Hermeticism continued this trend of naming aspects of the universe in terms of parts of the mind or self, or of human qualities.

Intuitional psychology sees the self as having levels or seats of consciousness which relate to levels of Reality as in the second of the two lists above. It relates to Being in the upper levels of the self, in soul, *nous* or higher mind, and universal being or intellect (levels (7)-(12)). Interestingly, Empirical philosophers operate at lower levels (levels (5)-(8)) as they are more centred on the rational, social ego which ruled the 18th century (in the poetry of Dryden and Pope, for example). The branches of philosophy followed by Empirical philosophers such as Locke and Hume dealt with ethics, language and politics; which were all to do with Existence rather than Being. The intuitional metaphysical philosopher operates on all 12 levels, implementing all the faculties within the self. He holds that identity links as much to Being as to Existence, and that culture should involve the bringing of Being to Existence in the form of works of art. Language should be extended to catch Being in images and reveal ontology, not be about social interaction at the level of Existence alone.

The intuitional multi-levelled self's connections with Being and Existence mean that as a philosophical movement Universalism has much in common with Existentialism, which is now defunct, having been superseded on the European continent by Phenomenology. Existentialists fell into two categories; those who approached Being through their soul (Kierkegaard, Heidegger, Jaspers and Marcel, and the Phenomenologist Husserl), and those who approached phenomenal Existence through their reason and rational, social ego (Sartre and Camus). Like Universalism, Existentialism had rational and intuitional wings. Intuitional Universalists in particular take issue with rational Existentialism's view of the arbitrariness of Existence. M. Roquentin in Sartre's *La Nausée* looks at a tree-root in a park and feels disgust, nausea at the gratuitousness of its existence. This nausea is a response of his rational ego, his reason, rather than his feeling soul, and, holding the same view of the world as Materialist scientists, lacks awareness of manifestation.

Wordsworth, looking at a similar tree-root, and connecting it esemplastically with the One, the Whole, the "Wisdom and Spirit of the Universe", would have felt joy rather than nausea. Heidegger, looking at his view of a mountain range from his hut near the Black Forest would have identified more with Wordsworth's perception than Sartre's. Heidegger asserted that the philosopher reveals the One using a poet's intuition and then rationalises it so that the One, already grasped by the intuition, can be understood by the reason.[9] Intuitional philosophy is like a reader's guide to a poet's works, rationalising a poetic perception of Being or Reality. It sympathises with van Gogh's "I am painting the infinite". Whitehead, who wrote "The purpose of philosophy is to rationalize mysticism" would have sympathised with Heidegger for in *Science and the Modern World* he devotes a chapter to praising Wordsworth and Shelley as deniers of scientific Materialism.

The intuitional self, residing outside the rational, social ego, is able to experience the workings of the Whole in a way that can elude the rational metaphysical philosopher. I have said (on p283) that the reason analyses and fragments the All or One into bits whereas intuition perceives the Whole, reunifies the fragments. Intuition's movement is from Existence to Being. The intuition perceives what the reason cannot perceive, that as it operates within a unity of Being the infinite order principle does not merely shape growth but can also shape our actions and fortunes if we put ourselves in harmony with its ordering force. The order principle works for good conditions (such as growth and development) to prevail in all life, the intuition perceives, and to the intuition order, which created the bio-friendly conditions in which we live, can make good things happen to us by detecting and advancing the yearnings in our cells. Our wishes and conduct influence the Whole and are influenced by it. That is the intuition's great hunch. To the All's 14 rational laws, intuition adds a complementary 14, which are all practical applications of the metaphysical perspective:[10]

the Law of Order – the infinite orders the finite but the finite cannot order the infinite;

the Law of Randomness – everything is random that has not been ordered and organised by the Law of Order;

the Law of Harmony – when the self is in harmony with Being it attracts happiness, as in Taoism;

the Law of Contradictions – all contradictions are ultimately reconciled within the overriding unity, just as $+A + -A = O$;

the Law of Unity – as Non-Being and Being have unity, all manifestation and multiplicity are contained within the unity, are interconnected and are one with the timeless infinite;

the Law of Multiplicity – multiplicity is contained within primordial unity and does not cease to be so contained when manifested into many.

the Law of Survival – as multiplicity does not cease to be contained within the unity of Being when it manifests, all beings who manifest do not cease to be contained within the unity of Being and return to it on death;

the Law of Illusion – those who do not know they are contained within the unityof Being may be under the illusion that no part of their being or consciousness will survive death;

the Law of Correspondence – as Being has unity, all phenomena of Existence manifest from the unity of Being in accordance with the Law of Unity, all phenomena share Being and are linked together and correspond to each other within the unity of Being in such a way that they all contribute to the universal harmony, and our outer world therefore reflects the thoughts, beliefs and attitudes of our inner world;

the Law of Symbolism – as all phenomena manifest from the One in accordance with the Law of Unity and the Law of Correspondence, the lower domain can always be taken to symbolise the reality of the higher order in which can be found its profoundest cause, or to put it more simply "as above, so below", and all phenomena can be intuitively perceived as symbols of aspects of the One and reveal Being;

the Law of Attraction or Thinking – as Being is a unity and the order principle shapes phenomena and therefore events, by working in harmony with the order principle the self can attract what it thinks (or prays for) and expects provided it takes initial concrete action to implement the new condition, while most things that happen to human beings have a cause within human thought;

the Law of Prosperity or Abundant Supply – as Being is a unity and the order-principle has dominion over events as well as phenomena, by working in harmony with the infinite order which means opening to the infinite, the self receives plenty to supply all needs provided it allows new energies to flow in which means cutting out dead wood so new growth can come through and giving readily to make room for what plenty will bring;

the Law of Sowing and Reaping – as the universe of Being is filled with both order and randomness, our actions either work with order or against it and have appropriate consequences, and what our being sows we will reap, either for ordered good or random ill;

the Law of the Good of the Whole – as the universe is a unity of infinite Being imbued with an order principle that operates for the good of the Whole, if we work with it and our being acts virtuously and selflessly for the good of the Whole, filled with the infinite our being maintains an equilibrium through feedback and receives good back for ourselves that we did not seek.

These practical laws are all more everyday applications of the intuition's perception of the infinite pervading the finite. Using their intuition in this way, the religious can claim that prayer, giving and rectitude all work in a practical sense, while those outside religion can claim equivalent benefits as their thought attracts desired circumstances to their lives.

The intuitional self achieves its perception of the infinite by the painstaking way it assembles its vision. Whereas the reason analyses the whole of Nature and fragments it into bits – as in the case of all the numbered data in chapters 7 and 8 – the intuition works in the opposite direction. It examines the bits, perceives the whole and reunifies the fragments as an archaeologist pieces together potsherds of a broken urn. We have seen on p171 that Coleridge referred to this faculty of the mind as "the esemplastic power of the imagination", the word "esemplastic" coming from the Greek "*eis en plattein*", "shape into one". The intuitional imagination takes the pieces and puts them back together again. The intuitional philosopher (of whom Coleridge himself in his philosophical reflections was an example) pieces science's fragmentation of Nature back into One Whole, as I did in Part Two where all the paragraphs form part of a whole structure like a stuck-together urn.

The intuitional philosopher does not rationalise – or extrapolate by reasoning – the Law of Order, but by esemplastically piecing the analysed bits back into a whole, intuitionally perceives the workings of order in the All or One, which includes the infinite and the finite in accordance with algebraic thinking ($+A + -A = O$). Receiving the metaphysical Light in his universal being, the intuitional philosopher intuitionally knows that it brings order into his *pneuma* and *psyche* which is reflected in his behaviour – a cleansing of the appetites, aversion to unnecessary clouding of the senses with alcohol or polluting them with irrelevant loud music or television programmes – and intuitionally locates the origin or order in the infinite which, his intuition tells him, manifested into metaphysical Light. He arrives at the same conclusion as the rationalist metaphysical philosopher, the difference being that whereas the rationalist metaphysical philosopher has arrived at his conclusions by using the concepts of his reason (the concept of the surfer and the data of chapters 7 and 8), the intuitional metaphysical philosopher reports on his direct experience.

Both the reason and intuition alike affirm a Law or Order and its opposite, a Law of Randomness; a Law of Harmony and its opposite, the Law of Contradictions; and a Law of Attraction or Thinking which, if used correctly, can lead to the Law of Abundance or Prosperity. Reason may acknowledge the other laws, including the Law of Unity, but as a more theoretical, analytical concept. The rational and intuitional philosophers use different faculties within the whole self to gain such insights. One of the difficulties with modern Analytic philosophy is that it operates theoretically and analytically and has lost touch with intuition.

Both the rational philosopher like Socrates and the intuitional philosopher like Whitehead agree that the way to live is to control the different levels or layers within the self at different times in such a way that the senses are controlled. Compare Plato's image of the charioteer controlling his horses through their harnesses, though the Universalist charioteer stands in level (7), objective consciousness, and controls a double quadriga: levels (4), (5), (6) and (10); and (8), (9), (11), and (12).[11] Unattached in its chariot, free from the demands of the sensual ego, inner serenity perceives the hidden harmony within the universe, that unity is behind all the surface contradictions that make the universe seem full of multiplicity and manifold difference, and that within this tranquillity comes an energy that orders one's life so it feels purposive and has direction. This energy works in harmony with the order principle and results in abundance within the self that is converted into circumstances of prosperity. Coleridge understood when he wrote, "I would make a pilgrimage to the deserts of Arabia to find the man who could make me understand how the *one can be many*."[12] He might have added, "And how the many can be One."

The intuitional metaphysical philosopher interprets all experience in terms of a system of general ideas that reflects *all* the parts of the self I listed on p299 – not just the reason and soul, but the spirit and universal being as well. The intuitional framework of ideas includes all 12 levels of consciousness and understands the totality of human experience in terms of *all* levels within the self. These correspond to layers and permutations of functioning cells in the brain (as outlined in chapter 8). All our experience can be understood in relation to the model Universalism offers of infinite energy manifesting into the universe and penetrating deep into universal being while a principle of order penetrates through eyes and chloroplasts to light-sensitive tissues and chlorophyll, sending signals to all creation. Humankind is thus deeply connected with the workings and operations of the universe.

Humans with a seat of consciousness in one of the higher levels perceive life to be more meaningful than those who perceive it from a seat of consciousness in one of the lower levels. The picture of a human feeling isolated and standing against the universe is wrong. Humans living in the higher levels perceive that all human beings are connected to the manifesting infinite by an invisible thread-like network of "ever-living Fire" or manifested metaphysical Light and are nourished by powers within the universe which reach their universal being. A stone carving of the Egyptian pharaoh Akhenaton demonstrates the caring, healing nature of these powers. Rays from the Aton (the sun-disc) end in hands which reach caringly towards the Pharaoh. In contrast, within the totality of the neural activity within and experience of the self, the rational, social ego is only in control at certain times when it needs to front human effort, in work situations or financial transactions for example. All the other parts of the self have their turn at different times, depending on the nature of the situation. Each part of the self takes charge when the demands of a situation make it the most appropriate part of the self to respond. Thus, ascending the hierarchy of levels:

in situations of danger, instinct controls;

in a new place, sense/impressions control;

in competitive sports, the ego and its link to the physical body control;

in work and social situations, and following up desires, the rational, social ego
    controls;

in the beauty of Nature, listening to beautiful music, looking at beautiful
    paintings, reading beautiful literature and loving a beautiful person, the soul
    or higher feeling controls;

in thinking a plan, the reason or higher thinking (*nous*) controls;

in intense experiences that go beyond our personality such as hearing messages
    from the dead or receiving memories of far lives, the spirit controls;

in opening to the "ever-living Fire" or metaphysical Light, the universal being or
    intellect controls.

The Intuitional Universe as a System of Light

Whereas in the rational universe, infinite Reality manifests and penetrates physical eyes and chloroplasts as an objective event, in the intuitional universe infinite Reality, the metaphysical Light, manifests and is received within the universal being or intellect, in the inner eye of contemplation, as a subjective experience.

In the Yogic practices of some forms of Hinduism and Buddhism the Light travels through the *chakras* or *cakras*, physic centres or centres of the invisible subtle body which, according to Hindu tradition, operates in conjunction with the body's plexuses near key parts of the spine. The top *chakra* is the crown *chakra*, which to Hindus is the seat of illumination. The seven Hindu *chakras* are from bottom up: root (base of spine), spleen, navel, heart, throat, brow and crown (which is associated with the pineal gland). In Buddhism there are four *chakras*. To Hindus (and Theosophists, who adopted the *chakra* system), the Light enters at the base of the spine and flows up the central nervous system from *chakra* to *chakra*, a process known as the rising of the serpentine fluid of Kundalini, and explodes into illumination in the crown *chakra*, which in Western thought is the intellect.[13]

In Western thought, the metaphysical Light does not travel *up* the spine. That is a Yogic phenomenon. By going into contemplation, the self shuts down many of the parts or centres within the self and brain I listed on p299: instincts, sense, ego. Detached from sense impressions and desires, it moves deeper from soul to spirit to universal being or intellect, which opens as a mirror to reflect the incoming rays of "ever-living Fire". What the contemplator sees is a reflection of a dazzling metaphysical Light that has manifested from the infinite, bringing energies from deep within the cosmos. These energies transform the self, turning it away from low desires and burning impurities which religions categorise as the accretions of "sin" from the pure soul beneath so that the recipient frequently loses the desire to smoke, drink alcohol or in some cases be promiscuous. The energies cause virtues to flourish and eliminate psychological problems and fear of death for they bring assurance that humans have an immortal spirit. The energies bring serenity and "the peace that passeth understanding", a sense of the unity of the universe and of human beings' oneness with all humankind and all creation.

The intuitional system contains a sea of energy or latent Light that has manifested into subtle, end-of-spectrum physical light, matter and consciousness. The same manifestational process, once out of Being and in Existence and directed by the universal order principle, shows itself to be a Light-inspired drive of organisms through matter to higher and higher levels of self-organising and ascending hierarchical wholes.

The intuitional universe is a perception of the intuitional self or consciousness. The received view of consciousness is that it is brain-dependent and ends when its brain stops functioning. However, according to the intuitional universe it is not impossible that each consciousness may be transmitted as light or photons through the brain rather than be produced by the brain. It may control and regulate the brain's electrical rhythms, regarding the finite brain as Shelley's "dome of many-coloured glass" which "stains the white radiance of eternity".[14] It may be that the white radiance of the universal being, spirit and soul come from outside and anchor and lodge within the many-coloured dome of the operating brain and appear to be dependent on the brain as brain function.

If consciousness is transmitted into the brain and the brain acts like a radio receiver, it is probably received by photons above the brain, in a "Bose-Einstein condensate".[15] This condensate, or substance produced by condensation, was proposed in a theory developed in 1924-25 by Einstein and the Indian physicist Satyendra Nath Bose on the statistical behaviour of a collection of photons, and it accounts for the streaming of laser light. Photons satisfying Bose-Einstein statistics are called bosons. Electrons satisfying Fermi-Dirac stastics are called fermions. These photons or bosons, including the Higgs field of bosons, have been undetected until now and were sought by CERN in its September 2008 experiment. If consciousness *is* received in the brain, the transmitted current of universal energy, perhaps arrives on carrier photons or bosons. The energy stimulates and drives the neurons and pushes their electrical impulses into interconnectedness and binding with each other. The activity of the neurons may thus be a consequence of the process of consciousness bosons, not its generator. And consciousness may be a tenant squatting in the home of the brain, which it deserts to go elsewhere, leaving the brain empty and lifeless.

To be completely clear, what I am suggesting is that consciousness may be composed of hitherto undetectable bosons (photons that obey Bose-Einstein statistics) and that the discovery of such bosons by CERN would make this proposal testable as a scientific fact.

I also want to make clear that metaphysical Light begins in the infinite, beyond physics, and, I submit, manifests into the finite, the universe after the Big Bang. At some stage, I submit, metaphysical Light manifests into finite bosons. CERN has sought this finite manifestation. Its particle chamber replicates the finite conditions *after* the Big Bang, by which time the infinite has manifested into the finite. It has tried to capture the moment of manifestation. I submit that the bosons of consciousness had an infinite origin, and have the infinite contained within their finite manifestation in accordance with the Law of Unity and the Law of Survival (see p303). According to metaphysics, in view of their infinite origin in Being, all finite phenomena contain within them that which is infinite. This was the point I argued with Bohm.

Evolution and brain function may have been spurred on and be consequences of the wave transmissions of the "ever-living Fire" or metaphysical Light which may manifest into physical light. As we have seen on pp246, 253-254 and 273, this may contain coded messages of order impelling all creatures to the next stage of their growth. Bohm was thinking along similar lines. He said that "Light can carry information about the entire universe"[16] and he held that information can move at superluminal speeds, speeds faster than the speed of light (which is just over 186,000 miles per second). When this happens as G.N. Lewis, a physical chemist, argued in the 1920s, there can be immediate contact at great distance as time slows down, distance is shortened and the two ends of a light-ray may have no time and no distance between them. This quantum non-locality is to physics what synchronicity is to psychology.

Thus, the invisible, latent, infinite "ever-living Fire" or metaphysical Light may travel faster than the speed of natural light, pouring order, truth and wisdom into consciousness and information into matter at speeds greater than the speed of light. Matter Bohm described as "condensed or frozen light". Matter forms when two photons collide. Matter is formed by light and there is a good scientific case for regarding matter as frozen light.

This is such an important idea that we should dwell on it. A moment's thought will reveal the connection between matter and light in Einstein's famous equation, $E(energy) = m(mass) c^2$ (the square of the speed of light in a vacuum). This equation tells us that mass is related to energy via the speed of light, that the speed of light fixes the "conversion factor" as to how much each kilogram of matter is worth in terms of energy.[17] Bohm, Einstein's *protégé*, took up this idea in conversation: "All matter is a condensation of light into patterns moving back and forth at average speeds which are less than the speed of light. Even Einstein had some hint of that idea. You could say that when we come to light we are coming to the fundamental activity in which existence has its ground, or at least coming close to it."[18] The transfer of order from light to living processes or organisms takes place via the absorption of sunlight into eyes and its transformation into disordered heat radiation.[19]

I would add that natural light is condensed metaphysical Light. And that the process of manifestation is a kind of condensation. This hidden network of the manifested Light may therefore be a network of "air waves". The Light is around humans and permeates them all the time. Just as neutrinos pass through human bodies as they have so little mass, so massless photon-like boson particles of manifested Light that may be tinier than neutrinos must also be able to pass through human bodies. Tinier-than-neutrino particles such as photinos/bosons, supersymmetric counterparts to photons just as gravitinos are counterparts to gravitons, may emerge from the hidden Reality beyond the quantum mechanic waves and space particles. All creatures are like sea-sponges living in a sea of Light. It is in fact a network of transcendent and immanent energy, manifesting from the infinite and flowing finitely into humans' intellects and into all chlorophyll.

The intuition brings its own intuitional slant to non-duality for it is deeply involved in the workings of consciousness. It will be helpful to list the metaphysical perspectives I identified on p281 alongside the various views of consciousness to make it absolutely clear what I am proposing and not proposing as regards the intuitional universe as a system of light:

Materialistic monism (that matter gave rise to mind or consciousness, scientific Materialism) – mind or consciousness is Materialistic brain-function within electrical neurons dependent on the brain, and dies when the brain dies;

dualism (that matter and mind are separate substances, Descartes' position) – mind or consciousness is composed of a substance separate from matter, and soul/spirit survives death whereas matter dies when the body dies and is recycled;

transcendental monism (that mind or consciousness gave rise to matter, that consciousness is the stuff of the universe, Idealism) – mind or consciousness is transmitted from manifesting metaphysical Reality and is perhaps composed of photon-like particles tinier than neutrinos such as photinos, but probably bosons, which form a halo-like system of light above the brain that uses brain-body, and soul, spirit and universal being survive death whereas matter is recycled;

metaphysical non-duality (that the infinite metaphysical Reality, a unity, gave rise to both matter and mind/consciousness, a diversity although from a common origin) – mind or consciousness is transmitted from manifesting metaphysical Reality and is perhaps composed of photon-like particles tinier than neutrinos such as photinos, but probably bosons, which form a halo-like system of light above the brain, a Bose-Einstein condensate, that uses brain-body, and soul, spirit and universal being and survives death whereas matter (atoms, cells and neurons) are recycled into a new form.

Universalism is a metaphysical non-duality, and this last perspective is the one I favour, as I said on p281. In intuitional metaphysical non-duality, consciousness appears ambivalently, depending on its level. Thus:

brain-body's feedback appears to control consciousness in levels (1)-(4), which seem to be dependent on the brain and appear as brain-function, suggesting that the autonomic consciousness dies when the brain dies;

consciousness appears to control brain-body's feedback in levels (5)-(12) but though it appears to be still dependent on brain and to be brain-function and die when the body dies in fact it may be transmitted from manifesting metaphysical Reality in photon-like particles tinier than neutrinos such as bosons which form a halo-like system of light, a Bose-Einstein condensate, that uses brain-body, and soul, spirit and universal being and may survive when the brain dies and matter (atoms, cells and neurons) are recycled into a new form;

both brain and consciousness interact at all levels, so in levels (1)-(4) brain-body appears to control consciousness but interacts and at times consciousness controls brain-body; and in levels (5)-(12) consciousness appears to control brain-body but interacts and at times brain-body appears to control consciousness.

In intuitional non-duality, consciousness knows Reality intuitionally by going into a meditative state (levels (6)-(8)) and passing from 30 cycles per second to below 4 cycles per second into universal being. As a general rule, the higher the level of consciousness towards level (12), the smaller the number of cycles per second in the brain. Consciousness receives an influx from the infinite metaphysical Light manifesting as photon-like particles tinier than neutrinos such as photinos, but probably bosons, beyond gamma rays on the electromagnetic spectrum, at the level of Being, which contains the ordering potentialities of Existence. Thus in metaphysical non-duality, consciousness forms as photon-like boson particles from the manifestation of metaphysical Light whereas brain-body form from the manifestation of the infinite as physical light, matter and other substances. Both consciousness and matter, in themselves diverse, share a common origin in the infinite Reality.

To state this common origin and unity even more clearly, in intuitional metaphysical non-duality there is:

body, brain, matter – nourished via the eyes by photons of natural light that, I submit, manifested from the bosons of infinite/metaphysical Light;

mind and consciousness – either transmitted and formed of photon-like particles tinier than neutrinos such as photinos, but probably bosons, that, I submit, manifested from the infinite/metaphysical Light and form a halo-like system of light above the brain, or – more of a compromise with Materialistic monism – lodging as an independent system of light within the electrical activity of the brain and nourished by photon-like particles tinier than neutrinos such as photinos, but probably bosons, that, I submit, manifested from metaphysical Light.

To be completely clear, within non-duality there are three possible views of consciousness/mind:

that all of it (levels (1)-(12)) is a consequence of brain-function and will die with the brain;

that part of it (levels (1)-(4)) is a consequence of brain-function that will die with the brain and part of it (levels (5)-(12)) is transmitted as bosons and lodges and will not die with the brain;

that none of it is a consequence of brain-function but is transmitted as bosons, and none of it (levels (1)-(12)) will die with the brain.

I submit that in a philosophical category either the second or the third proposition is true.

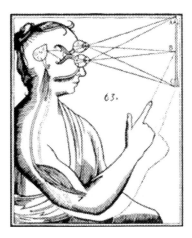

René Descartes' illustration of dualism.

Inputs are passed on by the sensory organs to the epiphysis (or pineal body or gland) in the brain and from there to the immaterial spirit.

The picture of consciousness/mind in intuitional non-duality is not unlike the picture Descartes literally used to illustrate dualism. (See illustration on p313.)[20] Inputs from the phenomenal world along with photons reach the eyes and are passed on through the sensory organs (photoreceptors) and neurons in the brain to the epiphysis, or pineal body or gland which Descartes called "the seat of the soul" in *Treatise of Man* (see Appendices 4 and 5), and from there, on this view, to the system of light which may include the soul, spirit, universal being, I-hood and consciousness. On this view, this system of light resembles a cloudy halo or aura of bosons that is tethered to the brain's pineal region and connected to the electrical network of neurons before birth and is released to leave the body shortly before death and then become a mist-like discarnate presence. It may continue to exist in a "body" composed of invisible bosons of light (an energy or frequency mediums can channel along with idiosyncratic language, behaviour traits and detailed memories) in new surroundings on "the other side" in Being and may reincarnate in the future just as it may have reincarnated several – or, indeed, many – times in the past. Accounts of near-death experiences report the feeling of the "I" hovering above the body on a kind of umbilical cord from the head. As I write this, my consciousness feels as if it is *outside* my skull – my I-hood does not feel as if it is locked away under my skull and behind my eyes – and it is possible that it is a field of bosons all round my head, a field of wispy boson light attached to my brain but all round "me". It may be an auric halo (a halo made of my aura) or a corona of light round my skull (like the small circle of light round the sun or moon known as a corona) or a crown (which derives from the Latin *corona*) of light, a wig of invisible boson light fastened within my brain and waving in air like hair under water as seaweed fastened to a rock waves in sea. Non-duality differs from dualism in holding that both matter/body and consciousness/mind (levels (5)-(12)) emerged from the one source, the infinite, and that the system of light that is consciousness/mind returns to the infinite whence it came while matter/body/the brain (levels (1)-(4)) return to the finite. It could be that the second proposition on p313 rather than the third proposition is true.

On this view, the conversion of photons into chemical energy for brain/body and the conversion of tiny neutrino-like or photino-like boson particles of manifested metaphysical Light into spiritual energy for soul/spirit/universal being/I-hood and consciousness both happen in accordance with Descartes' diagram on p313. Descartes' "new philosophy" in fact reached very near to the truth about consciousness but fell short by reporting a rift, a dualistic divide, between consciousness/mind and matter/body/brain, a divide that non-duality avoids by relating both bosons and photons to their common origin in having manifested from their common infinite source.

This is as close as it is possible to advance to the position of consciousness in body/brain. It is as far out as the surfer breasting the infinite. With these two views of tethered boson consciousness and of the infinite surrounding the universe I have penetrated as far as mind can reach.

In the intuitional universe, if consciousness is a transmitted system of light, then it is likely that the foetal brain has a potentiality or tendency to become conscious (the same "mechanism" we found at work in the inheritance of instinct, see pp181-182) and that its consciousness is lit on birth by photon-like bosons from the environment. The brain-body's feedback and maintenance of homeostasis are automatic and depend on "mechanisms" found in the brain and body whereas mind/consciousness is lit up like a television being turned on by a remote. The television has the potentiality to light up but only does so when it is primed by a remote. In terms of quantum mechanics or quantum cosmology (the application of quantum mechanics to the universe as a whole), a collection of indistinguishable photon-like bosons can appear non-locally a long distance away. We have seen that Bose-Einstein statistics measure such a collection of photon-like bosons and show how they occupy the available energy states, and that consciousness may be a Bose-Einstein condensate of bosons (see p309).

Transmitted consciousness may be a quantum phenomenon in which photon-like "consciousness bosons" tinier than neutrinos such as photinos travel instantly from intergalactic space, having manifested from infinity, and hover over brain cells in a halo-like ring of bosons, connecting and interacting with electrical neurons but fundamentally controlling them, forming integrative connections in rapid succession to make links with the brain's stored information. Transmitted consciousness bosons may hover above the brain and link with its multiplicity of information just as a computer operator bends over a computer while linking to its information. Bohm said in conversation, "The mind may have a structure similar to the universe, and in the underlying movement we call empty space there is actually a tremendous energy, a movement."[21] It may be that consciousness has the same structure as the universe in accordance with the Law of Correspondence which holds "as above, so below", and that the apparent emptiness of space round the brain is like empty intergalactic space in the universe – not empty but full of hitherto undetected energy.

It is likely that consciousness, mind and matter (Bohm's "frozen light") can be integrated into the scientific view by quantum mechanics, and that the relationship between consciousness, mind and matter will eventually be understood in terms of quantum mechanics or quantum cosmology (the application of quantum mechanics to the universe as a whole). Consciousness, in short, may be a system of high-frequency boson light, and whereas reductionism reduces the system of light to brain-functions Expansionism restores it to the system of light that it is.

Can we say that the origin of the universe and the origin of life took place with one end in view, the teleological creation of higher consciousness? No, for this would be a religious view, and philosophy operates in neutral terms. The intuitional Universalist philosopher echoes the rational Universalist's view of teleology (see pp279-280). He notes the ubiquity of the universal order principle and the development of higher consciousness which can "know" (in the sense of "directly experience") the infinite, but stops short of making assertions about purpose which cannot be evidenced or verified and belong to the terrain of religious belief.

Why, then, are we conscious? Looking back at our view of the origin of the universe, at the 40 contextual bio-friendly conditions and at the span of evolution, the Universalist philosopher can point out that within order as opposed to randomness, through self-organising evolution culminates in a higher (or deeper) consciousness and intellect which can know the surrounding infinite and he wonders if it has an aim, to throw up even higher (or deeper) consciousness and intellect in human beings. So what does it mean to be human? Darwinists, seeing humans in terms of survival, would say, "To compete and struggle to survive as a species." Universalists, seeing humans in terms of consciousness, would say, "To live through the higher virtues of intensified consciousness and intellect rather than through the lower vices, with compassion, tolerance and friendliness towards our fellow human beings." These were the virtues of the classical Greeks who inquired into, and were in quest of, how to live the good life and, aware of the infinite, founded Western philosophy.

I have said that Expansionist Universalism sees modern *Homo sapiens* as having evolved not by random chance but through an orderly drive by self-improvement or self-betterment, and through self-transcendence, to a higher (or deeper) consciousness that is beyond the consciousness of apes. The *genus Homo* has not reached the end of his evolution. We now are where we have reached, but there is a further development ahead. It can be expected that with further evolution, modern *Homo sapiens* can be seen as being on his way to a higher (or deeper) consciousness than we have now. This can be expected to increase our capacity to operate in the universal being (the highest or deepest consciousness) and to include experiences on level (12) on a more frequent and regular basis. In short, at some point in the future modern *Homo sapiens* can be expected to have a more developed transpersonal and universal consciousness, and will be more open to the hidden Reality or Light of Being known to the mystics.

I return to the idea that the drive for evolution may be a persistent drive within the genes for self-improvement, for the implementing of an inner aspiration for self-betterment. Every plant, tree, reptile, animal and human responds to its environment with a plan for improving itself, a persistent drive that realises its possibilities and helps bring in the next stage of its development. Every organism is in process, and just as we are more self-improved and self-bettered than Neanderthal and Cro-Magnon man, so in a few thousand years' time *Homo sapiens* will be more self-improved and self-bettered than we are, and will be further on in the process of self-development than we are in our intermediate state. Man is an unfinished process, an unfinished project whose aspirations are limitless and ongoing, for each time *Homo sapiens* reaches a new plateau in its ascent, new possibilities for self-improvement and self-betterment open ahead.

The brain's consciousness has developed enormously from the first cell. Reptiles' consciousness has been calculated as using 10^3 neurons, dogs 10^9 neurons[22] and humans (as we have seen on p231) 10^{11} and perhaps 10^{12} neurons – though we have seen on pp247-248 that the brain uses 3 times 10^{14} combinations of cells. If higher (or deeper) consciousness goes on developing, the human brain may develop and use 10^{13} neurons during the next phase of evolution. In a few million years' time mind can develop to a higher consciousness that is at present unimaginable. We now think of our computers as being advanced, but as we saw on p242 quantum computers are being spoken of which will make our computers appear backward in comparison. Future brains may make our present brains appear backward in comparison

It has been calculated that a human brain can process 100 trillion (10^{14}) instructions per second. Even at present the fastest supercomputer in the world, the IBM Blue Gene/l at Lawrence Livermore National Laboratory, is capable of handling 478.2 trillion instructions per second (against a 4-function calculator's 10 instructions per second). It is possible that the brain may be surpassed by normal personal computers in terms of instructions per second by 2030.[23]

However, we saw on p248 that the brain can undertake 10^{27} operations per second, many of which presumably do not involve instructions. There is of course more to brain power than instructions per second. While the human brain is doing a calculation that a calculator or desktop can do in a fraction of a second, it is also processing data from millions of nerve cells, monitoring visual input, aural input from both ears, sensory input from the body, regulating heartbeat and oxygen levels, monitoring hunger and thirst, controlling breathing and hundreds of other activities within the body. It surveys a vast interconnected network of cells for information that will solve a problem and simultaneously writes a legible solution in the correct symbols. The present range of supercomputers will not have these additional powers while handling instructions per second.

We have seen evolution as throwing up higher and higher (or deeper and deeper) consciousness, which exists in bio-friendly contextual conditions. This consciousness lodges within individual organisms – bodies and brains – and as we saw on pp252-253 whereas in each organism the brain controls and regulates the body's feedback "mechanism" and interacts with it so that at times the body also controls and regulates the brain, so the brain controls consciousness with the help of photons and interacts with it so that at times consciousness controls and regulates the brain's feedback "mechanism" with the help of photon-like bosons – as in the little finger movement where consciousness is monitored before the index finger is moved (see p249).

To sum up the intuitional system, it can be said that whereas the rational system is one of outer forces in which rays enter physical eyes and green leaves, the intuitional system is one of inner flow within an ever-dynamic process. Whereas in the rational system humans are integrated into the universe by forces, in the intuitional system of both East and West humans are integrated into the universe – fundamentally connected every second – by hidden or invisible, subtle, moving boson energies. In both cases, humans are inseparable parts of the Whole, and must be understood Expansionistically and integrationalistically rather than reductionistically.

Intuitional Approach to Metaphysics

The intuitional approach to metaphysics differs from the rational approach. The rational approach to metaphysics was based on a system of general ideas that interpreted every known concept and all experience from the perspective of the reason and the rational social ego. My rational system is based on non-duality, an infinite unity which manifested into becoming in the Form from Movement process and into a finite diversity of matter, manifested light and consciousness.

The intuitional approach to metaphysics, by contrast, is based on a system of general ideas which interprets all experience from the perspective of the universal being or intellect. My intuitional system is based on an *experience* of the infinite unity which manifests into becoming and into the finite as subtle, manifested, boson-like Light which enters intellects and chloroplasts and energises all creatures.

As in the case of the rational approach to metaphysics, a word of caution is necessary. The system of general ideas the intuition concocts explains intuition's place in the intuitive universe. Intuition's approach to metaphysics does not embrace *all* experiences encountered within the 12 levels of consciousness. It bypasses much rational experience. The intuition's approach to experience must be combined with the reason's approach, and will then encompass *all* the levels of activity in the all the 12 levels of consciousness, each of which has a centre that seems like a self. The self moves between levels of consciousness and centres, each of which relate to the universe in different ways. The intuition and reason, making separate approaches to aspects of the universe at different times and, at different times, engaging in all the experiences I listed under the 12 levels of consciousness, cumulatively have a very full knowledge of the universe and Nature's system.

We saw at the end of Part One and on p282 that today modern metaphysics can be said to have four subdivisions, which in descending hierarchical order are:

ontology (the study of Being);
psychology (what can be experienced of Being);
epistemology (what can be known about Being; and
cosmology (the structure of the universe).

I summed up the rational approach to these four subdivisions, and their integration with the manifestational process, on p292. Taking into account the

12 levels of consciousness, intuitional, Expansionist, Universalist metaphysical philosophy restates the four subdivisions of metaphysics in descending hierarchical order:

ontology – studying the infinite or timeless Being through experiences of Reality, " the over living Fire" or metaphysical Light, in the intellect of level (12) (source, All-One/Non-Being/Being);

transpersonal (or spiritual) psychology – how the spirit of level (11) can know the potentialities of Existence in Being, particularly how it can know psychic or paranormal phenomena and what seem to be far memories of distant lives (source, Being);

epistemology – how the mind knows Reality as an experience by transcending the reason and rational, social ego so that the top levels (7)-(12) of consciousness can free themselves from Existence – detach themselves from the world of the senses – and know Being (source, Being/Existence);

cosmology – the study and description of the structure, science and theory of the universe, see Part Two (source, Existence).

The intuitional multi-levelled self occupies different levels and operates within different seats of consciousness (see p299). The intuitional Universalist studies cosmology but experiences the infinite and knows what he studies from within, relating to it as an organic process rather than as a fixed system as perceived by the reason. Sensing the One intuitively from a seat of consciousness other than the reason, he feels at one with the One as did the Romantic poet Wordsworth (whereas Hegel, his inspiration, thought the One).

Intuitional metaphysical Universalism sees the purpose of language as not only to communicate socially within Existence and focus on linguistic analysis (rational metaphysical Universalism's view), but also to explore and reveal Being as well as Existence. To the intuitional philosopher, language is not an end in itself but a means to an end. It focuses on the universe and Nature, and serves the purpose of explaining their system of general ideas. To the intuitional philosopher, language is therefore outward-looking, not inward-looking.

The intuitional approach to metaphysics is thus one of direct experience from a deeper level of the self than the reason or rational, social ego. In fact, intuitional Universalism, as distinct from rational Universalism, has an existential dimension. For just as in Romanticism and Existentialism,

Universalism affects attitudes to living.

Romanticism's revolt against tradition, authority, reason and classical science brought with it certain attitudes that were lived: awareness of the infinite; a view of Nature as organic rather than mechanistic; a view of the individual as solitary and isolated rather than as a social being; trust in the shaping imagination; and a focus on extremes of feeling, including despair and the Romantic agony.

Existentialism's similar revolt against philosophical tradition, authority and rational and scientific systems in the name of subjectivity also brought attitudes that were lived: feeling one's own existence in solitude; freedom and free will; self-transcendence in projects; and extremes of emotion such as dread, care and anguish. Put like that, Existentialism seems a disguised re-run of Romantic attitudes. Existentialism went beyond Rationalism in rationalising the individual's intuitional sense of the human condition in concepts such as being-in-the-world and being-towards-death.

Intuitional Universalism's revolt against reason – and the tradition and authority of Logical Positivism and Linguistic Analysis – from the intellect, and against the fragmentation and reductionism of classical science, brought with it attitudes that are also lived: awareness of the infinite, awareness of the universe, of Nature's ecosystem and order; emphasis on the individual as world citizen; a global perspective in all disciplines; regarding humankind as one's brother (in accordance with the indications of DNA); living through all levels of the self; affirming the ultimate unity of the Whole, including the reunification of humankind. With such attitudes have come 14 universal experiences which are available to all humankind:

the universal experience of the unity and order of the infinite and the universe via intellect;

the universal experience of "ever-living Fire" or metaphysical Light, being-in-the-One;

the universal drive to self-improvement and self-betterment, which realises possibilities, leads to self-transcendence and helps bring in the next stage in one's development;

the universal order principle's advancement towards higher (or deeper) consciousness;

the universal freedom to break habit and choose one's future through self-transcendence, as the deep subconscious drive to self-improvement is reflected by willpower and acts out coded instructions;

the universal consciousness of purpose and awareness of meaning;

the universal self-transcendence into new levels of consciousness via projects;

the universal being's inner calm and serenity;

the universal inner awareness of one's own purpose;

the universal sense of the unity of all humankind;

the universal order principle's self-organising reflected in one's own life;

the universal order principle's readiness for the One (i.e. being in a state of
   readiness to experience Being);

the universal being-out-of-time (i.e. the mystical sense of the timeless unity of the
   universe);

the universal harmony within all humankind (a political idea of benevolent – not
   malevolent – world government).

To the Universalist, man is an unfinished being. Man is ever in process towards
his possibilities.

Philosophical Summary

What, then, is intuitional metaphysical philosophy or intuitional Universalism?
The intuitional Universalist experiences manifesting "ever-living Fire" or the
bosons of metaphysical Light that to rational Universalism is a concept rather
than a direct experience. He achieves this by vacating the rational, social ego for
the universal being or intellect. He opens to rays of light that are barely
discernible on the electromagnetic spectrum. Seeing Reality is a quasi-empirical
experience which has made a profound impact on those who experience it, such
as Augustine and Dante. The intuitional Universalist gets to know his multi-
levelled self and apprehends Reality directly. The intuitional Universalist
perceives laws of the universe through his intuition and similarly perceives that
the aim of evolution is to intensify consciousness. We have not reached the final
stage of evolution and our consciousness can be expected to develop.

We can now combine the Philosophical Summaries for chapters 9 and 10.
The rational Universalist understands the structure of the universe, a non-
duality, in terms of the concept of the infinite and the surfer and how the infinite
manifests into the finite through the order principle. The intuitional Universalist
experiences the infinite by moving to a different centre within his multi-levelled
self. The new philosophy of Universalism requires a mixture of these two

approaches: now rational, now intuitional.

In Part Three I have rationally stated a system of general ideas which can interpret all our experience of the universe – the experience I listed under 12 levels of consciousness. I have stated a four-tier manifestation process from the infinite to existence in which light condenses into and therefore forms matter, and I have followed it through with a philosophy of the origin and creation of the universe. I have tried to show how light within manifestation enters humans' and animals' eyes, brains and bodies, and plants' photosynthesising cells. Light controls the vast ecosystem of Nature and stimulates a coded drive to self-improvement in all organisms. I have shown that order and randomness co-exist, and that there are various laws within the metaphysical system. I have included our consciousness within the system and have shown how most of the traditional problems of philosophy are solved. These can be approached both rationally and intuitively.

The general system of ideas sees consciousness as split like the body, which is partly autonomic brain and partly bodily function. Consciousness is part finite, autonomic brain, and part infinite, a ring of light. It is a Bose-Einstein condensate, a form of "condensation" of light that is similar to the condensation of matter but it differs as it has an independent existence and survives death. A network of photons, also known as a field of bosons, *aither* or ether, links all living beings, who can absorb information from the infinite across its information-bearing particles. So where do we go when we "pass over"? Through a connecting "tunnel" into a level of Being that is otherwise separated from Existence. Looking into Existence from Being is like looking through a one-way mirror. We in Existence cannot see into Being, but departed spirits can see us in Existence. Spirits may continue in Being and their consciousness may survive.

If part of our consciousness is autonomic and another part can survive death, then the system we live in which includes the infinite, Nature's system and our everyday experience are all understandable and life did not begin as a pointless accident. Rather, the bio-friendly universe and evolution have been shaped by order. If part of consciousness survives death life has a meaning and purpose for it is related to an end. The metaphysical system leaves open the possibility, in a philosophical category, that part of consciousness *does* survive death and rejoin the infinite. This allows individuals to see that their lives have meaning and purpose as they advance through their projects and by self-transcendence towards the goal of life: knowledge of the metaphysical Light of Reality and of the possibility of an after-life – which they can embrace at the level of theological belief but which is nevertheless meaningful in philosophical terms. Metaphysics restores meaning to life and to the developing metaphysics of higher consciousness.

As Reality can be experienced intuitively, known in the intellect, as Light, our consciousness may be a high-frequency system of boson light and its development can be seen as an aspect of an orderly drive that is linked to the Light. The Universalist philosopher holds that the meaning of life among the species is the drive to high (or deep) consciousness, and that the meaning of an individual life is to see it in relation to the full context of life in the universe, as purposeful in relation to its projects. As Light conveys information to consciousness, it may be that we live in an *aither* of tides like sea-sponges in the sea, as filled with Light as sponges are filled with tides. Non-duality sees our bodies and consciousnesses as emerging from the same source, an ocean of Being and Light.

*

It is now time to pull chapters 9 and 10 (so far) together and focus on our aim of reconciling contradictions in modern philosophy.

Reunification of Rational and Intuitional Universalism

It must be remembered that intuitional Universalism is but one wing of Universalism. Universalism combines both rational and intuitional wings, thereby uniting the metaphysical and the scientific. Universalism therefore combines both Classicism, which we saw on p283 was an antecedent to rational metaphysics, *and* Romanticism, an antecedent to intuitional metaphysics. It therefore combines:

trust in the reason, emphasis on the finite universe, a view of Nature as order, clarity, balance, restraint, sense of beauty, a social view of humans via the rational, social ego, objectivity (Classicism/rational metaphysics); and

trust in the intuition and shaping imagination, awareness of the infinite behind the finite, revolt against reason, philosophical tradition and the fragmentation and reductionism of science, a view of Nature as organic dynamic process, a view of humans as individuals, attempting to live in all levels of the self, subjectivity (Romanticism/intuitional metaphysics).

In balancing the two, Universalism brings together conflicting traditions: reason and intuition; the metaphysical and the empirical/scientific; Classicism and Romanticism. In blending the rational and intuitional and in superimposing the intuitional on the rational, Universalism sees that they complement each other as a description of the whole from different parts of the self, one emphasising the finite universe, the other the manifesting infinite, in accordance with algebraic thinking $(+A + -A = O)$. Universalist philosophy thus implements the combination of rational and intuitional metaphysics.

The reunification of rational and intuitional Universalism works through 12 opposite pairs, the apparent dualism being reconciled within the reunification, each contradiction representing $+A + -A = O$:

trust in the reason – analysed fragments of science, and
trust in intuition and shaping imagination – making the universe/Nature whole;

emphasis on the finite universe, and
emphasis on the infinite behind the finite;

a view of Nature as an ordered ecosystem, and
a view of Nature as an ordered system of organic dynamic process;

emphasis on reason and mathematics, and
Expansionist revolt against reason, philosophical logic and language, and the
    fragmentation and reductionism of science;

rational clarity about what is, and
intuitional experience of what is;

sense of rational balance between the infinite and finite and all disciplines, and
revolt against past balance to return to the Whole;

restraint in being objective about Nature, and
emphasis on being subjective about Nature;

sense of the beauty of bio-friendliness and symmetry, and
sense of wonder at the phenomena of Nature;

view of humans as social in the concept of evolution, and
view of humans as individuals experiencing Reality;

rational, social, ego in social activities, and
universal being or intellect in solitary situations but at other times using all
    levels of self;

focus on objectivity, and
focus on subjectivity;

language used as an end, to communicate socially, and
language used as a means to an end, to reveal Being.

The reunification of rational and intuitional Universalism is an amalgamation that both wings can support. It makes possible a manifesto of Universalism, whose 15 main tenets are:

focus on the universe rather than logic and language;
focus on the universal order principle in the universe, a law which may act as a
    fifth force;
the universe/Nature manifested from the infinite/timelessness;
the universe/Nature and time began from a point and so everything is connected
    and one;
the infinite/timelessness can be known through universal being below the
    rational, social ego;
reunification of man and the universe/Nature and the infinite/timelessness;
reunification of fragmented thought and disciplines;
reunification of philosophy, science and religion;
focus on the bio-friendly universe, not a multiverse;
affirming order as being more influential than random accident;
affirming the structure of the universe as unique, its cause being the universal
    order principle from the infinite/timelessness/Void/Being/"*sea*" of energy;
affirming the eventual reunification of humankind;
affirming humankind as shaped by a self-organising principle so it is ordered and
    purposive;
affirming all history and culture as being connected, and one-world government

and religion;

affirming that life has a meaning.

Rational and intuitional Universalism can now restate metaphysics as a blend and unification of their two wings:

ontology – studying a static concept of Reality in the rational, social ego, and experiencing Reality as a dynamic, moving process through the universal being or intellect;

psychology – seeing the self in terms of the rational, social ego and seeing a multi-levelled self in a dynamic process whose variations can open to transpersonal spiritual experience;

epistemology – how the mind knows Reality as a concept through the reason and rational, social ego, *and* how the intellect experiences and knows Reality as an experience of metaphysical Light;

cosmology – the structure of the physical universe, and esemplastically shaping the fragments of scientific knowledge to return the fragments to a whole.

In this reunification intuition is fundamental and reason works in the service of intuition. Intuition sees but does not know how to say what it sees. Reason says but is limited in what it says, and is reinforced by the perceptions of intuition.

The intuitional approach can add experience to the 20 conceptual problems rational philosophers have pondered since the Presocratic Greeks (see pp288-289). For example: personal identity – to the intuitional philosopher, the full range of the multi-levelled self at whose deepest level is the experiencing universal being or intellect.

Reunification of Science and Metaphysics

Bergson in his *Introduction to Metaphysics*, 1903, called for "a much-desired union of science and metaphysics". Metaphysical science is science that connects to every possible concept, not just to every existing concept, and which therefore relates to the infinite. In the Middle Ages science and metaphysics were united.

The foundations of most modern objectivist, positivist and reductionist science have a Materialist ontology, for most science sees the phenomenal world as the only reality. Its epistemology is therefore sense data and, more recently, mathematics. The foundations of holistic science have a wholeness ontology, seeing the wholeness of the finite universe and its interconnectedness as the only reality. Its epistemology is intuitional inner knowing of interconnectedness within the finite universe. The foundations of a new metaphysical science, such as the one for which Bergson yearned, have a metaphysical ontology for it sees the infinite as the source of Reality, from which all else has manifested. Its epistemology is inner knowing of the metaphysical Light which manifests from the infinite. The foundations of a new metaphysical science, its ontology and epistemology are quite different from those of reductionist and holistic science.

A metaphysical science will question and challenge 150 years of reductionism and Materialism in the sciences and philosophy, and will carry metaphysics into many areas of science and philosophy. Metaphysical science should research, test, assemble evidence and do mathematics on the following 10 hypotheses:

(1) The origin of the universe: seeing the hot beginning (Big Bang) in relation to the "ever-living Fire" or metaphysical Light.

(2) Reality in the microworld: seeing Newton's expanding force, Einstein's cosmological constant, dark energy (if it exists), the principle of hidden variability and the origin of mass as effects of the manifesting *aither*-like metaphysical Light's varying intensities in different localities.

(3) Electromagnetic spectrum: seeing the high-frequency, gamma end of the spectrum in terms of the neutrino-like or photino-like bosons of manifested metaphysical Light.

(4) Brain physiology: seeing consciousness as a system of high-frequency boson light which connects to all rays on the electromagnetic spectrum in the region of 4 cycles per second.

(5) Mind-body problem: seeing the mind and consciousness as transmitted photon-like bosons bearing coded information that are not dependent upon physiological processes but which use the brain transmissively for their own metaphysical purposes, i.e. seeing the mind as independent of the body.

(6) Synchronicity: seeing coincidences as evidence of a unified interconnectedness of matter and consciousness through the *aither*-like manifesting Light.

(7) Mysticism: seeing religious experiences and mystical states of consciousness as contact with the metaphysical Light, which manifests into and pervades the universe, pouring universal energy into the universal being or intellect of humans.

(8) Order principle: seeing the infinite and Being (the metaphysical Light which can be known intuitionally or existentially through contemplation) as an ordering principle which sends photons into retinas and plants' chloroplasts, whence they pervade DNA.

(9) Evolution: within order as opposed to randomness, seeing all evolution in terms of the order principle and teleological evolutionary power of the neutrino-like, photino-like boson particles of manifested metaphysical Light, and its influence over a low-level "mechanism" that controls DNA exchange and chromosome division and evolution's drive for self-improvement and self-betterment.

(10) Philosophy: seeing philosophy as subject to metaphysical science, as framing a system of general ideas that include all possible concepts, the whole universe at all hierarchical levels and life as a whole, in terms of the manifesting infinite's metaphysical Light.

Through this agenda metaphysics returns science to its metaphysical foundations in the infinite. Metaphysical science will move science away from granular Materialism where visible matter occupies only 4 per cent of the matter in the universe, and away from reductionism. The discovery by CERN of a new ordering invisible Reality in the microworld would overthrow Heisenberg's uncertainty principle and explain how particles have their mass. It is an idea waiting to be discovered and verified, just as the philosophical idea of the atom waited 2,500 years after it was first proposed to be discovered and verified as a scientific entity in the 19th and 20th centuries.[24]

Metaphysical Revolution: Five Reunifications

Today's unifications seem new, but when seen in the historical context of the last 2,500 years they can be seen to be reunifications. The reunion of science and metaphysics will be a further step towards the reunion of science and philosophy; the reunion of the metaphysical and scientific traditions within philosophy; the reunion of the rational and intuitional philosophies within metaphysics; and the reunion of the linguistic and Phenomenological views within modern philosophy.

We have just been examining science and metaphysics. As regards science and philosophy, the Universalist philosopher, setting to one side new dimensions which cannot be proved outside mathematics but affirming the infinite for which there is scientific evidence, focuses on what science has not been able to prove as much as on what it has proved. As we saw in Part Two, the universe cannot be explained without an infinite singularity at the beginning – infinity – and the Universalist philosopher now sees the same unity behind the universe that the Greeks saw, which pre-existed the Big Bang in the initial point. Having surveyed the scientific evidence, the Universalist philosopher knows that a hidden Reality associated with universal order, which science has not yet measured but for which there is overwhelming evidence, is crucial to a new model of the universe which philosophy must reflect. I have endeavoured to import the scientific view of the universe set out in Part Two into my new philosophy of Universalism.

As regards the union of metaphysical and scientific traditions within philosophy, I have endeavoured to blend the metaphysical tradition of chapters 9 and 10 with the scientific view of the universe set out in Part Two to achieve that union.

It can now be seen that the scientific universe is a manifestation of the metaphysical "boundless", and the coming-together and union of the two divergent strands of the philosophical tradition can be seen as the metaphysical tradition's subsuming of the scientific tradition. It can also now be seen that the appropriate term for this union is "metaphysical" as metaphysical Reality manifests into scientific reality. It is therefore appropriate to describe this union in terms of "the metaphysical view of the universe".

Within this unification I have endeavoured to unite the rational and intuitional metaphysical philosophies, which I dealt with in chapters 9 and 10.

As regards the union within modern philosophy, I have endeavoured to bring together Linguistic Analysis and the Phenomenological view of Being. Language will focus on a precise description of Being's ordering of the universe and Nature, the phenomenal world of finite Existence. Language will therefore become outward-looking and will be a means to an end, to reveal infinite Being as well as finite Existence. Meaning and symbolism have layers that correspond to the infinite and finite levels and the 12 levels of consciousness. Literature, and in particular poetry, should contain levels of Being *and* Existence, as does Wordsworth's *Prelude*. Linguistic analysts will of course use language with their customary precision, but in an outward-looking way.

Universalism's aim is not to reject or refute but to bring out the one-sidedness of different approaches by integrating them into a more encompassing frame. It may appear to reject or refute one-sidedness but its aim is to reconcile and unify contradictions and opposing views within a systematic Grand Unified Theory of Everything.

To sum up, Universalism has reconciled five sets of contradictions:

science and metaphysics – through metaphysical science;

science and philosophy – by reconciling Part Two and chapters 9 and 10 (which are based on Part Two);

metaphysical and scientific traditions (Plato's and Aristotle's traditions) – by reconciling infinite Reality and the scientific view of the universe (chapters 9 and 10 and Part Two);

rational and intuitional philosophies within metaphysics – by stating rational metaphysics and overlaying intuitional metaphysics as if chapter 10 were laid over chapter 9;

linguistic and Phenomenological philosophies – by reconciling language and Being so language creates neologisms to define and describe the infinite perspective with precision and accuracy.

This process amounts to a Metaphysical Revolution. In so far as the reconciliations seem new, they constitute a Metaphysical Revolution, but when seen in the historical context of the last 2,500 years they can be seen to be a Metaphysical Restoration. I formulated a Metaphysical Revolution to return philosophy to metaphysics as long ago as 1980 and declared it in 1991. The Metaphysical Revolution will continue to work for the reconciliation of the five sets of contradictions.

Of movements, Whitehead wrote in *Introduction to Mathematics*, 1911[25]: "Operations of thought are like cavalry charges in a battle – they are strictly limited in number, they require fresh horses, and must only be made at decisive moments." The decisive moment is long overdue to effect a sudden and fundamental change in science and philosophy. The Universalist movement has begun at a decisive moment, and its Metaphysical Revolution is its crusading banner.

A New Discipline

Our philosophical view of the universe has reunified knowledge and scientific disciplines which have been separated since the Renaissance. All the disciplines in Part Two interconnect and should be treated as one interdisciplinary subject. They can no longer be taught in isolation from each other. Focusing on science today has brought philosophy (in all its branches) back to its original concern with the universe in the time of the Greeks. The new philosophy which has returned to the Presocratics at the origin of Western philosophy must be taught in conjunction with the disciplines in Part Two.

There should be a new discipline in the university curricula: the new philosophy of Universalism, which reunifies these five contradictions and has restored the universe and Nature to the foreground of philosophy. This new discipline will include the terrain of Part Two: cosmology, astrophysics, physics, biology, geology, plate tectonics, Ice Ages, ecology, biochemistry, physiology, psychology and most importantly, early Greek and Universalist philosophy. It will teach undergraduates about the origin of the universe and of life, the conditions in which evolution developed and the rise of intellectual consciousness, and – within order as opposed to randomness – all will be considered within a context of possible purpose.

This new course will deal with the contradictions in the universe and in disciplines, and with their reconciliation. It will focus on the unified vision. While it will balance enquiry with scepticism, its applications in wider disciplines can offer hope to new generations of students throughout the world.

PART FOUR

ORDER IN HUMAN AFFAIRS: APPLICATIONS OF UNIVERSALIST THINKING

"All are but parts of one stupendous whole,
Whose body nature is, and God the soul."
Pope, *An Essay on Criticism*, lines 267-8

11

New Understanding of the Past

This study would be incomplete if it did not briefly cover the applications of Universalism in non-scientific or environmental disciplines. For the scientific view of the universe and the perceptions of Universalist metaphysical philosophy alter our perception of several disciplines.

Universalism sees everything as having its origin in a microscopic pre-particle that became a particle and inflated following the Big Bang. Perceiving the universe of cosmology, physics and biology, all life and humankind as being interconnected and unified because they originated in one actual, literal point enables Universalists to see more recently developed areas of human life as interconnected and unified because they originated from a common metaphorical point.

From this Universalist perspective, history, culture and literature, through which we understand the past, must all be seen as having a common origin and therefore as interconnected and unified, as universal and global, as supra-national. History involves the entire history of humankind from the first recorded event to the present, and similarly culture and literature must be seen as wholes, covering the same space-time span. Universalism, seeing the whole, helps us understand the past in these three disciplines.

Historical Universalism

As all history emerged from one point that preceded the Big Bang and then from the first cell and the first human historical event, historical Universalism perceives history as one, an interconnected unity, a whole.

The history of a particular country cannot be studied in isolation from the history of the rest of the world as that would be partial history. History is the combined events that have befallen humankind as a whole from the recorded beginnings to the present. And so, history has a worldwide pattern.

To his great credit, Arnold Toynbee writing in the tradition of Edward Gibbon, who charted the decline of the Roman Empire, and of Oswald Spengler, who wrote of the decline of the West, saw this and wrote of world history as a whole. He was a giant who was before his time, perhaps because he worked for the British Royal Institute of International Affairs, an advocate of world government, and was a leading figure at Chatham House from 1925 to 1955.[1] He saw rising and falling civilisations in a very regular pattern, and attributed their motive force to "challenge and response". In 1954 he completely changed his 1934 scheme of civilisations, having lost confidence in his original scheme. He wrote: "I have been searching for the positive factor which within the last 5,000 years, has shaken part of Mankind...into the 'differentiation of civilization'....These manoeuvres have ended, one after another, in my drawing a blank."[2]

In my historical work,[3] I have examined the rise-and-fall pattern of 25 civilisations, and have found what I consider to be the law of history: that they go through 61 similar stages. Each begins as a culture in which there is a vision of Reality as the "ever-living Fire" or metaphysical Light, a vision which is received by a mystic (such as Mohammed, who saw the first page of the *Koran* written in Fire). People gather round the vision, and a religion forms round it. A new civilisation forms as the vision is taken abroad and the people expand. The civilisation grows as the vision is renewed (as in the quotations on pp295-296 of Augustine, Hildegard and Dante). So long as the civilisation follows its metaphysical central idea, its growth is strong. A civilisation is healthy so long as the vision of the metaphysical Light is seen and is central.

When the civilisation ceases to believe in its Reality, it turns secular and goes into a long decline which ends in conquest and occupation by foreign powers, or other civilisations. Eventually it passes within a successor civilisation, whose gods replace its own gods. The civilisation ends when it has passed into another civilisation (as the ancient Mesopotamian and Egyptian civilisations passed into the Arab civilisation).

In my study of history, living civilisations can be overlaid on the dead ones that have been through 61 stages. Overlaying reveals where the living civilisations are now. The European civilisation is found to be two-thirds through, about to enter a Union (the European Union). The Byzantine-Russian civilisation has just moved out of this stage (Communism) and is now in federalism. The North-American civilisation is a quarter-through in a globalist stage – the equivalent stage of Roman Empire when it ruled the world.

In all secular stages of all civilisations, cultures and times, human beings have encountered similar problems. In the early stages of civilisations, metaphysical visions of Being have been numerous (as in the monastic Dark Ages of the European civilisation). In the later stages of civilisations, Being tends not to be seen: the prevailing climate is one of scepticism, secular Materialism. In such times, thinkers keep their civilisation alive by reminding their fellow-citizens of the "ever-living Fire" or metaphysical Light that inspired it. They are against the trend of their day, but they ensure that the civilisation's fundamental vision is handed on to the next generation. In the long term, they are therefore fundamental to the health of their civilisation. There is no space here to develop this view, which can be read in other works.[4]

My philosophy of history is Universalist as it shows the vision of Being, or "ever-living Fire" or metaphysical Light, as being central to history. It is the motive force which Toynbee sought in vain. He looked for it within history and did not find it because it is outside history: the Reality perceived by intuitional lone mystics which is later followed by kings, generals and their economists, whose struggles for succession, wars and material prosperity form the basis of most history books. My philosophy of history is based on a movement away from partial history to "whole history". Its approach is both rational (in setting out the pattern in history) and intuitional (in sensing the whole, esemplastically piecing the fragments of civilisations back into a whole "urn"). In short, in my philosophy of history I have charted a historical approach to Being, which I see as central to religions and therefore to civilisations. On this view, history is the consequence of the vision of Being.

Civilisation's intellectual metaphysical vision becomes corrupted and turns secular. This is paralleled by a more debased vision of Utopias which begins revolutions. In revolutions one class or part of a civilisation tries to eliminate another class – as the French bourgeoisie and proletariat tried to eliminate the aristocracy. Revolutions end in massacres (purges and guillotinings).[5]

The rational progress of the Enlightenment, which came to an end, shocked by the barbarity of the French Revolution, was opposed to revelation, preferring reason – which is another way of saying that it was rational and not intuitional. The rational Enlightenment saw history as a rational progress towards perfection, a view neo-Darwinists took up and applied to evolution, seeing history as driven by natural selection and therefore accidental, chance, chaotic and purposeless.

The view that visions of Being cause civilisations to rise and that absence of this vision causes them to decline is anti-Enlightenment. This view sees history as having an underlying order, a rise-and-fall pattern in which can be detected the workings of the universal principle of order and pattern – and perhaps an underlying purpose and meaning: to embody the idea of Being and metaphysical

Light in religious buildings such as temples, cathedrals and mosques.

Universalist history, then, focuses on all humankind's history and therefore on all civilisations, their patterns of rise-and-fall and how their stages affect human beings in their lives. It is history in which the vision of the "ever-living Fire" or metaphysical Light is central to all cultures.

Being is seen in all cultures and ages, and historical Universalism focuses on the impact such visions of Being have had on stages in their civilisation's pattern. Each civilisation has local customs, traditions and differences while reflecting a universal global theme, and the whole of history has a richly unified and varied pattern that can reunify the fragmented discipline of history. In view of the coming emphasis on globalism, historical Universalism's time is coming.

Cultural Universalism

As all culture emerged from one point that preceded the Big Bang (via life and consciousness), and from the first cell and the first appearance of human culture, cultural Universalism perceives all cultures as one, an interconnected unity, a whole.

A culture is a people's common way of living. Edward Burnett Tylor wrote in 1871: "Culture…is that complete whole which includes knowledge, belief, arts, morals, law, custom, and any other capabilities and habits acquired by man as a member of society."[6] This complex whole grows out of primitive family groups and includes a people's behaviour: their language, ideas, attitudes, values, ideals, traditions, beliefs, customs, laws, codes, institutions, tools, technologies, techniques, material objects and works of art, and also their rituals and religious or sacred ceremonies. The complex whole includes writing, tool-making, weaving, potsherds, horror of incest and belief in spirits.[7]

The culture of a particular civilisation is its people's customs and artistic and intellectual achievements. It is a people's symbolic forms of expression, whereas civilisation includes buildings, technology and all the material things that contribute to a people's way of life. Culture cannot be studied in isolation from the cultures of the rest of the world. A particular civilisation's culture is part of one cultural impulse within civilisations, a one-world culture with common ground for all humankind and with local variations and differences. There is one supra-national culture throughout the world.

Just as we have seen on p265 that manifestation is in four tiers in a descending hierarchial order, so there is an ascending hierarchical order in the sciences which deal with different levels of human and social existence and of what exists. If the physical sciences focus on matter and especially at present on quantum gravity, and biology focuses on life and proposes an evolutionary synthesis while psychology focuses on mind and behaviour, the social sciences and humanities (the study of society, social relationships and human aspirations) focus on culture. Culture is therefore linked as a global force to the social sciences (such as economics, political science, sociology, social and cultural anthropology, social psychology and economic geography) and to the humanities. Although culture is a manifestation of a people's social organisation, it is also linked to individual aspiration treated as a social and cultural phenomenon. It is therefore closely identified with individual mystical experiences, the social organisation of religion, and with the expression of human culture in art, sculpture, music, literature and philosophy.

In each civilisation, during the growth phase when its vision of Reality or metaphysical Light is strong, its culture is unified round the central metaphysical idea like branches (to adapt Descartes' image) growing from a central trunk, which is the civilisation's religion – in the European civilisation, Christianity. During this early stage the trunk of the religion is fed with metaphysical sap and its branches – art, sculpture, music, literature and philosophy – all express the metaphysical vision of Being that is found in religion. Its leaves, individual works, express sap. There is unity of culture and "unity of Being".

In the European civilisation, during the Renaissance the philosopher Marsilio Ficino and the artist Sandro Botticelli shared Dante's metaphysical and religious vision of Being. The central idea of our European civilisation and culture, and of all civilisations and cultures, is the metaphysical vision of Reality as the "ever-living Fire" or Light (expressed in European art as the halo) which is beyond the world of the senses but knowable within the universal Whole.

When the metaphysical sap stops, the branches grow brittle and the culture is fractured and fragmented. Art, religion and philosophy cease to be filled with metaphysical sap and, deprived of natural vigour, start falling apart. When the sap fails, the branches turn dry, sere leaves fall and the civilisation and its culture decline.

The state of a local culture depends on the state of its civilisation. When civilisations rise and are healthy, they embody the One, Being, which is reflected first in the mystical vision and later in rational philosophy, as was the case with the Ionian Presocratics like Anaximander and Parmenides, whose works were on one of the branches from the Iranian tradition of the religion of Ahura Mazda. When civilisations lose contact with the One, with the metaphysical Light, there is cultural decline, and art, religion and philosophy turn secular, deal with surfaces and no longer embody the sap in the civilisation's religious trunk. The culture becomes shallow and unhealthy, and declines.

The process of secularisation has taken place in European civilisation today. The metaphysical sap of Christianity began to fail with the Renaissance and was drying up at the end of the 17th century. Since that time 50 "isms" or doctrinal movements have arisen, representing secular, philosophical and political traditions that demonstrate fragmentation, loss of contact with the One and disunity within the declining European civilisation. They are:

humanism	accidentalism
scientific	nihilism
revolution/reductionism	communism
mechanism	conservatism
Rosicrucianism	imperialism
Rationalism	totalitarianism
Empiricism	Nazism
scepticism	fascism
Atomism/Materialism	Stalinism
Enlightenment/deism	pragmatism
Idealism	progressivism
Realism	Phenomenology/Existentialism
liberalism	stoicism
capitalism	vitalism
individualism	intuitionism
egoism	modernism/post-modernism
atheism	secularism
radicalism	objectivism
utilitarianism	positivism
determinism	analytic and linguistic
historicism	philosophy/logical empiricism
nationalism	ethical relativism
socialism	republicanism
Marxism	hedonism/Epicureanism

anarchism/syndicalism   structuralism/post-structuralism

Darwinism   holism

These "isms" or doctrinal movements were unthinkable before the Renaissance because Christendom was unified as the medieval philosophy curriculum demonstrated. The diversity of the "isms" or doctrines reveals the multiplicity within which humans now live. They are all ways of doing things for all people or attitudes based on one particular faculty of consciousness's many faculties and multi-levelled self. Few offer ways back to Reality, the One. Indeed, the majority suggest that today, in Eliot's words, "human kind/Cannot bear very much reality" ('Burnt Norton'). Deism, Rationalism, Idealism and Existentialism are all flawed because they engage parts of consciousness, not the whole of consciousness, as we have seen. Holism applies to the universe and Nature, but not to the infinite or timelessness.

I have not included fundamentalism or unitarianism, which may provide routes back to the One. I have not included Classicism and Romanticism, which are reunified in, and therefore aspects of, Universalism and provide rational and intuitional ways back to Reality. I have not included Expansionism for the same reason, and of course Universalism itself, which was an "ism" before its new meaning in this book of "focusing on the universe, establishing the universal order principle, reunifying disciplines, focusing on mankind as a whole and, through universal being or intellect, contacting the One". In its quest to restore contact with the universe, Nature, the One and unity it counteracts many of the other "isms", and is the "ism" that most offers a way back to Reality.

The Universalist counteracts the "isms" by conducting a "Universalist Revolution in Thought and Culture". The Metaphysical Revolution is a restoration of the metaphysical, mystical tradition alongside science and in opposition anti-metaphysical humanism, Materialism, Rationalism, Empiricism, scepticism, mechanism, positivism and reductionism. The Universalist is discontented with the existing cultural order and opposes it. He calls for support to start a necessary change, a revolution in thought and culture, so that true values are reflected in the metaphysical-secular spectrum of our culture, which should show symptoms of cultural health rather than terminal disease.

The Universalist Revolution declares:

(1) The phenomenal world of the senses is not ultimate Reality. The infinite

is behind the finite universe. This perception challenges and sweeps aside humanism, Materialism, Rationalism, Empiricism, scepticism, mechanism, positivism and reductionism in all disciplines.

(2) We can know infinite Being, or the One, beyond and behind the phenomenal world of the senses – rationally via what the surfer breasts, and intuitively via the vision of Reality, of the "ever-living Fire" or metaphysical Light.

(3) Each one of us is therefore not a reductionist collection of atoms on a dunghill whose mind is mere brain function or a solely social ego, but a being with 12 levels of consciousness and a possibly immortal spirit (or invisible body) within his or her visible body.

(4) Universalism states the universe in terms of the infinite and science, and particularly in terms of metaphysical Being, or "ever-living Fire" or Light, which manifests into the structure of the universe. It is a science that studies the structure of the universal Whole, which includes every possible concept of the mind and the metaphysical layers of manifesting Being. It also studies perception of the One, the intellect's experience of the Fire or Light which can be reported phenomenologically (through a study of consciousness or perception) in "self-reports" which are quasi-empirical.

It is also a practical, contemplative, intuitional philosophy that "existentialises" metaphysical Reality, whose universal energy can be known existentially in the contemplative vision. It contacts the universal energy of the "ever-living Fire" or metaphysical Light, Reality, and applies its consequences to all disciplines.

(5) The central idea of all civilisations and cultures is the vision of Reality as "ever-living Fire" or metaphysical Light, which is beyond the world of the senses but knowable within the universal Whole.

(6) This documented vision of Reality should be reinstated in philosophy to reconcile, unite and move beyond Logical Positivism, Linguistic Analysis and Phenomenology.

(7) The vision of Reality has inspired the growth of all civilisations in history. History studies the universal Whole and should have a global perspective. The metaphysical vision which is common to all civilisations is the best basis for a common world culture.

(8) This vision of Reality should have a place in the spectrum of literature – novels, plays and poems – and exist alongside and challenge secular, technique-oriented literature. Literature misleads if it conveys an exclusively

surface view of life, if it assumes that appearances are all and does not hint at Being or Reality. Literature should be truth-bearing and glimpse or reveal metaphysical Being or Reality.

Universalist literature is neo-Baroque as it combines the metaphysical and secular and unites the world of the senses and spirit by seeking the sunburst experience and unity It reveals harmony between apparent opposites: sacred and profane, regular and irregular, order and disorder, stillness and movement, Being and Becoming, eternity and time. Universalist literature combines Classicism and Romanticism, statement and image, social situations and sublime metaphysical vision, traditional and organic form.

(9) Universalist philosophy, history and literature offer a vision of harmony, meaning and order in relation to the universal Whole after the *angst* and anxiety of 20th-century thought, particularly in modernist literature and Existentialist philosophy. Universalism emphasises the contemplative gaze, union with the universal Whole and the rustic pursuit of reflection amid tranquillity, the vision of mystic writers, artists and sculptors.

(10) European artists – practitioners in painting, music, architecture and sculpture as well as in literature – should transmit the sap of the cross-disciplinary vision of the One "ever-living Fire" or Light in their works and connect themselves to the Universalist Revolution's revitalisation of the European civilisation's central idea. (Vincent van Gogh's swirling starry night skies attempted to do this in his own way: "I am painting the infinite." See p302.) They will thereby contribute to the return of a common, unified European culture. This will affect the North-American culture and the whole of Western and global culture.

(11) European culture needs to be re-formed to restore (to use Matthew Arnold's words) the most perfect works from the best self, which constitute the highest expressions of culture, and to reconnect philosophy, history and literature to the tradition of the unity of vision of the "ever-living Fire" or metaphysical Light. The Revolution should have a hearing, which will spread awareness of the consequences for European culture of the Revolution's fundamental shift in perception. This will affect the North-American culture and the whole of Western and global culture.

(12) On the Universalist principle that all the metaphysical ideas of all civilisations are essentially the same vision, a revival of the common metaphysical vision in the European civilisation and culture will be essentially the same as corresponding revivals in all civilisations and cultures, which,

however, are at different stages of advancement. The revival within the European culture can therefore inspire an international Universalist movement to focus on the common metaphysical ground of all cultures and create a world culture, which can be a force for world peace. This will affect the North-American culture and the whole of Western and global culture.

The implementation of this 12-point Declaration will have the following seven consequences in modern European thought and culture:

(1) Renewal within our secularised, humanistic culture, which is in terminal decay.

(2) Restoration of a vision reflected in all the European arts over hundreds of years, round which all civilisations have grown and whose renewal revives our culture and revitalises our civilisation. This will affect the North-American culture and the whole of Western and global culture.

(3) Identification of the unifying principle in the universe of physics and the formulation of a Form from Movement Theory (or a full Grand Unifying Theory).

(4) Renewal of philosophy through a new metaphysical philosophy, Universalism.

(5) Introduction of a new global perspective in history through a new history which takes account of all civilisations and cultures, not just slices of nation-state history, and identifies the unifying principle in all civilisations.

(6) Restoration of the essential European vision in literature through a new literature which mixes the metaphysical and secular as did Baroque art (thus neo-Baroque literature), which draws on many disciplines and reflects our Age. This will affect the North-American culture and the whole of Western and global culture.

(7) Revival of culture by a group of practitioners who revive the essential European vision in their work, acting like a Pre-Raphaelite Group. This will affect the North-American culture and the whole of Western and global culture.

My Declaration exists both at the level of contemplation and at the level of action. It is a programme for change which demands reflection and requires action.

The Universalist restores the culture of his own civilisation, and in so doing makes a contribution to restoring the health of the world culture. The vision of Being in all cultures and ages is a combined tradition that brings a sense of meaning and purpose to cultural improvement today.

Universalism restores a one-world culture as the common ground for humankind. By conducting a "Revolution in Thought and Culture" against modern secularised culture, with a view to making the metaphysical sap flow again in the trunks, branches and leaves of all the particular civilisations, Universalism restores cultures to health they once enjoyed. The common ground cultures enjoy is: the concept and experience of Reality; the infinite; the scientific view of the universe and Nature's ecosystem, which includes the metaphysical and scientific philosophical traditions; and the shared experience of world history, its civilisations and cultures.

Just as Universalism sees world history as a unity, so it sees world culture as a unity, a combination of all the civilisations' religions, philosophies and arts – a one-world culture. Its method of proceeding is both rational (in setting out a world culture of all civilisations) and intuitional (in sensing the whole, esemplastically piecing together the fragmented cultures into one world culture). Universalism's reunification of philosophy and science at five levels and its restoration of the metaphysical that surrounds and permeates the universe are themes of all cultures.

Literary Universalism

As all literature emerged from one point that preceded the Big Bang, from the first cell and from the first instance of human literature, literary Universalism sees all literature, the literature of all countries – as one, an interconnected unity, one supra-national literature. The poetry of each country is part of all nations' poetry, and all nations' poetry is one universal poetry that, at its deepest, reflects the infinite and timeless One.

In our time literature has been reduced to a materialist and social-humanist outlook with few aspirations or great insights. In poems, plays, short stories and novels, intuitional and rational Universalist *literati* express *experiences* of Being intermingled with rational interpretation. Revelations of Being are inevitably occasional. Much of Wordsworth's *Prelude* is about everyday life and social situations rather than about epiphanies of Being in mountain scenery, and there is a considerable amount of rational interpretation. Nevertheless, Universalist writers bring Being back into human life and focus on the existential contact between the individual and Being amid rational narrative, as distinct from rational philosophers' *thinking* about Being. Universalist literature focuses on experiences and revelations of Being in literary works, experiences which give added meaning to lives and suggest that human lives have purpose.

I have said that the intuition esemplastically approaches the One and perceives unity whereas the reason's analysis sifts and makes distinctions, disunites and separates into parts. Universalist literature, including poetry, appeals to the intuitive faculty and reflects the oneness of the universe, but it also appeals to the reason as it interprets experience rationally in a blend of the intuitional and rational, Romanticism and Classicism, and the opposites of sense and spirit in the Baroque Age. Universalist literature is the successor to Vitalism and Existentialism on the intuitional side, and to Empiricism on the rational side as it emphasises the empirical nature of the experience of Reality and of the scientific view of the universe.

The Universalist poet reshapes and restructures the apparently chaotic universe into a structure of order in the act of writing a poem. It could be said that he lets the universal order principle into literature. In ordering the universe, he approaches, or puts himself in readiness for, the One and receives back from it a glimpse or reassurance that the universe is an ordered one, with a purpose and a meaning. The Universalist poet restates the order in the universe in both intuitional (experiential) and rational (interpretative) terms.

Universalist poets who are imbued with this sense of order include Homer, Virgil, Dante, Shakespeare, Marlowe, Wordsworth, Shelley and Eliot. The worlds of Greek and Roman epic are presided over and ordered by the gods, and the world of Dante by God. Shakespeare has a profound sense of order, the disruption of which affects the "great chain of being". Marlowe sees order in terms of Heaven, Wordsworth in terms of the "Wisdom and Spirit of the universe". Shelley's sense of order can be found in 'Adonais' (which is about the death of Keats) – "the One remains, the many change and pass" – and Eliot's sense of order can be found in *The Four Quartets*, where (in 'East Coker') he wrote of "the still point of the turning world". Dostoevsky dealt with the nihilists who attempted to overthrow order, and their defeat reinstates the idea of order. Tolstoy relates war and peace to the fundamental order in society.

The forms or genres of a Universalist man of letters all reveal Being. Universalist poems investigate the universe with precise language and catch intimations of unity, of the presence of the One, Being. They offer sudden revelations of Being and capture Being in the moment. Epic relates the extremities of war and peace to notions of Heaven and Hell. Verse plays link order to the Being behind the divine right of kings, and to awareness of Being. Autobiography, diaries and short stories link the everyday to growing awareness of Being and contact with Being.

All these forms can be used to probe and investigate the universe. Just as miners find one drill is useful for one kind of coalface, and another for another, so different literary forms or genres are appropriate depending on the particular coalface of the universe the writer drills.

As the Nothingness, or All or One, manifests through Non-Being to Being and Existence, there are layers or tiers or multiple states of Being and of consciousness which reflect them. Humankind is presented against a complex, layered, tiered universe. Universalism ranges over all lives and cultures and focuses on the inward Being of the "ever-living Fire" or Light.

In conclusion, Universalism sees human order and organisation creating one-world history, one-world culture and one-world literature. Each member of humankind lives within one of the surviving civilisations that combine local differences with global unity. Universalists hold that the arts and literature should reflect the reunification of humankind. Besides being a new view that makes possible a new understanding of the past, Universalism also makes possible a new view of the pressing future which our world is bringing in.

12

SHAPING THE FUTURE

Universalism has applications in contemporary global issues which will shape the future. Increasingly governments are seeing the need for a global solution to the earth's problems, all of which are interconnected. Action by only a few governments on such matters as climate change and the distribution of the earth's resources to include the poor will always be ineffectual and inadequate, and only a pan-world effort will make a difference.

This view accords with Universalism's view of every human as a global citizen. Just as all cultures collectively form a world culture, so all religions collectively form a world religion, and a pan-world solution to conflicting religions would make some headway in finding a solution to the problem of religious strife, which is behind many of the problems of the present time. Universalism regards all religions as being potentially one religion. Similarly, Universalism regards all governments as potentially collectively forming one world government that will deal with the world's most immediate practical problem: its response to climate change.

Environmental Universalism
As the earth and our environment emerged from one point that preceded the Big Bang, environmental Universalism sees all environmental developments as being aspects of one process and interconnected. Perceiving the earth's environment as one, Universalists affirm a one-world, or to put it in its wider context, a one-universe environmental movement.

An astronaut in space sees our planet as one. We have seen from the 40 bio-friendly conditions in chapter 5 that the earth cannot be said to have an exclusively accidental, chance environment. I have already referred to the idea (on pp172-173) that the earth might have an equilibrium and an environment that regulates itself by homeostasis. Such a process would be an environmental equivalent of the homeostasis that regulates the body and brain.

This equilibrium is hard to detect or corroborate in a time when the ozone layer has been depleted by CFCs, when greenhouse gases have caused global warming, when there has been a rapid growth in human population and when short-term warming conceals long-term cooling. Much of this obfuscation is due to human contamination.

The depletion of the ozone layer by CFCs was discovered in 1985,[1] when the British base at Halley in the Antarctic Peninsula discovered that the ozone layer above the Antarctic had a hole in it the size of the United States, Canada and Mexico.[2] This hole has continued to grow. The depletion in the stratosphere has been found to have been caused by artificial chemicals such as chlorofluorocarbons (CFCs) and halons in the northern hemisphere.

CFCs were first marketed by the DuPont corporation in refrigerators in 1929,[3] and they were used as aerosol propellants for hair, paint, insect sprays, air fresheners, foam, packing and insulation after the Second World War. Although forming less than one per cent of the atmosphere,[4] the ozone in the stratosphere filters out and restricts the amount of ultraviolet-B (UV-B) radiation reaching the surface of the earth. High levels of UV-B have been saturating Antarctica and the Southern Ocean in spring and early summer, threatening phytoplankton on which the Antarctic marine ecosystem depends. A band of shorter wavelength UV-C has also appeared.[5]

Just under one quarter of the earth's atmosphere is oxygen.[6] The CFC molecules rose to the stratosphere and dislodged ozone (which is three atoms of oxygen bound together). The atmospheric radiation separates chlorine from a CFC molecule. Tiny ice crystals form in the stratosphere. These crystals provide places where chlorine is separated to destroy ozone. The chlorine captures one of the three oxygen atoms, turning ozone into oxygen. The chlorine-oxygen alliance is swiftly broken when another single oxygen atom takes the oxygen atom and forms another oxygen molecule. The chlorine then looks to destroy another oxygen molecule and this pattern is repeated over and over again.[7]

Under the Montreal Protocol of 1987, an international treaty, the production of CFCs has been eliminated.[8] However, as a result of the backlog of CFCs in the atmosphere about 50 per cent of ozone is missing in the Antarctic each spring whereas only 5 per cent is missing in the Arctic. Scientists have found that the protective ozone over the South Pole will take until 2068 to recover.[9] It is expected that there will be an ozone hole every spring for the next hundred years, and that the global effects of the ozone hole will remain significant in a thousand years' time. The ozone problem in the Antarctic may cause the southern hemisphere to have a new Ice Age – or rather glacial.

That greenhouse gases cause global warming has been suspected since the 1940s. The sun's radiance and radiation warms the earth, and the earth's temperature allows it to radiate away the radiant energy it receives from the sun. About 50 per cent of solar energy – light and heat – reaches the ground, of which 70 per cent is absorbed by rocks, the vegetation on continents and the oceans, warming the earth, while 30 per cent is reflected back into space by bipolar ice.[10]

When water vapour, carbon dioxide, methane and other gases which together make up under 0.1 per cent of the atmosphere are exposed to infra-red radiation, they form greenhouse gases, trapping the earth's radiating-out of part of the sun's energy. About 60 per cent of global warming is caused by the greenhouse effect and 16 per cent by methane. Carbon dioxide, which formed 98 per cent of the early earth when the air temperature was 290°C, only forms 0.038 per cent today when the air temperature is 16°C. The missing carbon dioxide has been pulled out of water during the making of reefs and limestone.[11]

The Intergovernmental Panel on Climate Change (IPCC) has said that greenhouse gases have caused global warming, which is greatest round the north and south polar regions. Since the beginning of the 20th century the average temperature just above the earth's surface has increased by about 0.6°C. According to British data since the 1940s, there has been a sustained temperature increase in the earth's atmosphere of around 2.5°C in the Antarctic Peninsula. The National Climatic Data Centre and NASA (both in the US) and the Climate Research Group of the University of East Anglia (in the UK) concur.[12]

A report for the European Commission on climate change published in 2007 made dire predictions. By 2071 climate change will cause droughts and floods that will kill 90,000 people a year. Droughts in southern Europe will cause fires and reduce soil fertility and crop yields will fall by a fifth. There will be floods in Hungary's Danube basin, and sea levels will rise by up to a metre, causing damage that will cost tens of billions of euros. Oceans will acidify, and fish will migrate northwards.[13]

In Antarctica ice shelves are breaking up and glaciers are cracking, blocks of broken-away ice are falling into the sea and drifting as icebergs. Since 1995 the Larsen Ice Shelf, the Filchner-Ronne Shelf and the Ross Ice Shelf have all lost ice the size of Rhode Island (2002), larger than Delaware (1998) and a third of the size of Switzerland (2000) respectively. Winter ice seems to be contracting in West Antarctica but not in East Antarctica, where small shelves are stable.

Across the world, mountain glaciers are in retreat: in the Alps, Andes, Himalayas and Rocky Mountains; and in New Zealand, East Africa and Greenland. The US Glacier National Park may have no glaciers after 2050. There is later freezing and earlier melting from Alaska to the UK, where spring is coming earlier with flowers blooming and birds breeding earlier. As a result of melting glacier ice, the sea level has risen 4 inches since 1900.[14]

In Antarctica the annual air temperature averages −4.5°C, and in the winter when the sun disappears for six months the temperature can reach −80°C. The water temperature is warmer. About 55 million years ago the surface temperature of the Southern Ocean was 15°C. Now it is a maximum of 3°C, because surface waters have been carried down into deeper waters.[15] About 8 per cent of Antarctic fish only live in Antarctica. Having no blood cells as there is oxygen in sea water, they live in a water temperature of −1.8°C and if it rises above +1.5°C (a low-temperature tolerance of 3 per cent) they die. Coral reefs live at a temperature of 1°C and die if the water rises above 3°C. Melting ice and warmer waters affect krill, the staple food of whales, seals, dolphins, penguins and albatrosses, and there has been a reduction of 6 per cent to 12 per cent in marine stocks. Warmer-water species now invade the warmer cold waters, threatening the habitat of colder-water species and endangering their survival. As ice melts the sea level rises, and mangroves, coral reefs and sea-grass meadows will not survive. If, as has been proposed, global warming increases by 1.5°C to 5.5°C during the 21st century sea levels will rise, South Florida may be inundated and cities near the sea may be flooded, such as New York, New Orleans, Amsterdam, Copenhagen, Tokyo and Buenos Aires. Water vapour will increase and there may be a reduction in water levels of lakes, though increased rain may keep levels stable.[16]

Natural phenomena may have contributed to these atmospheric-based changes. The sun has cycles when small dark sunspots appear on its face, the last two sunspot cycles falling between 1979 and 2001.[17] Radiation from the sun produces chemical isotopes such as carbon-14 and beryllium-10. An estimate can be made of solar variability for the last few thousand years by examining these two in rocks and fossils. Measuring greenhouse gases began in the International Geophysical Year Programme of 1957-8. In the 40 years since then the carbon dioxide in the atmosphere has gone up by 20 per cent, and is now up 0.5 per cent each year on what it was in 1750. Russian drilling in Antarctica has found cores that show the concentration of carbon dioxide exceeds that at any time since 420 million years ago. During this time high carbon-dioxide levels coincide with high temperatures, and low carbon-dioxide levels with low temperatures. We can assume there will be high temperatures – and high carbon-dioxide levels – for the rest of the 21st century.[18] Methane has doubled since 1750. These gases will decline as a result of the Montreal Protocol, but not yet. Microscopic particles and droplets known as aerosols caused by fossil fuels are also in the atmosphere but these do reflect some of the earth's sunshine back into space before it reaches the earth's surface. Volcanic eruptions have also spewed volcanic aerosols and ash into the atmosphere.

Despite these natural contributions to the atmosphere, in 2001 the IPCC scientists concluded that the increase in temperature cannot be explained by sunspots and volcanic eruptions alone, and that global warming was mainly due to human activity. The consequences may be that there is more plant growth due to the fertilization of carbon dioxide, but soil will become more desiccated.[19]

Universalist thinking on global warming is very much in tune with the environmental movement which brings a global perspective to conserving energy, developing alternatives to fossil fuels and preventing greenhouse gases from escaping to the atmosphere. The globalist Kyoto Protocol of 1997 began the reduction of carbon-dioxide emissions.

The rapid growth of the human population is an essential part of the problem, as Universalist thinking realises. The growth of the human population was a flat line from 160,000 years ago to shortly before AD1800, after which it began rising dramatically. The human population reached 1 billion in 1804, 2 billion in 1927 (123 years later), 3 billion in 1960 (33 years later), 4 billion in 1974 (14 years later), 5 billion in 1987 (13 years later) and 6 billion in 1999 (12 years later). It is projected to reach 7 billion in 2013 (15 years later), 8 billion in 2028 (15 years later) and 9 billion in 2054 (26 years later).[20]

Man has had a huge impact on his environment, covering green lands with roofs and roads and filling the night sky with electric lights. Man has exceeded the earth's carrying capacity. It has been estimated[21] that humans are consuming 1,000 times more than other animals of our size; are producing 100,000 times more carbon dioxide than other animals of our size; and have 1,000 times the human population than other animals of our size. (We should be 6 million on this comparison and in 2008 we were well over 6 billion.) At the present rate of expansion, 750 years from now humans will cover the entire earth, and in 2,000 years' time the volume of human flesh will equal the volume of the earth.[22] Animals forage in 2 per cent of their area, man forages in 80 per cent and eats meals whose ingredients have come from many countries.

The human population has nearly trebled in my lifetime and resources have not kept up with this increase. There is a problem with water in some regions, for example in Indonesia which is over-populated and where human faeces are fertilizing rice fields and polluting Djakarta bay. The human population is the problem behind global warming, with its expectations of an ever higher and higher life style, demand for water and production of effluents.

How should mankind deal with the problem of human population? There has already been an attempt to control it through the *Global 2000 Report to the President*.[23] Accepted by President Carter in 1980 when the human population was just over 4 billion, this project has failed in its attempt to cap and reduce the human population. Now, despite the project's having persuaded China to legislate and confine families to one child per family, the world's population is 6.7 billion and approaching 7 billion. Solutions that reduce the population back towards the 1980 figure are fundamentally neo-Hitlerian and racist. Many of the countries whose populations have dramatically increased recently are in the Third World, and the *Global 2000 Report*'s commentators referred to these as "useless eaters", meaning people who do not work but eat the earth's food. To put the genie back into the 1980 bottle would require the human population to be reduced by nearly 3 billion, a thought that cannot be countenanced.

Humans will resist attempts to reduce the human population. Many (including the Pope) would say that modern *Homo sapiens* should not be measured biologically by being given a status equivalent to the status of penguins, seals and whales; that modern *Homo sapiens* dominates Nature as the only creature with self-awareness, self-consciousness, reason, intellect, and a perhaps immortal soul and spirit; and that modern *Homo sapiens* should have a privileged position and not be required to reduce his population. Some are scornful about global warming, seeing it as an excuse to make us buy new cars, televisions and kitchen goods that comply with anti-warming criteria and strengthen demand, which the capitalist system needs to remain healthy.

Many would say that Nature will solve the problem of human population by adjusting from overpopulation to underpopulation – perhaps by nuclear warfare or perhaps through disease. Modern *Homo sapiens* is, as we have observed, one species which can interbreed and is therefore susceptible to a new strain of a virus which could decimate our species, indeed wipe humanity out.

At this point Universalist thinking demands that we should stand back and take a long-term view for long-term cooling is the context of our short-term warming. We have found that within our Ice Age which began 40/33.7 million years ago (see p137), there are meteorological, geological and man-made influences which may accelerate or reverse this trend in the short term or in the long term.

From one point of view, there is a prospect of a long-term period of global cooling as greenhouse gases shield the earth from the sun. Within our interglacial period, a lesser, recent "mini"-glacial stage known (in terms of our definition, wrongly) as the "Little Ice Age" began c.1500 and reached its maximum advance c.1750. It receded until 1850 when warming began with the Industrial Revolution and lasted until 1950. Global warming in our time may therefore be to some extent part of a cyclical pattern, a continuation and intensification of the warming of 1850-1950.[24]

Predictions based on Milankovitch's work in palaeoclimatology suggests that, as we have seen (see pp141-142) eccentric orbiting is now the most dominant influence, and that this and the axis wobble combined suggest that the ice sheet is in retreat, whereas the tilt of the axis suggests that it is advancing. From 20,000 years ago North America was under ice, and the ice sheet has been in retreat for the last 11,500-10,000 years. The sea level around Europe has risen, and there are predictions that if Greenland's ice melts, the sea will rise 6 metres, which would mean that Florida and Bangladesh would be flooded. Some predictions say this could happen in our lifetime, others that it will not happen until long after our time. If the Antarctic continent were to melt it has been estimated that the sea level would rise 66 metres – which, with the 6 metres from Greenland would make 72 metres in all – but no one is suggesting that this is imminent.[25] There has been a temperature change in Europe, which is now warmer than it was in the Middle Ages. It seems that the Age of Aquarius, the water-pourer in astrology, may indeed be ahead.

However, from this point of view we are still in an Ice Age, and though we must act to prevent global warming to ameliorate the conditions which our generation and our children's and grandchildren's generations will experience, it could be that in as little as a thousand years' time people will lament that what we have done to reduce global warming and to avoid enlarging the hole in the ozone layer will have contributed to the global *cooling* future generations will experience when the ice sheet advances again.

From another point of view, there is a prospect of a short-term period of global warming as our carbon emissions harm the earth's atmosphere and cause the earth to grow warmer, further melting ice sheets and raising sea levels. It is a measure of this global warming in our interglacial period that the Arctic's North-West

Passage, which traverses the polar ice in a 3,200-mile waterway connecting the Atlantic and Pacific Oceans across the top of Canada and Alaska and which has defeated virtually all previous attempts to sail it until now, is now navigable. It may be regularly open by 2027, even by 2012. This is because satellite images have shown that the ice sheet has shrunk 38,600 square miles (100,000 sq kms) per year over the past decade, and ten times as much in 2006 alone.[26] Polar bears may soon be threatened with extinction as the ice sheet is breaking up, sometimes leaving them stranded on floating bits of ice. Some are drowning, some are dying of starvation due to climate change.

The same pattern of breaking ice can be detected in the Antarctic. Everywhere in Antarctica the glaciers are retreating. The large glacier, Risting glacier, filled the end of the Drygalski fjord in South Georgia in 1957, showing that in 50 years a huge amount has melted. The entire fjord was once the glacier, and as the glacier melts, so the fjord increases. Losses of ice due to melting in West Antarctica increased 59 per cent from 83 billion tons in 1996 to 132 billion in tons in 2006, and 60 billion tons of ice melted from the Antarctic Peninsula in 2006, an increase of 140 per cent since 1996. In November 2007 I saw huge cracks in the glaciers on either side of the disused Argentinean station in Paradise Bay on the Antarctic Peninsula. Within the next hour part of one of the glaciers broke off and left several hundred yards of the bay under pack ice, which our Zodiac crunched through. I was told that calving glaciers often cause waves on the stony beach where I had landed. I encountered similar fracturing near the British Base A in Port Lockroy, so my own observation confirmed that the retreat of glaciers in these two places is typical of Antarctica. It looks as if more parts of glaciers may calve, and it was no surprise that in early 2008 a chunk of the Wilkins Ice Shelf 25.5 by 1.5 miles broke off, triggering the disintegration of 405 square kilometres of the shelf's interior.[27]

Short-term phenomena, then, such as the emission of greenhouse gases may affect the earth's atmosphere and climate, but these have to be seen against the background of long-term cycles, and against the prospect of a new glacial within our continuing Ice Age in perhaps 1,000-1,500 or 16,500-18,000 years' time. As the earth has a continent over the South Pole (Antarctica) and an almost land-locked polar sea over the North Pole (the Arctic Ocean), geologists believe that the earth will continue to endure glacial periods for the foreseeable future. To generations a thousand years from now I say, "I am sorry our generation compounded your problems, but we had to act in the best interests of *our* time to prevent the deaths of many millions from starvation and flooding."

The environmental movement has championed global warming but is caught within this glacial-interglacial cycle. Its findings that greenhouse gases contribute to changes in the atmosphere which can change climate and environmental conditions have to be acted on for our generation and for the coming generations. These findings about gases have to be balanced with the underlying pattern that climate changes may be consequences of the glacial-interglacial cycles of our Ice Age, which can change the earth's atmosphere.

In its approach, the environmental movement looks at human contributions to a deterioration in the earth's atmosphere and seeks to restrict and rectify the damage caused by emissions. In doing this it sees with a global perspective that focuses on the whole planet rather than regions and national boundaries. Whether it realises this or not, the environmental movement has adopted Universalist thinking.

Has the earth a self-regulating principle? If we forget about the label Gaia, is it a living self-organising unity? Are we on one self-regulating planet? If the earth has an equilibrium, its own version of the homeostasis we have found in the body and brain, its own version of negative feedback, it can be expected to adjust the imbalances of the depleted ozone layer, the greenhouse effect, the growth in human population, the damage to our oceans and atmosphere. Universalist thinking expects solutions to these problems.

Political (World-Government) Universalism

As all politics emerged from one point that preceded the Big Bang, from the first cell and from the first human political organisation, political Universalism sees the whole world as being ordered as one political entity, an interconnected unity, a whole. As all humans are world citizens they have human rights, which include a human right to live under a world government that has abolished war, famine and disease. The world order's world government must be democratic so every member of humankind has a democratic vote. Political Universalism affirms a world government under the UN that is not totalitarian but allows each human being the maximum freedom and attacks poverty.

Our global identity is reinforced by the web created by Tim Berners-Lee, which theoretically allows each world citizen to communicate with all others (provided they are on email). The web symbolises the interconnectedness of humankind just as the distinctive DNA signature of all world citizens symbolises the uniqueness of each individual.

Political Universalism minimises the conflicts that divide people and eliminates divisions by negotiation. Ultimately it must logically believe that national borders do not matter, that mass migration – emigration and immigration – is a neutral, indeed a good, thing, and that natural identity should yield to multi-culturalism because that will follow when a world government is in place.

There have been strides towards a world government, with many trade institutions founded and moves towards American, European and Pacific Unions.[28] It is intended that they will eventually merge and that there will be a one-world political structure and currency. In September 2000, 149 participating heads of government and officials from 31 other nations attended the UN Millennium Assembly and Summit of world leaders in New York, and, with only 14 world leaders missing or unrepresented, ushered in the age of global governance by adopting a revised version of the UN Charter, which was known as the Charter for Global Democracy.[29] They all appeared for a photo, suggesting the unofficial inauguration of a world government. There is now a transitional period before a world government is formally established, and this new thinking is causing a considerable amount of pain as old national attitudes give way to new internationalist, global priorities.

However, the establishment of a world government – a United States of the World or Universal World State – is not as easy as it seems. All the living civilisations of historical Universalism, each of which is at a different stage of its development, have to come together and voluntarily place themselves under a world government. Some civilisations are clearly reluctant to do this, and old-style nationalistic attitudes remain a problem. Russia, for example, is not behaving as co-operatively as world-government enthusiasts might wish, and officially-communist China is determined to put China's interests before world interests in matters such as the world response to climate change and the future of Tibet. Furthermore, there are still dictatorships which oppose the world order, and Islamic fundamentalism is anti-Western and opposes democratisation and Westernization by resistance movements and suicide bombings.

A world-government form of political Universalism has never happened before. World empires such as the Roman Empire have always been within a civilisation, and it is possible that political Universalism may prove to be an imperial phase within the North-American civilisation.

The integrity of the world order that is endeavouring to form a world government is an issue. There have been several attempts to form a world government during the last 100 years, and the same mega-rich families have been involved dynastically over several generations.[30] They are in league with the multinational companies and have a business outlook on acquiring the world's resources, including oil.[31] The Iraq operation was not untypical. The US and British military occupied Iraq, handed the oil over to newly set-up bodies and moved multinationals in to reconstruct Iraq out of Iraq's oil wealth with relatively small financial assistance from the US. The beneficiaries have been the self-interested businessmen operating behind the military.

A world order led by benevolent world rulers of integrity can spread freedom and democracy and bring paradise to earth, abolish war, famine and disease and liberate the poor from their poverty. If, however, the rulers are self-interested, then the world's resources will not be fairly distributed.[32] A large share will be acquired by the West, which is eager to continue its dominance in the face of competition from the East.

We have seen (on p351) that there is a problem with the rapid growth of the human population, which is forecast to exceed the capacity of the earth's resources that can sustain it. There have been reports on whether there should be population control or reduction, and it has been suggested that contrived wars, famines and indiscriminate man-made diseases such as AIDS have been used to reduce the rising population. A world government managing population reduction in a way that lacks integrity would be a disaster for millions of people. Clearly, if a world order were ruled by a dictatorship such as Hitler's or Stalin's, and had a population-reduction agenda, the world would not be a pleasant place for humans.

Political Universalism is Utopian in wanting to improve the lot of humankind and bringing universal freedom, democracy and relief from poverty, war, famine and disease. It is ameliorist, and has nothing to do with the activities and policies of those who do not have humankind's best interests at heart. It looks ahead to a united world beyond the war against terrorism, and endeavours to make the earth a Paradise – so long as the operators of the world order are up to Universalism's standards. Universalism will be the philosophy of a good globalist world government – if there ever is one.

Religious Universalism

As all religion emerged from one point that preceded the Big Bang, from the first cell and from the first appearance of organized religion, religious Universalism affirms that all humankind will eventually be saved from the finite universe of time into infinite timelessness, not just members of one particular religion. Religious Universalism sees the prospect of all religions becoming merged into a one-world religion based on what all religions have in common: the infinite, timeless Being that has been intuitionally glimpsed by mystics of all religions as the "Divine" or inner Light. All human beings are equal in relation to the metaphysical Light that can fill their souls and order Nature's ecosystem.

I have dealt fully with the tradition of this common denominator in all religions elsewhere.[33] The vision of metaphysical Light or Being is found in every generation and culture. We have seen that it is a universal experience received in the universal being or intellect. This experience can bring together Catholic, Protestant and Orthodox Christianity, Islam, Judaism, Hinduism, Buddhism, Jainism, Sikhism and Taoism and many other religions such as Zoroastrianism, which all share the experience. There can be regional and local variety round this universal experience as in the Greek time when Zeus became identified with the local gods of religions in Asia Minor, and in the Roman time when Jupiter became identified with local gods such as Zeus and Sol who had taken over from Mithras in Asia Minor. In this one-world religion (if it happens) there will be one experience of the "ever-living Fire" or metaphysical Light which all religions have in common and which has been stated slightly differently within each religion, and there will be a merging of God, Allah, Yahweh, Brahman, the Enlightened Buddha, Nirvana, Om Kar, the Tao and Ahura Mazda. The words of their prophets will be regional and local variations of the universal God: Christ, Mohammed, Moses, Siva, the Buddha, Mahavira, Guru Nanak, Lao-Tze (or Lao-Tzu).

The posture of the world religion will be Unitarian Universalist – "Unitarian" in the sense of "believing that God is not a Trinity but one person", a posture that will enable there to be common ground between the Gods or all the higher religions. All religions have views on how humans should lead a good life in order to be rewarded in Heaven or Paradise (the infinite timelessness) or by Nirvana (extinction). They all have codes of conduct such as the Ten Commandments, which broadly share the same principles but differ in detail from religion to religion.

A one-world religion can cover every region and historical period in a United States of the World (or Universal World State). It will include shamanism which has been practised in much of Siberia and Central Asia since c.50,000BC and whose central experience was the vision of metaphysical Light which could be "seen at midnight".[34] Religion began in pre-history, and during our Ice Age's last glacial, Stone-Age man lived out of reach of the frozen ice in the mouths of caves. The interiors of the caves were pitch-black and frightening, and cave art illustrates their shamanistic rituals. Neanderthal man, who died out c.29,000BC in Spain and c.27,000BC in the Caucasus,[35] conducted ceremonies over Neanderthal dead. The one-world religion will include the most ancient attempts at religion, including all the Middle-Eastern cults, as well as the higher religions for, doctrine aside, they all focused on the same essential religious experience of the "ever-living Fire" or Light. Religious Universalism's one-world religion can therefore now reconcile and unite all cults and faiths during the last 50,000 years.

Such a one-world religion will spread with the spread of political Universalism. A world government would introduce a world religion, and each region would be allowed its own cultural and religious variations so it can practise its own tradition in relation to the universal experience and the whole. It is the connection of local traditions with the whole that would be different – and difficult. A one-world religion would have to be imposed from the centre, for Islamic fundamentalism (one of the more prominent dissenters) would take exception to it and for a while the two would merely co-exist. Religion would be more syncretistic, with faiths blending into each other. Particular faiths would lose purity, as would nations, in a cosmopolitanised world where all citizens feel at home in many countries, which they would regard as regions. The universal one-world religion would be a *koiné* or melting-pot in which many religions are melted down like metals. Religion may be more shallow by becoming a mixture of faiths.

Religious Universalism – a religion composed of all religions – would be challenged by atheists and sceptics, and would simply co-exist alongside their disbelief. Atheists and those opposed to all religion such as Richard Dawkins adopt a position akin to that of the medieval Church: Darwinism (Dawkins' God)[36] is right and one must not speculate beyond it in the way that Galileo speculated beyond the Christian universe in the Middle Ages. Atheists have a Materialistic view of human life and death, and deny that metaphysical Light is received in the universal being or intellect, or that the spirit will survive death. Their view is interpretative, they believe that an after-life is scientifically irrelevant in terms of the discoveries of recent physics. However, we have seen that there may be "radio transmissions" between spirits of the dead that occupy a part of the infinite and living consciousnesses on earth. There is a view that such an "after-life" is in a fifth dimension. In philosophical terms it can be located in Being rather than Existence, and if Being is a fifth dimension then the Universalist philosopher will not disagree with this view.

Life on earth, we have seen, involves a mixture of order and randomness. The religious view depends on the quasi-empirical experience of Being manifesting in metaphysical Light and order. Atheists exaggerate the prevalence of randomness at the expense of order. If there were no order, there would be chaos and nothing would exist, live or develop. Religion tends to attribute providential importance to order, and religion's problem is to cope with randomness.

Universalism, then, is a one-world and one-universe philosophy that should be applied to the big issues of today: the health of the planet; good government; and the spread of perennial religion. Universalism sees a one-world environmental movement, a political one-world government and a one-world religion as well as a one-world history, culture and literature. Using "culture" in its widest sense as the intellectual achievement produced by civilisations, Universalism helps us to understand the culture of the past, present and near-future.

The Universalist perspective of the Metaphysical Revolution has introduced a new paradigm, a new view of man which releases him from Materialism and reductionism and offers hope of contact with Being. This can make possible a new Golden Age in literature and the arts.

An ideal of the Renaissance was the "many-sided man" at home in many disciplines: Leonardo, Michelangelo and Bacon, an ideal later perpetuated by Goethe and others. Today, it is vital that there should be "many-sided men" who can view philosophy, science, history and literature in relation to each other and all disciplines. It is very easy for a specialist scientist, operating within the sciences and no other disciplines, to assume that Materialism is the norm. He needs ontological philosophy and historical and literary thinking to inform his scientific specialism and connect him with the Whole, the One or All, to demonstrate that this is not so. The Whole perspective brings a very different slant on, and approach to, the universe.

Epilogue

I began my *Prolegomena* with the School's Quadrangle at Oxford's Bodleian library (see p1), which I often walked through on my way to the Radcliffe Camera. At one end stands the black statue of the 3rd Earl of Pembroke, William Herbert of Wilton, which Shakespeare visited in 1603 when *As You Like It* was performed before James I. As patron of a group of artists, writers and musicians including Shakespeare and Ben Jonson, he may have been the Mr W.H. to whom Shakespeare dedicated his *Sonnets* (though an Earl would never have been addressed as "Mr" without good reason). Facing him at the other end of the Quadrangle, sitting on a throne on high, is King James I, shown a few years after the settling of America at Jamestown. In the second decade of the 17th century philosophy was still reasonably united, as was the European civilisation: Greek roots, Christian trunk and branches – philosophy, art, sculpture, music and literature – filled with metaphysical sap. It was a time of harmony and unification, the time of Sir Francis Bacon before the Civil War in England broke things up, shortly before Descartes split the late Renaissance unity with his dualism.

That Arcadian harmony and unification can be revived if philosophy can be successfully reunified. The aim is to carry a very modern view of the universe into philosophy throughout the Western world so that a latter-day William Herbert can be patron to metaphysical philosophers, thinkers, writers, artists and musicians in peaceful countries throughout the Western world.

Summary

We have come a long way. The theme of this book has been that Western philosophy must return to its origin in the Presocratic Greeks. Its rightful focus is on the universe and Nature's ecosystem. It should restore the metaphysical view of the infinite, and bring together the metaphysical and scientific traditions (Part One). I have done this by importing the scientific view of the universe into philosophy (Part Two), by uniting science and metaphysics and by uniting the rational and intuitional approaches to metaphysics (Part Three). I have also united the metaphysical and scientific traditions and the linguistic and Phenomenological philosophies. I have applied Universalism practically to the history, culture and literature of the past and to the environmental movement, political world government and one-world religion in the near-future (Part Four).

In the interests of clarity what I have done in the course of this book is:

(1) to return philosophy to its Greek roots in the boundless *apeiron* or infinite;

(2) to restore the infinite to philosophy;

(3) to restore the finite universe and Nature to philosophy;

(4) to restore the scientific view of the universe to philosophy;

(5) to restore metaphysics to philosophy;

(6) to propose a new Form from Movement Theory to account for the origin and creation of the universe;

(7) to reconcile and unite science and metaphysics through metaphysical science which will investigate the universal order principle;

(8) to reconcile and reunify science and philosophy;

(9) to reconcile and unite the metaphysical and scientific traditions in the history of philosophy;

(10) to unite the rational and intuitional metaphysical tradition;

(11) to reconcile and unite linguistic and Phenomenological philosophies;

(12) to propose that the multi-levelled self has 12 levels of consciousness;

(13) to state a new university curriculum for a new discipline, Universalism;

(14) to demonstrate how Universalism can transform studies of the past and movements in the present and near-future;

(15) to launch a Metaphysical Revolution to bring about these changes in philosophy;

(16) to launch a Revolution in Thought and Culture to bring about these changes in culture.

In these 16 restorations, reconciliations, unifications, proposals and revolutions, I have reconciled the contradictions of philosophy by using algebraic thinking: $+A + -A = O$. Universalism has clear answers to the ultimate questions of metaphysics, seeing them in terms of the infinite and science. I have effected a Metaphysical Revolution to restore metaphysics alongside science in a redefinition of what philosophy should be doing, in a move away from logic and language. It is time for a change of emphasis in philosophy.

I have defined "Universalism" as focusing: on the universe; on a universal order principle; on humankind's place in the universe; on universal cosmic energies which stimulate the growth of organisms and plants; and the universal being or intellect where these energies or rays are received.

I have described how metaphysical ontology approaches Reality rationally *and* intuitionally; how metaphysical psychology is transpersonal, by occupying a level of consciousness different from that of everyday consciousness within the multi-levelled self; how metaphysical epistemology describes knowing the metaphysical Light in the universal being or intellect; and how cosmology is part of the scientific view of the universe. All these branches of contemporary metaphysics are bound together by the model of Reality manifesting from Nothingness to Non-Being, into Being and finally into Existence, being received quasi-empirically in the experiencing intellect and being thought about in the reason and rational, social ego. Human beings are a part of, and within, Nature's ecosystem and are nourished and nurtured by forces that may be carried by natural light.

In a philosophical category, the universe is as fundamentally orderly as it is also fundamentally a thing of chance. In algebraic thinking, $+A + -A = O$: the Law of Order + the Law of Randomness = Being in Existence. (The Law of Order begins in timeless, infinite Being, the Law of Randomness begins in finite chaotic Existence, which is prone to accident and chance.) To turn it round, Being in Existence is the Law of Order + the Law of Randomness. I can also express the reconciliation as: the One Being's Order + Existence's Randomness = the whole system. In other words, the contradictions are reconciled in fundamental unity to the benefit of the whole.

Reductionism assumes that the universe, earth and man are exclusively accidental and without purpose. I have demonstrated that there is an alternative interpretation. Universalist Expansionist philosophy reconnects humankind with the universe and Nature through the workings of the universal order principle. It offers an interpretation of the universe as orderly and purposive in terms of testable scientific disciplines and their cultural applications. We are all brothers and sisters on one perhaps self-regulated planet in which a self-organising principle has taken us through evolution to self-improvement, to higher consciousness and ability to sense the One.

Reductionism and Materialism are severely dented by the 40 bio-friendly conditions (chapter 5) and by the invisibility of 96 per cent of the matter in the universe when visible matter constitutes only 4 per cent of this matter. The probability of such a universe's being accidental and these conditions being coincidental is, as we have seen on p124, 10^{120}, which in my view delivers reductionism and Materialism a knock-out blow.

In this account, I have been careful to distinguish categories, to keep philosophical categories separate from religious categories and testable scientific categories. The new philosophy brings traditional philosophy, testable scientific data and perennial religion together again in a form that does not violate testable science.

If the universe has a self-organising system on a self-regulating planet – self-regulating on account of the universal order principle which permeates the finite universe – it offers order and purpose in one's life and a connection with the oneness of humankind and the infinite, timeless One. Within a "sea" of cosmic energy the universe began from a point which was, according to my Form from Movement Theory, self-aware. Is the point of the finite universe to become aware – conscious of itself again through the evolution of human consciousness to new heights, or rather depths – over the next few million years?

This work has introduced a philosophy of the order within the universe. In a metaphysical sense it returns philosophy to around 1910 when the metaphysical emphasis was in the hands of Bergson, whose ambition was to unite metaphysics and science through his process philosophy; T.E. Hulme; Husserl; Whitehead, whose process philosophy of organism looked back to Darwin and Bergson; and William James. All of these saw the problem confronting philosophy. They belonged to a time before philosophers withdrew from the universe, bemused by the complexity of Einstein's revolutionary view of it, and also from metaphysics into comforting logic and language, rational security.

At the end of *A Brief History of Time*, Hawking wrote correctly: "Philosophers reduced the scope of their inquiries so much that Wittgenstein, the most famous philosopher of this century, said, 'The sole remaining task for philosophy is the analysis of language.' What a comedown from the great tradition of philosophy from Aristotle to Kant!"[1] I could not agree more. In our time, the philosophers have abdicated and cosmologists and thinkers about the universe, scientists practising what had been known before the 18th century as "natural philosophy" (see p3), have moved onto their ground and have filled the vacuum they vacated. The successes of Newton and Einstein propelled science onto the philosophers' ground. In our time, the role of the philosopher has been assumed by the thoughtful cosmologist or interpreter of science and reflecter on cosmology. In pre-18th-century terms, the "natural philosophers" have occupied the ground of philosophy, including the ground once occupied by "metaphysics".

Thus, Hawking and I occupy the ground which has traditionally been occupied by the true philosophers. We may fundamentally disagree about the universe's origin, which he seeks in the finite and which I find in the infinite, and about the cause of the weakness of gravity, which he seeks in superstrings and many dimensions and which I find in the expanding force of light; but in the past philosophers with different emphases often disagreed. Roger Penrose, John Barrow, Paul Davies, Martin Rees, John Gribbin and others who have attempted to describe the universe as accurately as possible – rather than merely interpret science – occupy the same ground of "natural philosophy". There is a public thirst for accurate description of the universe, which is why Hawking's *A Brief History of Time* sold 10 million copies. The public are not receiving *any* description of the universe from the philosophers and regard cosmologists and thinkers about the universe as surrogate philosophers.

The new "cosmologist philosophers" have gone some way towards fulfilling the philosophers' ancient task of thinking deeply about the universe and offering a consistent system but they have significantly failed to address philosophy's need to return to its metaphysical roots and to reunite the metaphysical and scientific. This was only to be expected as they have no interest in the ground of philosophy beyond the confines of "natural philosophy". Though occupying – squatting on – the ground of philosophy they have worked within cosmology without regard for the philosophical tradition, and so they have not contributed to the Metaphysical Revolution.

Closing Vision

Universalism is an "everythingism" in the sense that it is a Theory of Everything in so far as it explains everything in terms of fifth force that operates via a field of "*aither*" CERN has sought and pervades everything including the four forces with which it was once united, and stimulates all Nature's phenomena to grow and develop in accordance with coded information in photons and DNA. (See Postscript.)

My hope is that others will take up this theme in a more comprehensive way, produce thousands more examples than I have produced and demonstrate that the Law of Order is not merely a philosophical concept but is indeed a scientific energy that pervades the universe and acts as a fifth fundamental force. If what I have been saying in this book resonates, then others should swell my theme. This book has been my *Prolegomena*, an introduction, to a new discipline which I now entrust to a new generation of thinkers in the hope that they will make sweeping changes to philosophy so that, as in the time of the early Greeks, it can again assist the young of coming generations in understanding the operation of the universe and the purpose of the upward thrust of evolution through twelve levels of consciousness.

A lot of work, in the form of papers and essays, needs to be done to achieve the programme of the Metaphysical Revolution. A lot of books need to be written about the Revolution's fundamental shift in perception. Universalist philosophers need leisure to reflect, complete independence from preferment at work, for promotion may affect judgements, and a sufficient income to buy a library.

Philosophers, historians and writers must come forward to co-create the work of the Revolution. Metaphysics is now at the centre of philosophy, and ontology – the study and science of Being – is at the centre of metaphysics.

In this work I have laid out a jumbled thousand-piece jigsaw whose interlocking bits wear fragments of pictures from different disciplines, and with esemplastic intuition I have pieced it together to give an interlocking picture of the universe. My picture of the universe is up-to-date. Details of the science will change. There will be new discoveries. Picturing the universe is like photographing a river. *Panta rhei*, the moving flux flows on. There is soon a new river before the camera, but it is in shape, structure and essence the same river. The universe ever changes but ever remains the same. And so my picture of it will remain true forever. My picture of the universe will be updated but its principles will endure so long as there is time and humankind contemplates what is, and what is forever.

With that thought I turn away from the work to be done and concentrate on the figure of the lone surfer on the sea of invisible energy, endlessly crouching in his astronaut's spacesuit on the crest of the perpetual wave of the universe as it ceaselessly accelerates into the infinite, endlessly breasting the infinite, the boundless (*to apeiron*), which most of modern philosophy and much of modern cosmology believes does not exist. The surfer is out as far as mind can reach and is calmly still as, his consciousness tethered to the pineal-body region of his brain, in intuitional trance, he crouches, surges, races, gazes into and experiences the Reality his peers so indignantly deny. With a supreme effort of the imagination he has placed himself at the cutting-edge of truth and is intent on increasing his hard-won Faustian knowledge.

In his consciousness, moving between 12 levels or seats of his self, he knows that the aim of life is to arrive at the truth of why we are here in the universe, to replace himself by leaving behind children and to live in harmony with Reality.

He knows that the purpose of life is the *genus Homo*'s thrust, by self-improvement, self-betterment and self-transcendence, from ape to superconsciousness which can experience Reality, the metaphysical Light, and to cleanse and heighten his own consciousness to prepare for the survival of part of his self after death. He knows that the meaning of life is to know with certainty and profound joy that life is not a chance, random accident but is the careful work of a caring order principle that is bio-friendly and imbues human activity with a profound meaning as humans journey towards the survival of part of their consciousness beyond this life, to a new form of existence which is only a breath away. In his multi-levelled consciousness he knows that the aim, purpose and meaning of *his* life mirror the aim, purpose and meaning of life in his projects and self-appointed tasks, his domestic harmony and his existential – or rather, universal and Universalist – oneness with the All or One, with Being, within the scheme of manifestation. Now within his universal being, his intellect, and now within his soul breaks the metaphysical Light, filling his consciousness and his limbs with a profound meaning that brings with it "a peace that passeth understanding". He feels wonderfully serene and accepts the conditions of existence. He loves life and *his* life, and though solitary and aware that he can fall off the crest of his wave before his next breath, he is confident that when he departs this life he will live on unseen in infinite Being (in accordance with the Law of Survival which states that all beings do not cease to be contained within the unity of Being) and return to it on death, and he counts himself happy.

Surf on, brave surfer, surf on – but when you are ready, return and tell us exactly what the infinite is like. Is the dark relieved by glimmers of "ever-living Fire" or Light?

POSTSCRIPT

THE 2008/9 CERN EXPERIMENT AND THE ORDER PRINCIPLE

On 10 September 2008, amid groundless fears that it would create an enormous black hole that would swallow the earth or possibly suck the earth inside out, the first beam of subatomic particles in the Large Hadron Collider (LHC) at CERN was successfully steered around the full 27 kilometres (17 miles) of the world's most powerful particle accelerator 330 feet below the Swiss-French border. Both clockwise and counter-clockwise beams worked, and all was ready to simulate, and replicate, what happened a trillionth of a second after the Big Bang. The wires and 9,300 magnets were cooled by liquid nitrogen to achieve a temperature of −271.25C (1.9 degrees above absolute zero) as at 271C helium becomes "super-fluid" (flowing with virtually no viscosity) and can therefore conduct away heat. The heat that would eventually be generated by beam collisions would be 14TeV (tera-electronvolts, tera denoting a factor of 10^{12}).

Unfortunately the LHC broke down on 19 September due to a faulty connection between two magnets, which melted, leaking a ton of helium and causing the temperature of 100 magnets to rise by around 100C. The fault affected part of the accelerator. Beam collisions would be delayed until the spring of 2009.

The beams would be accelerated to 99.9999991 per cent of the speed of light, and would complete 11,245 17-mile laps in a single second. Beams were to smash together up to 600 million times a second. The first collision of beams, and the smashing of protons and lead ions against each other, would be captured by detectors. The detectors were so vast that one was housed in a cavern that could enclose the nave of Westminster Abbey. The information would then be analysed by the LHC Computing Grid, a global computing grid designed to organise information extremely quickly and handle 15 million gigabytes (or 21.4 million CDs) of LHC-related data every year.

The detectors would trace the subatomic *débris* thrown off by the collisions to reveal new particles and effects. The analysis would be looking for the creation of new particles present in the first trillionth of a second after the Big Bang. The data produced and analysed during the two or three years after the experiment was expected to throw light on the origin of mass; the working of gravity; the question of the existence (or non-existence) of extra dimensions; and the nature of the 96 per cent of the universe that cannot be seen, i.e. of dark matter (23 per cent) and dark energy (73 per cent). (Some hold that dark matter fills 25 per cent of the universe and dark energy 71 per cent.) The analysis would also reveal the evidence (or otherwise) for the Higgs boson, a new type of tiny particle called a scalar and vector boson which has been dubbed the "God particle". This was proposed by Peter Higgs in 1964 as an explanation of why matter has mass (which we perceive as weight) – how a massless particle could turn into a massive particle – and how it can coalesce to stars, planets and people.

Previous atom-smashers had not found this boson. The LHC's predecessor, the Large Electron-Positron Collider, LEP, which occupied the same tunnel in the 1980s, was not sufficiently powerful to find the Higgs boson. It had a practical application as it produced the world-wide web, the internet, which was invented by Tim Berners-Lee to help researchers share information generated by the LEP. It was the means in the late 1980s of analysing CERN's data, and there was now the prospect that the LHC's computing grid would have a similar practical application – that we could all have access to supercomputers capable of handling a large volume of data from our desks: lightning-fast computers linked to a new world grid. (Practical applications of Grid technology are already being explored in the medical world, to share information about mammograms and to kill cancer tumours with hadron therapy or particle beams, and it is possible that nuclear waste can be broken down by firing a beam of protons to generate a shower of neutrons. Also, a better understanding of the photons in sunlight will advance knowledge of climate change.)

It was anticipated – or rather, hoped – that when the data of the LHC CERN experiment had been analysed there would be a definitive view on whether Higgs bosons exist or whether there are supersymmetrical partner particles, which at the time of the experiment had no existence outside mathematical theory. Thus the Alice detector would attempt to create quark-gluon plasma; the CMS detector would look for new particles and phenomena; and the LHCb detector would study the relationship between matter and antimatter. It was the Atlas detector that would search for new physical phenomena, including the Higgs boson, supersymmetry and extra dimensions.

There was some rivalry between the theory of the Higgs boson and the theory of supersymmetry. The LHC had been set up to check the Higgs theory. (It was ironical that Higgs' first paper containing a short mathematical proof of his theory was rejected by the CERN-based editor of the *European Journal of Physics Letters* in the summer of 1964.) The day before the beginning of the LHC CERN experiment Stephen Hawking jokingly said[1] that it would be "more exciting" if the experiment did not find the "God particle": "That will show something is wrong and we will need to think again." He was championing the theory of "partner particles" of supersymmetry and extra dimensions. Hawking had once placed a $100 bet that the Higgs boson does not exist. Higgs was irritated, and at a subsequent press conference in Edinburgh said of a previous paper that Hawking had written (echoing Penrose's criticism of Hawking, see p62): "Finally, I don't think the way he does it is good enough. He puts together theories in particle physics with gravity...in a way which no theoretical physicist would believe is the correct theory. I am very doubtful about his calculations." The two men were both hoping to be vindicated by the LHC CERN experiment, and both were contenders for the Nobel Prize, which was sure to be awarded to the one who was proved right.

Both theories – the Higgs boson and supersymmetric particles theories – have implications for my order principle. The Higgs theory suggests that all particles that are not Higgs bosons travel through a field of (at the time of writing theoretical, non-evidential) Higgs bosons, which, like the Greek *aither* that was never found, slow the particles down, interact with them and give rise to their mass by attaching themselves to them. Hadrons are particles with mass, the best-known being protons and neutrons, all of which are composed of smaller units called quarks that have been bound together, and gluons which stick on quarks. Smashing protons and hadrons, particles with mass, could therefore separate Higgs bosons. The Higgs boson, if it exists, would be a "mechanism" of the order principle as it would order matter, and as bosons include photons they would provide a physical basis for the Universalist view expressed in the foregoing pages that photons have a role in ordering matter and consciousness, and the universe. Supersymmetry, on the other hand, a theoretical, non-evidential hypothesis at the time of writing, suggests that all particles have an accompanying partner known as a superparticle or sparticle. This symmetry at the subatomic level could explain why 96 per cent of the universe is invisible, and could provide an alternative "mechanism" for the ordering of matter and consciousness.

The final outcome of the LHC CERN experiment raises the prospect that the new philosophy of Universalism, which admits the universe back into philosophy, will soon have a physical, evidential basis in so far as its order principle may soon have a scientifically confirmed "mechanism" in either the Higgs boson or supersymmetry, which would reveal how particles, including photons, organise and order matter, biological forms, consciousness, Nature and the universe. Would CERN find a new ordering invisible Reality or ordering principle in the microworld?

APPENDICES

Appendix 1: Constants and Ratios

A. 326 Fundamental Physical Constants – Complete Listing in Alphabetical Order

326 constants and ratios are listed together with their values, uncertainty rating and units of measurement.[1] In many cases, the values of these constants and ratios are just right for life, and if a particular value were slightly higher or lower then the physical world would be different and would not be able to sustain life. For 326 constants and ratios to be just right for life is extraordinary. Each of the 326 values also has to be right in relation to the other 325 values. The scale and degree of the complexity and order required to achieve a balance between all 326 constants and ratios almost defies belief. These constants and ratios are listed here to show the immensity and complexity of the system which balances so many different values in such a way that they are in harmony with each other and do not conflict. It is beyond belief that such a fine balance could be a random accident. One value could just be a random event, but for 326 interacting and interlocking values to be in balance suggests a degree of order way beyond the random.

Quantity	Value	Uncertainty	Unit
alpha particle-electron mass ratio	7294.299 5365	0.000 0031	
alpha particle mass	6.644 656 20 e-27	0.000 000 33 e-27	kg
alpha particle mass energy equivalent	5.971 919 17 e-10	0.000 000 30 e-10	J
alpha particle mass energy equivalent in MeV	3727.379 109	0.000 093	MeV
alpha particle mass in u	4.001 506 179 127	0.000 000 000 062	u
alpha particle molar mass	4.001 506 179 127 e-3	0.000 000 000 062 e-3	kg mol^-1
alpha particle-proton mass ratio	3.972 599 689 51	0.000 000 000 41	
Angstrom star	1.000 014 98 e-10	0.000 000 000 90 e-10	m

atomic mass constant	1.660 538 782 e-27	0.000 000 083 e-27	kg
atomic mass constant energy equivalent	1.492 417 830 e-10	0.000 000 074 e-10	J
atomic mass constant energy equivalent in MeV	931.494 028	0.000 023	MeV
atomic mass unit-electron volt relationship	931.494 028 e6	0.000 023 e6	Ev
atomic mass unit-hartree relationship	3.423 177 7149 e7	0.000 000 0049 e7	E h
atomic mass unit-hertz relationship	2.252 342 7369 e23	0.000 000 0032 e23	Hz
atomic mass unit-inverse meter relationship	7.513 006 671 e14	0.000 000 011 e14	m^-1
atomic mass unit-joule relationship	1.492 417 830 e-10	0.000 000 074 e-10	J
atomic mass unit-kelvin relationship	1.080 9527 e13	0.000 0019 e13	K
atomic mass unit-kilogram relationship	1.660 538 782 e-27	0.000 000 083 e-27	kg
atomic unit of 1st hyperpolarizablity	3.206 361 533 e-53	0.000 000 081 e-53	C^3 m^3 J^-2
atomic unit of action	1.054 571 628 e-34	0.000 000 053 e-34	J s
atomic unit of 2nd hyperpolarizablity	6.235 380 95 e-65	0.000 000 31 e-65	C^4 m^4 J^-3
atomic unit of charge	1.602 176 487 e-19	0.000 000 040 e-19	C
atomic unit of charge density	1.081 202 300 e12	0.000 000 027 e12	C m^-3
atomic unit of current	6.623 617 63 e-3	0.000 000 17 e-3	A
atomic unit of electric dipole mom.	8.478 352 81 e-30	0.000 000 21 e-30	C m
atomic unit of electric field	5.142 206 32 e11	0.000 000 13 e11	V m^-1
atomic unit of electric field gradient	9.717 361 66 e21	0.000 000 24 e21	V m^-2
atomic unit of electric polarizablity	1.648 777 2536 e-41	0.000 000 0034 e-41	C^2 m^2 J^-1
atomic unit of electric potential	27.211 383 86	0.000 000 68	V
atomic unit of electric quadrupole mom.	4.486 551 07 e-40	0.000 000 11 e-40	C m^2

Quantity	Value	Uncertainty	Unit
atomic unit of energy	4.359 743 94 e-18	0.000 000 22 e-18	J
atomic unit of force	8.238 722 06 e-8	0.000 000 41 e-8	N
atomic unit of length	0.529 177 208 59 e-10	0.000 000 000 36 e-10	m
atomic unit of mag. Dipole mom.	1.854 801 830 e-23	0.000 000 046 e-23	J T^-1
atomic unit of mag. Flux density	2.350 517 382 e5	0.000 000 059 e5	T
atomic unit of magnetizability	7.891 036 433 e-29	0.000 000 027 e-29	J T^-2
atomic unit of mass	9.109 382 15 e-31	0.000 000 45 e-31	kg
atomic unit of momentum	1.992 851 565 e-24	0.000 000 099 e-24	kg m s^-1
atomic unit of permittivity	1.112 650 056... e-10	(exact)	F m^-1
atomic unit of time	2.418 884 326 505 e-17	0.000 000 000 016 e-17	s
atomic unit of velocity	2.187 691 2541 e6	0.000 000 0015 e6	m s^-1
Avogadro constant	6.022 141 79 e23	0.000 000 30 e23	mol^-1
Bohr magneton	927.400 915 e-26	0.000 023 e-26	J T^-1
Bohr magneton in Ev/T	5.788 381 7555 e-5	0.000 000 0079 e-5	Ev T^-1
Bohr magneton in Hz/T	13.996 246 04 e9	0.000 000 35 e9	Hz T^-1
Bohr magneton in inverse meters per tesla	46.686 4515	0.0000 0012	m^-1 T^-1
Bohr magneton in K/T	0.671 7131	0.0000 0012	K T^-1
Bohr radius	0.529 177 208 59 e-10	0.000 000 000 36 e-10	m
Boltzmann constant	1.380 6504 e-23	0.000 0024 e-23	J K^-1
Boltzmann constant in Ev/K	8.617 343 e-5	0.000 015 e-5	Ev K^-1

Quantity	Value	Uncertainty	Unit
Boltzmann constant in Hz/K	2.083 6644 e10	0.000 0036 e10	Hz K^-1
Boltzmann constant in inverse meters per kelvin	69.503 56	0.000 12	m^-1 K^-1
characteristic impedance of vacuum	376.730 313 461...	(exact)	ohm
classical electron radius	2.817 940 2894 e-15	0.000 000 0058 e-15	m
Compton wavelength	2.426 310 2175 e-12	0.000 000 0033 e-12	m
Compton wavelength over 2 pi	386.159 264 59 e-15	0.000 000 53 e-15	m
conductance quantum	7.748 091 7004 e-5	0.000 000 0053 e-5	S
conventional value of Josephson constant	483 597.9 e9	(exact)	Hz V^-1
conventional value of von Klitzing constant	25 812.807	(exact)	ohm
Cu x unit	1.002 076 99 e-13	0.000 000 28 e-13	m
deuteron-electron mag. Mom. ratio	-4.664 345 537 e-4	0.000 000 039 e-4	
deuteron-electron mass ratio	3670.482 9654	0.000 0016	
deuteron g factor	0.857 438 2308	0.000 000 0072	
deuteron mag. Mom.	0.433 073 465 e-26	0.000 000 011 e-26	J T^-1
deuteron mag. Mom. to Bohr magneton ratio	0.466 975 4556 e-3	0.000 000 0039 e-3	
deuteron mag. Mom. to nuclear magneton ratio	0.857 438 2308	0.000 000 0072	
deuteron mass	3.343 583 20 e-27	0.000 000 17 e-27	kg
deuteron mass energy equivalent	3.005 062 72 e-10	0.000 000 15 e-10	J
deuteron mass energy equivalent in MeV	1875.612 793	0.000 047	MeV
deuteron mass in u	2.013 553 212 724	0.000 000 000 078	u
deuteron molar mass	2.013 553 212 724 e-3	0.000 000 000 078 e-3	kg mol^-
deuteron-neutron mag. Mom. ratio	-0.448 206 52	0.000 000 11	

Quantity	Value	Uncertainty	Unit
deuteron-proton mag. Mom. ratio	0.307 012 2070	0.000 000 0024	
deuteron-proton mass ratio	1.999 007 501 08	0.000 000 000 22	
deuteron rms charge radius	2.1402 e-15	0.0028 e-15	m
electric constant	8.854 187 817... e-12	(exact)	F m^-1
electron charge to mass quotient	-1.758 820 150 e11	0.000 000 044 e11	C kg^-1
electron-deuteron mag. Mom. ratio	-2143.923 498	0.000 018	
electron-deuteron mass ratio	2.724 437 1093 e-4	0.000 000 0012 e-4	
electron g factor	-2.002 319 304 3622	0.000 000 000 0015	
electron gyromag. ratio	1.760 859 770 e11	0.000 000 044 e11	s^-1 T^-1
electron gyromag. ratio over 2 pi	28 024.953 64	0.000 70	MHz T^-1
electron mag. Mom.	-928.476 377 e-26	0.000 023 e-26	J T^-1
electron mag. Mom. anomaly	1.159 652 181 11 e-3	0.000 000 000 74 e-3	
electron mag. Mom. to Bohr magneton ratio	-1.001 159 652 181 11	0.000 000 000 000 74	
electron mag. Mom. to nuclear magneton ratio	-1838.281 970 92	0.000 000 80	
electron mass	9.109 382 15 e-31	0.000 000 45 e-31	kg
electron mass energy equivalent	8.187 104 38 e-14	0.000 000 41 e-14	J
electron mass energy equivalent in MeV	0.510 998 910	0.000 000 013	MeV
electron mass in u	5.485 799 0943 e-4	0.000 000 0023 e-4	u
electron molar mass	5.485 799 0943 e-7	0.000 000 0023 e-7	kg mol^-1
electron-muon mag. Mom. ratio	206.766 9877	0.000 0052	
electron-muon mass ratio	4.836 331 71 e-3	0.000 000 12 e-3	

electron-neutron mag. Mom. ratio	960.920 50	0.000 23	
electron-neutron mass ratio	5.438 673 4459 e-4	0.000 000 0033 e-4	
electron-proton mag. Mom. ratio	-658.210 6848	0.0000 0054	
electron-proton mass ratio	5.446 170 2177 e-4	0.000 000 0024 e-4	
electron-tau mass ratio	2.875 64 e-4	0.000 47 e-4	
electron to alpha particle mass ratio	1.370 933 555 70 e-4	0.000 000 000 58 e-4	
electron to shielded helion mag. Mom. ratio	864.058 257	0.000 010	
electron to shielded proton mag. Mom. ratio	-658.227 5971	0.0000 0072	
electron volt	1.602 176 487 e-19	0.000 000 040 e-19	J
electron volt-atomic mass unit relationship	1.073 544 188 e-9	0.000 000 027 e-9	u
electron volt-hartree relationship	3.674 932 540 e-2	0.000 000 092 e-2	E_h
electron volt-hertz relationship	2.417 989 454 e14	0.000 000 060 e14	Hz
electron volt-inverse meter relationship	8.065 544 65 e5	0.000 000 20 e5	m^{-1}
electron volt-joule relationship	1.602 176 487 e-19	0.000 000 040 e-19	J
electron volt-kelvin relationship	1.160 4505 e4	0.000 0020 e4	K
electron volt-kilogram relationship	1.782 661 758 e-36	0.000 000 044 e-36	kg
elementary charge	1.602 176 487 e-19	0.000 000 040 e-19	C
elementary charge over h	2.417 989 454 e14	0.000 000 060 e14	$A\ J^{-1}$
Faraday constant	96 485.3399	0.0024	$C\ mol^{-1}$
Faraday constant for conventional electric current	96 485.3401	0.0048	$C\_90\ mol^{-1}$
Fermi coupling constant	1.166 37 e-5	0.000 01 e-5	GeV^{-2}
fine-structure constant	7.297 352 5376 e-3	0.000 000 0050 e-3	

Quantity	Value	Uncertainty	Unit
first radiation constant	3.741 771 18 e-16	0.000 000 19 e-16	W m^2
first radiation constant for spectral radiance	1.191 042 759 e-16	0.000 000 059 e-16	W m^2 sr^-1
hartree-atomic mass unit relationship	2.921 262 2986 e-8	0.000 000 0042 e-8	u
hartree-electron volt relationship	27.211 383 86	0.000 000 68	Ev
Hartree energy	4.359 743 94 e-18	0.000 000 22 e-18	J
Hartree energy in Ev	27.211 383 86	0.000 000 68	Ev
hartree-hertz relationship	6.579 683 920 722 e15	0.000 000 000 044 e15	Hz
hartree-inverse meter relationship	2.194 746 313 705 e7	0.000 000 000 015 e7	m^-1
hartree-joule relationship	4.359 743 94 e-18	0.000 000 22 e-18	J
hartree-kelvin relationship	3.157 7465 e5	0.000 0055 e5	K
hartree-kilogram relationship	4.850 869 34 e-35	0.000 000 24 e-35	kg
helion-electron mass ratio	5495.885 2765	0.000 0052	
helion mass	5.006 411 92 e-27	0.000 000 25 e-27	kg
helion mass energy equivalent	4.499 538 64 e-10	0.000 000 22 e-10	J
helion mass energy equivalent in MeV	2808.391 383	0.000 070	MeV
helion mass in u	3.014 932 2473	0.000 000 0026	u
helion molar mass	3.014 932 2473 e-3	0.000 000 0026 e-3	kg mol^-1
helion-proton mass ratio	2.993 152 6713	0.000 000 0026	
hertz-atomic mass unit relationship	4.439 821 6294 e-24	0.000 000 0064 e-24	u
hertz-electron volt relationship	4.135 667 33 e-15	0.000 000 10 e-15	Ev
hertz-hartree relationship	1.519 829 846 006 e-16	0.000 000 000 010 e-16	E_h

hertz-inverse meter relationship	$3.335\ 640\ 951\ldots$ e-9	(exact)	m^-1
hertz-joule relationship	$6.626\ 068\ 96$ e-34	$0.000\ 000\ 33$ e-34	J
hertz-kelvin relationship	$4.799\ 2374$ e-11	$0.000\ 0084$ e-11	K
hertz-kilogram relationship	$7.372\ 496\ 00$ e-51	$0.000\ 000\ 37$ e-51	kg
inverse fine-structure constant	$137.035\ 999\ 679$	$0.000\ 000\ 094$	
inverse meter-atomic mass unit relationship	$1.331\ 025\ 0394$ e-15	$0.000\ 000\ 0019$ e-15	u
inverse meter-electron volt relationship	$1.239\ 841\ 875$ e-6	$0.000\ 000\ 031$ e-6	Ev
inverse meter-hartree relationship	$4.556\ 335\ 252\ 760$ e-8	$0.000\ 000\ 000\ 030$ e-8	E_h
inverse meter-hertz relationship	$299\ 792\ 458$	(exact)	Hz
inverse meter-joule relationship	$1.986\ 445\ 501$ e-25	$0.000\ 000\ 099$ e-25	J
inverse meter-kelvin relationship	$1.438\ 7752$ e-2	$0.000\ 0025$ e-2	K
inverse meter-kilogram relationship	$2.210\ 218\ 70$ e-42	$0.000\ 000\ 11$ e-42	kg
inverse of conductance quantum	$12\ 906.403\ 7787$	$0.000\ 0088$	ohm
Josephson constant	$483\ 597.891$ e9	0.012 e9	Hz V^-1
joule-atomic mass unit relationship	$6.700\ 536\ 41$ e9	$0.000\ 000\ 33$ e9	u
joule-electron volt relationship	$6.241\ 509\ 65$ e18	$0.000\ 000\ 16$ e18	Ev
joule-hartree relationship	$2.293\ 712\ 69$ e17	$0.000\ 000\ 11$ e17	E_h
joule-hertz relationship	$1.509\ 190\ 450$ e33	$0.000\ 000\ 075$ e33	Hz
joule-inverse meter relationship	$5.034\ 117\ 47$ e24	$0.000\ 000\ 25$ e24	m^-1
joule-kelvin relationship	$7.242\ 963$ e22	$0.000\ 013$ e22	K
joule-kilogram relationship	$1.112\ 650\ 056\ldots$ e-17	(exact)	kg
kelvin-atomic mass unit relationship	$9.251\ 098$ e-14	$0.000\ 016$ e-14	u

Quantity	Value	Uncertainty	Unit
kelvin-electron volt relationship	8.617 343 e-5	0.000 015 e-5	Ev
kelvin-hartree relationship	3.166 8153 e-6	0.000 0055 e-6	E_h
kelvin-hertz relationship	2.083 6644 e10	0.000 0036 e10	Hz
kelvin-inverse meter relationship	69.503 56	0.000 12	m^-1
kelvin-joule relationship	1.380 6504 e-23	0.000 0024 e-23	J
kelvin-kilogram relationship	1.536 1807 e-40	0.000 0027 e-40	kg
kilogram-atomic mass unit relationship	6.022 141 79 e26	0.000 000 30 e26	u
kilogram-electron volt relationship	5.609 589 12 e35	0.000 000 14 e35	Ev
kilogram-hartree relationship	2.061 486 16 e34	0.000 000 10 e34	E_h
kilogram-hertz relationship	1.356 392 733 e50	0.000 000 068 e50	Hz
kilogram-inverse meter relationship	4.524 439 15 e41	0.000 000 23 e41	m^-1
kilogram-joule relationship	8.987 551 787... e16	(exact)	J
kilogram-kelvin relationship	6.509 651 e39	0.000 011 e39	K
lattice parameter of silicon	543.102 064 e-12	0.000 014 e-12	m
lattice spacing of silicon	192.015 5762 e-12	0.000 0050 e-12	m
Loschmidt constant (273.15 K, 101.325 kPa)	2.686 7774 e25	0.000 0047 e25	m^-3
mag. Constant	12.566 370 614... e-7	(exact)	N A^-2
mag. Flux quantum	2.067 833 667 e-15	0.000 000 052 e-15	Wb
molar gas constant	8.314 472	0.000 015	J mol^-1 K^-1
molar mass constant	1 e-3	(exact)	kg mol^-1
molar mass of carbon-12	12 e-3	(exact)	kg mol^-1

molar Planck constant	3.990 312 6821 e-10	0.000 000 0057 e-10	J s mol^-1
molar Planck constant times c	0.119 626 564 72	0.000 000 000 17	J m mol^-1
molar volume of ideal gas (273.15 K, 100 kPa)	22.710 981 e-3	0.000 040 e-3	m^3 mol^-1
molar volume of ideal gas (273.15 K, 101.325 kPa)	22.413 996 e-3	0.000 039 e-3	m^3 mol^-1
molar volume of silicon	12.058 8349 e-6	0.000 0011 e-6	m^3 mol^-1
Mo x unit	1.002 099 55 e-13	0.000 000 53 e-13	m
muon Compton wavelength	11.734 441 04 e-15	0.000 000 30 e-15	m
muon Compton wavelength over 2 pi	1.867 594 295 e-15	0.000 000 047 e-15	m
muon-electron mass ratio	206.768 2823	0.000 0052	
muon g factor	-2.002 331 8414	0.000 000 0012	
muon mag. Mom.	-4.490 447 86 e-26	0.000 000 16 e-26	J T^-1
muon mag. Mom. anomaly	1.165 920 69 e-3	0.000 000 60 e-3	
muon mag. Mom. to Bohr magneton ratio	-4.841 970 49 e-3	0.000 000 12 e-3	
muon mag. Mom. to nuclear magneton ratio	-8.890 597 05	0.000 000 23	
muon mass	1.883 531 30 e-28	0.000 000 11 e-28	kg
muon mass energy equivalent	1.692 833 510 e-11	0.000 000 095 e-11	J
muon mass energy equivalent in MeV	105.658 3668	0.000 0038	MeV
muon mass in u	0.113 428 9256	0.000 000 0029	u
muon molar mass	0.113 428 9256 e-3	0.000 000 0029 e-3	kg mol^-1
muon-neutron mass ratio	0.112 454 5167	0.000 000 0029	
muon-proton mag. Mom. ratio	-3.183 345 137	0.000 000 085	

Quantity	Value	Uncertainty	Unit
muon-proton mass ratio	0.112 609 5261	0.000 000 0029	
muon-tau mass ratio	5.945 92 e-2	0.000 97 e-2	
natural unit of action	1.054 571 628 e-34	0.000 000 053 e-34	J s
natural unit of action in Ev s	6.582 118 99 e-16	0.000 000 16 e-16	Ev s
natural unit of energy	8.187 104 38 e-14	0.000 000 41 e-14	J
natural unit of energy in MeV	0.510 998 910	0.000 000 013	MeV
natural unit of length	386.159 264 59 e-15	0.000 000 53 e-15	m
natural unit of mass	9.109 382 15 e-31	0.000 000 45 e-31	kg
natural unit of momentum	2.730 924 06 e-22	0.000 000 14 e-22	kg m s^-1
natural unit of momentum in MeV/c	0.510 998 910	0.000 000 013	MeV/c
natural unit of time	1.288 088 6570 e-21	0.000 000 0018 e-21	s
natural unit of velocity	299 792 458	(exact)	m s^-1
neutron Compton wavelength	1.319 590 8951 e-15	0.000 000 0020 e-15	m
neutron Compton wavelength over 2 pi	0.210 019 413 82 e-15	0.000 000 000 31 e-15	m
neutron-electron mag. Mom. ratio	1.040 668 82 e-3	0.000 000 25 e-3	
neutron-electron mass ratio	1838.683 6605	0.000 0011	
neutron g factor	-3.826 085 45	0.000 000 90	
neutron gyromag. Ratio	1.832 471 85 e8	0.000 000 43 e8	s^-1 T^-1
neutron gyromag. ratio over 2 pi	29.164 6954	0.000 0069	MHz T^-1
neutron mag. Mom.	-0.966 236 41 e-26	0.000 000 23 e-26	J T^-1
neutron mag. Mom. to Bohr magneton ratio	-1.041 875 63 e-3	0.000 000 25 e-3	

neutron mag. Mom. to nuclear magneton ratio	-1.913 042 73	0.000 000 45	
neutron mass	1.674 927 211 e-27	0.000 000 084 e-27	kg
neutron mass energy equivalent	1.505 349 505 e-10	0.000 000 075 e-10	J
neutron mass energy equivalent in MeV	939.565 346	0.000 023	MeV
neutron mass in u	1.008 664 915 97	0.000 000 000 43	u
neutron molar mass	1.008 664 915 97 e-3	0.000 000 000 43 e-3	kg mol^-1
neutron-muon mass ratio	8.892 484 09	0.000 000 23	
neutron-proton mag. Mom. ratio	-0.684 979 34	0.000 000 16	
neutron-proton mass ratio	1.001 378 419 18	0.000 000 000 46	
neutron-tau mass ratio	0.528 740	0.000 086	
neutron to shielded proton mag. Mom. ratio	-0.684 996 94	0.000 000 16	
Newtonian constant of gravitation	6.674 28 e-11	0.000 67 e-11	m^3 kg^-1 s^-2
Newtonian constant of gravitation over h-bar c	6.708 81 e-39	0.000 67 e-39	(GeV/c^2)^-2
nuclear magneton	5.050 783 24 e-27	0.000 000 13 e-27	J T^-1
nuclear magneton in Ev/T	3.152 451 2326 e-8	0.000 000 0045 e-8	Ev T^-1
nuclear magneton in inverse meters per tesla	2.542 623 616 e-2	0.000 000 064 e-2	m^-1 T^-1
nuclear magneton in K/T	3.658 2637 e-4	0.000 0064 e-4	K T^-1
nuclear magneton in MHz/T	7.622 593 84	0.000 000 19	MHz T^-1
Planck constant	6.626 068 96 e-34	0.000 000 33 e-34	J s
Planck constant in Ev s	4.135 667 33 e-15	0.000 000 10 e-15	Ev s
Planck constant over 2 pi	1.054 571 628 e-34	0.000 000 053 e-34	J s
Planck constant over 2 pi in Ev s	6.582 118 99 e-16	0.000 000 16 e-16	Ev s

Quantity	Value	Uncertainty	Unit
Planck constant over 2 pi times c in MeV fm	197.326 9631	0.000 0049	MeV fm
Planck length	1.616 252 e-35	0.000 081 e-35	m
Planck mass	2.176 44 e-8	0.000 11 e-8	kg
Planck mass energy equivalent in GeV	1.220 892 e19	0.000 061 e19	GeV
Planck temperature	1.416 785 e32	0.000 071 e32	K
Planck time	5.391 24 e-44	0.000 27 e-44	s
proton charge to mass quotient	9.578 833 92 e7	0.000 000 24 e7	C kg^-1
proton Compton wavelength	1.321 409 8446 e-15	0.000 000 0019 e-15	m
proton Compton wavelength over 2 pi	0.210 308 908 61 e-15	0.000 000 000 30 e-15	m
proton-electron mass ratio	1836.152 672 47	0.000 000 80	
proton g factor	5.585 694 713	0.000 000 046	
proton gyromag. ratio	2.675 222 099 e8	0.000 000 070 e8	s^-1 T^-1
proton gyromag. ratio over 2 pi	42.577 4821	0.000 0011	MHz T^-1
proton mag. Mom.	1.410 606 662 e-26	0.000 000 037 e-26	J T^-1
proton mag. Mom. to Bohr magneton ratio	1.521 032 209 e-3	0.000 000 012 e-3	
proton mag. Mom. to nuclear magneton ratio	2.792 847 356	0.000 000 023	
proton mag. Shielding correction	25.694 e-6	0.014 e-6	
proton mass	1.672 621 637 e-27	0.000 000 083 e-27	kg
proton mass energy equivalent	1.503 277 359 e-10	0.000 000 075 e-10	J
proton mass energy equivalent in MeV	938.272 013	0.000 023	MeV
proton mass in u	1.007 276 466 77	0.000 000 000 10	u

proton molar mass	1.007 276 466 77 e-3	0.000 000 000 10 e-3	kg mol^-1
proton-muon mass ratio	8.880 243 39	0.000 000 23	
proton-neutron mag. Mom. ratio	-1.459 898 06	0.000 000 34	
proton-neutron mass ratio	0.998 623 478 24	0.000 000 000 46	
proton rms charge radius	0.8768 e-15	0.0069 e-15	m
proton-tau mass ratio	0.528 012	0.000 086	
quantum of circulation	3.636 947 5199 e-4	0.000 000 0050 e-4	m^2 s^-1
quantum of circulation times 2	7.273 895 040 e-4	0.000 000 010 e-4	m^2 s^-1
Rydberg constant	10 973 731.568 527	0.000 073	m^-1
Rydberg constant times c in Hz	3.289 841 960 361 e15	0.000 000 000 022 e15	Hz
Rydberg constant times hc in Ev	13.605 691 93	0.000 000 34	Ev
Rydberg constant times hc in J	2.179 871 97 e-18	0.000 000 11 e-18	J
Sackur-Tetrode constant (1 K, 100 kPa)	-1.151 7047	0.000 0044	
Sackur-Tetrode constant (1 K, 101.325 kPa)	-1.164 8677	0.000 0044	
second radiation constant	1.438 7752 e-2	0.000 0025 e-2	m K
shielded helion gyromag. ratio	2.037 894 730 e8	0.000 000 056 e8	s^-1 T^-1
shielded helion gyromag. ratio over 2 pi	32.434 101 98	0.000 000 90	MHz T^-1
shielded helion mag. Mom.	-1.074 552 982 e-26	0.000 000 030 e-26	J T^-1
shielded helion mag. Mom. to Bohr magneton ratio	-1.158 671 471 e-3	0.000 000 014 e-3	
shielded helion mag. Mom. to nuclear magneton ratio	-2.127 497 718	0.000 000 025	
shielded helion to proton mag. Mom. ratio	-0.761 766 558	0.000 000 011	
shielded helion to shielded proton mag. Mom. ratio	-0.761 786 1313	0.000 000 0033	

Quantity	Value	Uncertainty	Unit
shielded proton gyromag. ratio	2.675 153 362 e8	0.000 000 073 e8	s^-1 T^-1
shielded proton gyromag. ratio over 2 pi	42.576 3881	0.000 0012	MHz T^-1
shielded proton mag. Mom.	1.410 570 419 e-26	0.000 000 038 e-26	J T^-1
shielded proton mag. Mom. to Bohr magneton ratio	1.520 993 128 e-3	0.000 000 017 e-3	
shielded proton mag. Mom. to nuclear magneton ratio	2.792 775 598	0.000 000 030	
speed of light in vacuum	299 792 458	(exact)	m s^-1
standard acceleration of gravity	9.806 65	(exact)	m s^-2
standard atmosphere	101 325	(exact)	Pa
Stefan-Boltzmann constant	5.670 400 e-8	0.000 040 e-8	W m^-2 K^-4
tau Compton wavelength	0.697 72 e-15	0.000 11 e-15	m
tau Compton wavelength over 2 pi	0.111 046 e-15	0.000 018 e-15	m
tau-electron mass ratio	3477.48	0.57	
tau mass	3.167 77 e-27	0.000 52 e-27	kg
tau mass energy equivalent	2.847 05 e-10	0.000 46 e-10	J
tau mass energy equivalent in MeV	1776.99	0.29	MeV
tau mass in u	1.907 68	0.000 31	u
tau molar mass	1.907 68 e-3	0.000 31 e-3	kg mol^-1
tau-muon mass ratio	16.8183	0.0027	
tau-neutron mass ratio	1.891 29	0.000 31	
tau-proton mass ratio	1.893 90	0.000 31	
Thomson cross section	0.665 245 8558 e-28	0.000 000 0027 e-28	m^2

Quantity	Value	Uncertainty	Unit
triton-electron mag. Mom. ratio	-1.620 514 423 e-3	0.000 000 021 e-3	
triton-electron mass ratio	5496.921 5269	0.000 0051	
triton g factor	5.957 924 896	0.000 000 076	
triton mag. Mom.	1.504 609 361 e-26	0.000 000 042 e-26	J T^-1
triton mag. Mom. to Bohr magneton ratio	1.622 393 657 e-3	0.000 000 021 e-3	
triton mag. Mom. to nuclear magneton ratio	2.978 962 448	0.000 000 038	
triton mass	5.007 355 88 e-27	0.000 000 25 e-27	kg
triton mass energy equivalent	4.500 387 03 e-10	0.000 000 22 e-10	J
triton mass energy equivalent in MeV	2808.920 906	0.000 070	MeV
triton mass in u	3.015 500 7134	0.000 000 0025	u
triton molar mass	3.015 500 7134 e-3	0.000 000 0025 e-3	kg mol^-1
triton-neutron mag. Mom. ratio	-1.557 185 53	0.000 000 37	
triton-proton mag. Mom. ratio	1.066 639 908	0.000 000 010	
triton-proton mass ratio	2.993 717 0309	0.000 000 0025	
unified atomic mass unit	1.660 538 782 e-27	0.000 000 083 e-27	kg
von Klitzing constant	25 812.807 557	0.000 018	ohm
weak mixing angle	0.222 55	0.000 56	
Wien frequency displacement law constant	5.878 933 e10	0.000 010 e10	Hz K^-1
Wien wavelength displacement law constant	2.897 7685 e-3	0.000 0051 e-3	m K

B. 152 FUNDAMENTAL PHYSICAL CONSTANTS — GROUPED UNDER FIELDS

The fields the main constants are in can immediately be grasped from the headings below:[2]

UNIVERSAL

speed of light in vacuum

magnetic constant

electric constant

characteristic impedance of vacuum

Newtonian constant of gravitation

Planck constant

Planck mass

Planck temperature

Planck length

Planck time

ELECTROMAGNETIC

elementary charge

magnetic flux quantum

conductance quantum

    inverse of conductance quantum

Josephson constant

von Klitzing constant

Bohr magneton

nuclear magneton

ATOMIC AND NUCLEAR

General

fine-structure constant

    inverse fine-structure constant

Rydberg constant

Bohr radius

Hartree energy

quantum of circulation

Electroweak

Fermi coupling constant

weak mixing angle

Electron

electron mass

   energy equivalent

electron-muon mass ratio

electron-tau mass ratio

electron-proton mass ratio

electron-deuteron mass ratio

electron to alpha particle mass ratio

electron charge to mass quotient

electron molar mass

Compton wavelength

classical electron radius

Thomson cross section

electron magnetic moment

   to Bohr magneton ratio

   to nuclear magneton ratio

electron magnetic moment anomaly

electron g-factor

electron-muon magnetic moment ratio

electron-proton magnetic moment ratio

electron to shielded proton magnetic moment ratio

electron-neutron magnetic moment ratio

electron-deuteron magnetic moment ratio

electron to shielded helion magnetic moment ratio

electron gyromagnetic ratio

Muon

muon mass

energy equivalent

muon-electron mass ratio

muon-tau mass ratio

muon-proton mass ratio

muon-neutron mass ratio

muon molar mass

muon Compton wavelength

muon magnetic moment

to Bohr magneton ratio

to nuclear magneton ratio

muon magnetic moment anomaly

muon *g*-factor

muon-proton magnetic moment ration

Tau

tau mass

energy equivalent

tau-electron mass ratio

tau-muon mass ratio

tau-proton mass ratio

tau-neutron mass ratio

tau molar mass

tau Compton wavelength

Proton

proton mass

energy equivalent

proton-electron mass ratio

proton-muon mass ratio

proton-tau mass ratio

proton-neutron mass ratio

proton charge to mass quotient

proton molar mass

proton Compton wavelength

proton rms charge radius

proton magnetic moment

to Bohr magneton ratio

to nuclear magneton ratio

proton g-factor

proton-neutron magnetic moment ratio

shielded proton magnetic moment

to Bohr magneton ratio

to nuclear magneton ratio

proton magnetic shielding

proton gyromagnetic ratio

shielded gyromagnetic ratio

Neutron

neutron mass

energy equivalent

neutron-electron mass ratio

neutron-muon mass ratio

neutron-tau mass ratio

neutron-proton mass ratio

neutron molar mass

neutron Compton wavelength

neutron magnetic moment

to Bohr magneton ratio

to nuclear magneton ratio

neutron g-factor

neutron-electron magnetic moment ratio

neutron-proton magnetic moment ratio

neutron to shielded proton magnetic moment ratio

neutron gyromagnetic ratio

Deuteron

deuteron mass

energy equivalent

deuteron-electron mass ratio

deuteron-proton mass ratio

deuteron molar mass

deuteron rms charge radius

deuteron magnetic moment
  to Bohr magneton ratio
  to nuclear magneton ratio
deuteron-electron magnetic moment ratio
deuteron-proton magnetic moment ratio
deuteron-neutron magnetic moment ratio

Helion

helion mass
  energy equivalent
helion-electron mass ratio
helion-proton mass ratio
helion molar mass
shielded helion magnetic moment
  to Bohr magneton ratio
  to nuclear magneton ratio
shielded helion to proton magnetic moment ratio
shielded helion to shielded proton magnetic moment ratio
shielded helion gyromagnetic ratio

Alpha particle

alpha particle mass
  energy equivalent
alpha particle to electron mass ratio
alpha particle to proton mass ratio
alpha particle molar mass

PHYSICO-CHEMICAL

Avogadro constant
atomic mass constant
  energy equivalent
Faraday constant
molar Planck constant
molar gas constant
Boltzmann constant

molar volume of ideal gas

    Loschmidt constant

Sackur-Tetrode constant

Stefan-Boltzmann constant

first radiation constant

first radiation constant for spectral radiance

second radiation constant

Wien displacement law constant

C. 75 RATIOS – EXTRACTED FROM A.[3]

alpha particle-electron mass ratio
alpha particle-proton mass ratio
deuteron magnetic moment to Bohrmagenton ratio
deuteron magnetic moment to nuclear magneton ratio
deuteron-electron magnetic moment ratio
deuteron-electron mass ratio
deuteron-neutron magnetic moment ratio
deuteron-proton magnetic moment ratio
deuteron-proton mass ratio
electron gyromagnetic ratio
electron gyromagnetic ratio over 2 pi
electron magnetic moment to Bohr magneton ratio
electron magnetic moment to nuclear magneton ratio
electron to alpha particle mass ratio
electron to shielded helion magnetic moment ratio
electron to shielded proton magnetic moment ratio
electron-deuteron magnetic moment ratio
electron-deuteron mass ratio
electron-muon magnetic moment ratio
electron-muon mass ratio
electron-neutron magnetic moment ratio
electron-neutron mass ratio
electron-proton magnetic moment ratio
electron-proton mass ratio
electron-tau mass ratio
helion-electron mass ratio
helion-proton mass ratio
muon magnetic moment to Bohr magneton ratio
muon magnetic moment to nuclear magneton ratio
muon-electron mass ratio
muon-neutron mass ratio
muon-proton magnetic moment ratio
muon-proton mass ratio
muon-tau mass ratio
neutron gyromagnetic ratio
neutron gyromagnetic ratio over 2 pi
neutron magnetic moment to Bohr magneton ratio
neutron magnetic moment to nuclear magneton ratio
neutron to shielded proton magnetic moment ratio

neutron-electron magnetic moment ratio
neutron-electron mass ratio
neutron-muon mass ratio
neutron-proton magnetic moment ratio
neutron-proton mass ratio
neutron-tau mass ratio
proton gyromagnetic ratio
proton gyromagnetic ratio over 2 pi
proton magnetic moment to Bohr magneton ratio
proton magnetic moment to nuclear magneton ratio
proton-electron mass ratio
proton-muon mass ratio
proton-neutron magnetic moment ratio
proton-neutron mass ratio
proton-tau mass ratio
shielded helion gyromagnetic ratio
shielded helion gyromagnetic ratio over 2 pi
shielded helion magnetic moment to Bohr magneton ratio
shielded helion magnetic moment to nuclear magneton ratio
shielded helion to proton magnetic moment ratio
shielded helion to shielded proton magnetic moment ratio
shielded proton gyromagnetic ratio
shielded proton gyromagnetic ratio over 2 pi
shielded proton magnetic moment to Bohr magneton ratio
shielded proton magnetic moment to nuclear magneton ratio
tau-electron mass ratio
tau-muon mass ratio
tau-neutron mass ratio
tau-proton mass ratio
triton magnetic moment to Bohr magneton ratio
triton magnetic moment to nuclear magneton ratio
triton-electron magnetic moment ratio
trion-electron mass ratio
triton-neutron magnetic moment ratio
triton-proton magnetic moment ratio
triton-proton mass ratio

APPENDIX 2: BIOLOGY'S BIG BANG: THE CAMBRIAN EXPLOSION

Taxa, the units of biological classification, are arranged in a hierarchy from kingdom to subspecies. In protists, plants and animals the descending order is: kingdom, phylum (in plants, division), class, order, family, genus, species, subspecies or race.

The animal kingdom has about 40 phyla, six of which are divided into subphyla. The phyla are arranged in alphabetical order within each geological period. The subphyla are indented and are also arranged alphabetically within their phyla.

Evolution may be expected to throw up new phyla gradually, in dribs and drabs, over the 3.8 billion years since life began. It is therefore somewhat extraordinary that at least 19, and possibly as many as 35, out of 40 new phyla were thrown up within a period (the Cambrian period) that spanned only 70 million years. Many of these new phyla[4] are thought to have emerged about 540 million years ago.

1. Precambrian (before 570 million years ago): 3 Phyla (possibly 4)

Cnidaria	Porifera
Kimberella	Track-making worm
Gelatinosa (Late Cambrian)	
Nuda	

2. Cambrian (570-500 million years ago): 19 Phyla

Annelida	Echinodermata
Arthropoda	Asterozoa (Late Ordovician)
Anomalocariida	Blastozoa
Chelicerata	Crinozoa
Crustacea	Echinozoa
Hexapoda (Upper Devonian)	Homalozoa
Myriapoda (Late Siluran)	Halkieriida
Trilobita	Hemichordata
Brachiopoda	Hyolitha
Craniiformea	Mollusca

Linguliformea

Rhynchonelliformea

Chaetognatha

Chordata

   Cephalochordata

   Craniata

   Urochordata

Coeloscleritophora

Ctenophora

Amphineura

Cyrtosoma

Diasoma

Nematoda

Onychophora

Phoronida

Pogonophora

Priapula

Tardigrada

Tardipolypoda

3. Later Geological Periods: 6 Phyla

Ordovician

(500-440 million years ago)

   Bryozoa/Ectoprocta

Devonian

(395-345 million years ago)

   Echiura

Pennsylvanian

325-280 million years ago)

   Nemertina

Pennsylvanian

(195-136 million years ago)

   Entoprocta

Tertiary

(65-2 million years ago)

   Nematomorpha

   Rotifera

4. No Known Fossils: 12 Phyla

Acanthocephala

Cycliophora

Dicyemida

Gastrotricha

Gnathostomulida

Kinoryncha

Loricifera

Orthonectida

Pentastoma

Placozoa

Platyhelminthes

Sipuncula

APPENDIX 3: NATURAL LIGHT'S EFFECT ON GROWTH

The intensity of natural light has an effect on the light sensitivity of photobiological processes. The photons of natural light stimulate photosynthesis, flowering, seed germination, chlorophyll formation and phototropism (plants bending towards light), and human colour vision, black-and-white vision and vision of starlight, at different intensities in watts per square metre.[5] Some of these intensities of photons are so tiny that until the last century it was not known that they existed. Even Newton would be amazed at some of the findings in this table.

Light intensity in watts per square metre	Effects on growth
1,000/10^3/ one thousand	Clear sunlight at noon in June, 00000 lux, light saturation for photosynthesis in wheat, 20000 lux
100/10^2/ one hundred	Daylight, overcast sky, 1000-10000 lux
10/10^1/ ten	Compensation point of photosynthesis (photosynthesis and respiration equal), 100-1000 lux
1/10^0/ one	Photoperiodic control of flowering, about 100 lux
0.1/10^{-1}/ one tenth	Late twilight stimulation of seed germination (8 minutes of red light)
0.01/10^{-2}/ one hundredth	Moonlight, max 0-2 lux
0.001/10^{-3}/ one thousandth	Human colour vision, phototaxis in flagellates

$0.0001/10^{-4}/$ Chlorophyll formation
one ten-thousandth   (perceptible greening, red light)

$0.00001/10^{-5}/$ Phototropism (i.e. bending towards light)
one hundred-thousandth   in oat seedling (blue light)

$0.000001/10^{-6}/$ Phototropism in fungus
one millionth   (20 minutes of blue light)

$0.0000001/10^{-7}/$ Black and white image perception in man
one ten-millionth   (the intensity value refers to the
                                                illumination of the objects)

$0.000000001/10^{-9}/$ Light from a bright star (Sirius)
one billionth

$0.0000000001/10^{-10}/$ Effect on plant growth (inhibition of
one ten-billionth   growth in lower part of stem of oat
                                                seedling, red light)

$0.000000000001/10^{-12}/$ Light from a star of the 6th magnitude,
one trillionth   barely visible by the unaided eye
                                                (0.000000002 lux at the pupil)

$0.000000000000000001/10^{-18}/$ Light from the weakest stars that can
one quintillionth   be recorded with the largest telescopes

For red, blue and green light, see the bottom enlargement of the
electromagnetic spectrum on p400.

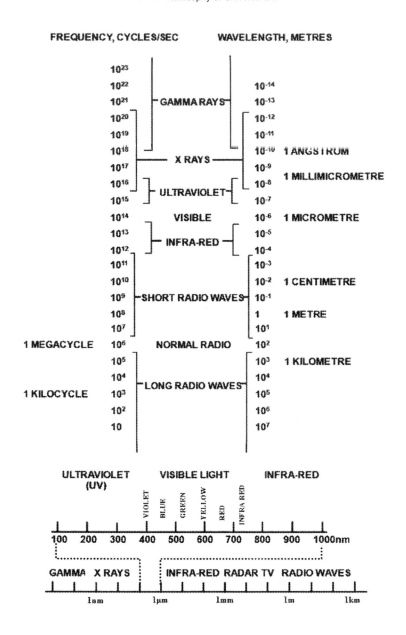

The electromagnetic spectrum with enlargement for light.

APPENDIX 4: THE HUMAN BRAIN

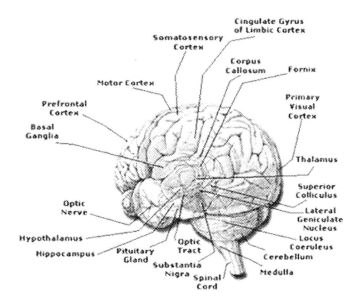

General view of the human brain.[6]

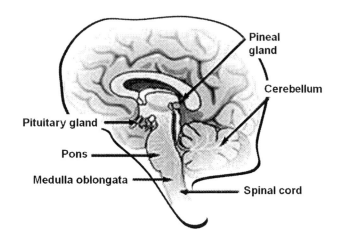

The pineal gland or pineal body, or epiphysis, which Descartes called the "seat of the soul" in *Treatise of Man*. The claustrum is very near it.[7]

The human brain showing the claustrum and amygdala
(The piriform cortex can be designated paleocortex,
and the pallidum can be designated the palaeostriatum.
The "palaeo" suggests a very old, primitive area of the brain.)[8]

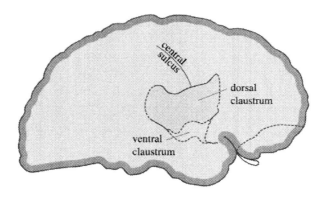

Francis Crick's illustration of the claustrum in his last paper,
which he was "refining" the day he died.[9]

APPENDIX 5: THE HUMAN EYE AND RECEPTION OF LIGHT

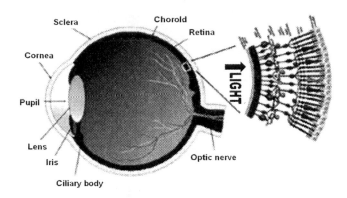

General view of the human eye.[10]

How photons are conveyed to the brain.[11]

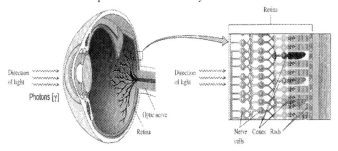

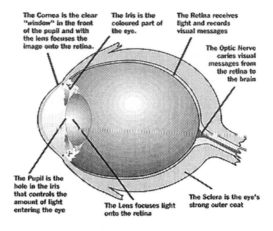

The Cornea is the clear "window" in the front of the pupil and with the lens focuses the image onto the retina.

The Iris is the coloured part of the eye.

The Retina receives light and records visual messages

The Optic Nerve caries visual messages from the retina to the brain

The Pupil is the hole in the iris that controls the amount of light entering the eye

The Lens focuses light onto the retina

The Sclera is the eye's strong outer coat

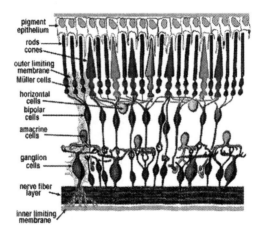

pigment epithelium
rods
cones
outer limiting membrane
Müller cells
horizontal cells
bipolar cells
amacrine cells
ganglion cells
nerve fiber layer
inner limiting membrane

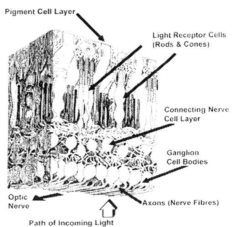

Pigment Cell Layer

Light Receptor Cells (Rods & Cones)

Connecting Nerve Cell Layer

Ganglion Cell Bodies

Optic Nerve

Axons (Nerve Fibres)

Path of Incoming Light

Light-brain transfer in close-up.[1]

Richard Dawkins' illustration of the reception of light.[13]

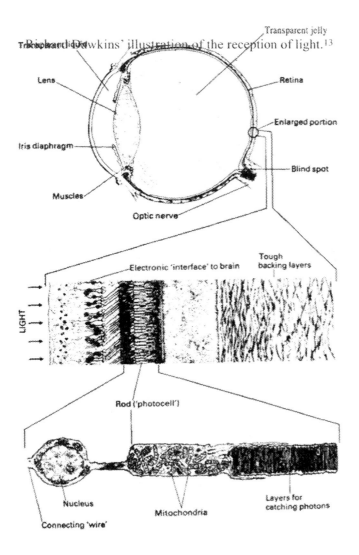

APPENDIX 6: THE SPREAD OF PHILOSOPHY FROM IONIA TO ITALY AND GREECE

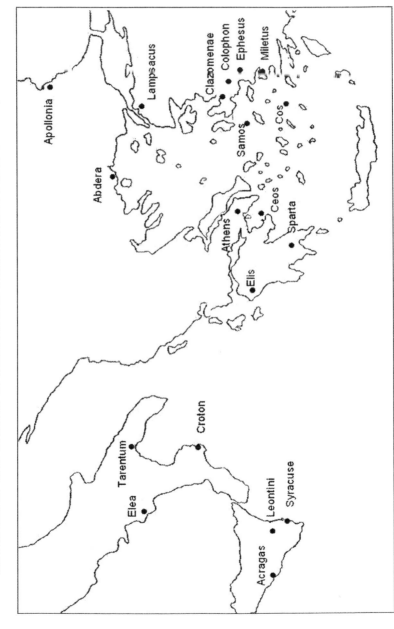

City-states showing where philosophers lived in Ionia (to the east) and where they moved to in Italy and Sicily (to the west).

NOTES AND REFERENCES TO SOURCES

Philosophical works do not usually contain Notes. However, as one aspect of Universalism rests on a science of the universe it is important that the scientific information at least should have the capacity to be confirmed as accurate, and so this work is an exception. Gone are the days when a philosopher can write a "system of general ideas" and metaphysical scheme about the universe out of his head without close reference to the world about him and without giving precise information regarding the universe he describes.

A reader needs to be able to corroborate factual information I have used. There are often many corroborating sources. I have confined myself to an accessible text or website which a reader can consult relatively easily to check my data. This does not mean that I have taken my information from that particular text or website. There may be more authoritative sources for a particular piece of information, but these may be hard for a reader to access.

Sometimes my sources are in out-of-print manuscripts or texts which are hard to locate. In such cases I have added "quoted in" and provided a reliable, in-print, widely-available work and page number where my source can readily be found. All facts can be checked in the 15th edition of the *Encyclopaedia Britannica*, the edition most found in libraries.

PRELIMINARIES

1. The image is based on an image that appeared on the front page of the English newspaper *The Independent* on 24 April 1992 under the headline "How the universe began". The image of the astronaut-surfer on top, an enlargement of which appears on pxvi, was designed by Alexander Kirk.

PROLOGUE

1. Some useful introductions to metaphysics are: Brian Carr, *Metaphysics*; D.W. Hamlyn, *Metaphysics*; George N. Schlesinger, *Metaphysics*; and Alan R. White, *Methods of Metaphysics*.
2. See the *Routledge Encyclopaedia of Philosophy*, file://C:\DOCUME~1\user\LOCALS~1\Temp\HQl6N4C.htm

PART ONE
THE STORY OF PHILOSOPHY

1. Aristotle, *Physics,* trans. by R.P. Hardie and R.K. Gaye, Table of Contents, Book III, Part IV, written c.350BC. See http://classics.mit.edu/Aristotle/physics.3.iii.html

1. The Origins of Western Philosophy

1. Robin Waterfield, *The First Philosophers*, p3. The eclipse may have been that of 21 September 582BC.

2. A.A. Long, ed., *The Cambridge Companion to Early Greek Philosophy.* See pp47, 79 for *to apeiron* and *gonimon.*

3. Sextus Empiricus, *Against the Mathematicians*, 22.26-23.20; quoted in Jonathan Barnes, *Early Greek Philosophy*, p45. Several translations distilled.

4. Clement, *Stromateis*, v 109,1, and *Simplicius, Commentary on Aristotle's Physics,* 23.11, 23.20; quoted in G.S. Kirk, J.E. Raven and M. Schofield, *The Presocratic Philosophers*, p169; and in Waterfield, *op. cit.*, pp26-27.

5. Aristotle, *Metaphysics* 986b 21-5, Ross, quoted in Kirk, Raven and Schofield, *op. cit.*, p172 and in Waterfield, *op. cit.* p28.

6. Simplicius, *Commentary on Aristotle's Physics*, 22.26-23.20; quoted in Barnes, *Early Greek Philosophy,* pp43-44. For a discussion on Xenophanes' relation to Eleatic monism, see Xenophanes of Colophon, *Fragments*, trans. by J.H. Lesher, pp190-193.

7. Aristotle, *Metaphysics*, 987b 29-30; see also *Metaphysics*, 987b 11; 990a 12, 30; 1080b 16-22; 1083b 8; 1090a 20-35.

8. Heracleitus, *On the Universe*, in *Hippocrates*, trans. by W.H.S. Jones, Loeb Classical Library, vol. IV, p47; Kirk, Raven and Schofield, *op. cit.*, p198; Long, ed., *op. cit.*, p99. For the soul consisting of eternal Fire, see Charles H. Kahn, *The Art and Thought of* Heraclitus, pp128, 238-243 and 248ff.

9. Plato, *Cratylus*, 401d, 411b, 436e, 440. Also *Philebus,* 43a; *Symposium,* 207d; *Theaetetus*, 160d, 177c, 179d, 181d, 183a. Quoted by Kirk, Raven and Schofield, *op. cit.*, p186.

10. Heracleitus, *op. cit.*, p495 and Kirk, Raven and Schofield, *op. cit.*, p195.

11. Heracleitus, *op. cit.*, p493. See Kahn, *op. cit.*, p171: "The order of the universe is…understood as a work of cognition and intention."

12. Plato, *Parmenides*, 128a, b.

13. Kirk, Raven and Schofield, *op. cit.*, p240.

14. Simplicius, *Commentary on Aristotle's Physics*, 78,5; 145,1 and 5, 144,29; quoted in Kirk, Raven and Schofield, *op. cit.*, pp248-251; Parmenides of Elea, *Fragments*, trans. by David Gallop, p65.

15. Kirk, Raven and Schofield, *op. cit.*, p251.

16. Simplicius, *Commentary on Aristotle's Physics*, 145,27; quoted in Kirk, Raven and Schofield, *op. cit.*, p251.

17. Waterfield, *op. cit.*, p53-55. For space as a plenum, see David Bohm, *Wholeness and the Implicate Order*, p191.

18. Kirk, Raven and Schofield, *op. cit.*, pp249-251.

19. Barnes, *Early Greek Philosophy*, pxxxix; Waterfield, *op. cit.*, p64; Mike Lucas, *Antarctica*, p20. Xenophanes said that the whole is one, spherical according to Theodoretus, *Treatment of Greek Afflictions*, 4.5, quoted in Xenophanes of Colophon, *Fragments, op. cit.*, p216. Parmenides may have derived his spherical earth from Xenophanes.

20. Long, *op. cit.*, p119.

21. Long, *op. cit.*, pp126-127.

22. Long, *op. cit.*, pp125-127. Also, Kirk, Raven and Schofield, *op. cit.*, pp269, 277.

23. Jonathan Barnes, *The Presocratic Philosophers*, pp203-204, 220-222.

24. John Barrow, *The Infinite Book*, p20.

25. Kirk, Raven and Schofield, *op. cit.*, p191.

26. Barnes, *Early Greek Philosophy*, pxli.

27. Barrow, *The Infinite Book*, pp178-9.

28. Plato, *Sophist*, 242d.

29. Plato, *Republic*, 515a, 532b.

30. Long, *op. cit.*, p47.

31. A summary of Aristotle's remark that Xenophanes "with his eye on the whole world (or universe) said that the One was god (or God)", quoted in Kirk, Raven and Schofield, *op. cit.*, p172 and referred to in Note 5.

32. For shamanism and Bon-po, see Nicholas Hagger, *The Light of Civilization*, pp18-25.

33. For the Kurgans, see Hagger, *The Light of Civilization*, pp29-38; and *The Rise and Fall of Civilizations*, civilisation no.1, *passim*. The three waves were in c.4400-4300BC; c.2600BC; and c.2400-2000BC.

34. For Utu and Shamash, see Hagger, *The Light of Civilization*, pp39-49.
35. For Ra, see Hagger, *The Light of Civilization*, pp50-57.
36. For Dyaeus Pitar and Zeus, see Hagger, *The Light of Civilization*, pp36, 69-70.
37. For Mitra, see Hagger, *The Light of Civilization*, pp109-111, 278-279.
38. For *rta* as cosmic order, see Jeanine Miller, *The Vision of Cosmic Order in the Vedas*, pp13-28, 231, 241.
39. For Vedism, Hinduism and Buddhism, see Hagger, *The Light of Civilization*, pp278-289.
40. For the Iranian Zoroaster and Ahura Mazda, see Hagger, *The Light of Civilization*, pp106-112; and *The Last Tourist in Iran*, pp62-67, 101-111.
41. For the Hittite Storm-and-Weather god Baal and Yahweh, see Hagger, *The Light of Civilization*, pp84-96.
42. For the Eleusinian mysteries, see Hagger, *The Light of Civilization*, pp73-75.
43. For *T'ien* and the *Tao*, see Hagger, *The Light of Civilization*, pp330-341.
44. For the Celtic Du-w, see Hagger, *The Light of Civilization*, pp97-100.
45. For Hermetic and Neoplatonist texts, see Hagger, *The Light of Civilization*, pp401-404 and 406-409.
46. For the Clear Light of the Void, see Hagger, *The Light of Civilization*, pp343-350.
47. For Sufism and al-Arabi, see Hagger, *The Light of Civilization*, pp259-274.
48. Shelley, 'Adonais', L11.

2. The Decline of Western Philosophy and the Way Forward

1. A paraphrase of excerpts from Plato, *Letter VII* to Dion, 326b and *Republic* V. 473d; both in Plato, *op. cit.*
2. The sources for this tradition are all late: Joannes Philoponus, a late Neoplatonic Christian philosopher who lived in Alexandria in the 6th century AD, in his commentary on Aristotle's *De Anima, Comm. In Arist. Graeca*, XV, ed. M. Hayduck, Berlin 1897, pp117,29; and Elias, who also lived in Alexandria in the 6th century AD, in his commentary on Aristotle's *Analytics, Cat., Comm. In Arist. Graeca*, XVII, pars 1, ed. A. Busse, Berlin, 1900, pp118,18. Joannes Tzetzes, a Byzantine author from the early 12th century AD, in *Chiliades*, VII, 973, gives a fuller version: "Let no one ignorant of geometry come under my roof."

3. See *Encyclopaedia Britannica*, 14. 531-538, for Plato.

4. See *Encyclopaedia Britannica*, 1. 1162-1171, for Aristotle.

5. See Hagger, *The Light of Civilization*, pp406-409, for Plotinus.

6. See *Encyclopaedia Britannica*, 14. 539-545, for Platonism and Neoplatonism until the end of the section.

7. See Hagger, *The Light of Civilization*, pp128-133, for Augustine.

8. Augustine, *Confessions*, 8.7.

9. Quoted in many sources, including David W. Kling, *The Bible in History*, Oxford University Press, n/d.

10. See *Encyclopaedia Britannica*, pp1155-1161 for Aristotelianism until the end of the section.

11. *The Notebooks of Leonardo Da Vinci*, ch. I, Prolegomena and General Introduction to the Book on Painting, 10, second Introduction for experience; and ch. XIX, Philosophical Maxims, 1154, for instrumental or mechanical science, and 1158 for mathematical sciences – see http://en.wikisource.org/wiki/The_Notebooks_of_Leonardo_Da_Vinci

12. See *Encyclopaedia Britannica*, 6. 766-770, for Empiricism.

13. See *Encyclopaedia Britannica*, 15. 527-532, for Rationalism.

14. T.S. Eliot, 'The Metaphysical Poets' (1921), in T.S. Eliot, *Selected Prose*, pp111-120. Eliot writes: "In the seventeenth century a dissociation of sensibility set in, from which we have never recovered; and this dissociation, as is natural, was aggravated by the influence of the two most powerful poets of the century, Milton and Dryden." In 'Milton' (1947) Eliot reconsidered laying the blame at the door of Milton and Dryden, correctly for the responsibility was Descartes'.

15. James Boswell, *The Life of Samuel Johnson*, abridged version, *Everybody's Boswell*, p99.

16. See *Encyclopaedia Britannica*, 9. 189-193, for Idealism.

17. See *Encyclopaedia Britannica*, 15. 539-542, for Realism.

18. Alfred North Whitehead, *An Anthology*, selected by F.S.C. Northrop and Mason W. Gross, p567.

19. See *Encyclopaedia Britannica*, 1. 799-807, for Analytic and Linguistic Philosophy.

20. A transcript of the 1948 broadcast was reprinted in Bertrand Russell, *Why I Am Not a Christian*, p155.

21. See *Encyclopaedia Britannica*, 14. 210-215, for Phenomenology.

22.   See *Encyclopaedia Britannica*, 7. 73-79, for Existentialism.

23.   Alfred North Whitehead, *op. cit.,* p567.

24.   Alfred North Whitehead, *op. cit.,* p845.

25.   See http://plato.stanford.edu/entries/wolff-christian/#The

PART TWO
THE SCIENTIFIC VIEW OF THE UNIVERSE
AND THE ORDER PRINCIPLE

3. Cosmology and Astrophysics:
The Big Bang and the Origin of the Universe

1.   See Julian Barbour, *The End of Time*, p12, for the originality of this idea.

2.   John Barrow, *The Infinite Book*, pp122-123.

3.   Simon Singh, *Big Bang*, pp156-161.

4.   Singh, *op. cit.,* pp149-156.

5.   Singh, *op. cit.,* pp 214-229, 248-261. The discovery in 1998 that the universe's expansion is accelerating means that some calculations based on "Hubble time" are now unsafe. The oldest stars may not now be older than the universe.

6.   Singh, *op. cit.,* pp341-347, 418-421, 438-440.

7.   Singh, *op. cit.,* pp422-437.

8.   Singh, *op. cit.,* pp453-461, 471-473.

9.   Roger Penrose, *The Emperor's New Mind*, p543. For Dante's infinitesimal "point", see *Paradiso*, canto XXVIII.

10.   Penrose, *The Emperor's New Mind,* pp435-447.

11.   Martin Rees, *Just Six Numbers: The Deep Forces That Shape The Universe*, p44.

12.   Penrose, *The Emperor's New Mind,* pp541-544.

13.   Penrose, *The Emperor's New Mind,* pp436, 470-471; Stephen Hawking, *A Brief History of Time,* pp88-92, 133-136.

14.   Georg Cantor, *Gesammelte Abhandlungen*, 3378; quoted in Rudy Rucker, *Infinity and The Mind*, p9 and in Barrow, *The Infinite Book*, pp93-94.

15.   Barrow, *The Infinite Book*, p94.

16.   Barrow, *The Infinite Book*, p38.

17.   Singh, *op. cit.,* p489.

18.   Hawking, *op. cit.,* pp137-138.

19. Hawking, *op. cit.*, pp116, 133-136.

20. Hawking, *op. cit.*, pp140-141.

21. Penrose, *The Emperor's New Mind*, p471.

22. Barrow, *The Infinite Book*, pp103-109.

23. John Gribbin, *In the Beginning*, pxii.

24. Gribbin, *In the Beginning*, p229.

25. Singh, *op. cit.*, p482. Details of the Pamela probe and discovery were revealed by Mirko Boezio at a cosmology conference in Stockholm and reported in *The Sunday Times*, 21 September 2008. Boezio emphasised that the results were preliminary and had not been subjected to peer review.

26. To check facts and timings in this section, see Tom Duncan with Heather Kennett, *Advanced Physics*, pp493-507.

27. Penrose, *The Emperor's New Mind*, p419.

28. Gribbin, *In the Beginning*, p11. For inflation by 10^{54}, faster than the speed of light, in the next paragraph see http://abyss.uoregon.edu/~js/21st_century_science/lectures/lec25.html

29. For 10^{23} stars see Fred Adams, *Origins of Existence*, p66.

30. For 10^{23} grains of sand see Fred Adams and Greg Laughlin, *The Five Ages of the Universe*, pxxiii. And http://en.wikipedia.org/wiki/1000000000000_(number)

31. The acceleration was discovered from studies of distant supernovae and results announced by two international teams of researchers. For acceleration, see Martin Rees, *Our Cosmic Habitat*, pp104-108

32. For acceleration beginning 8 billion years after inflation, 5 billion years ago, see Barrow, *The Infinite Book*, p147. Barrow claims that dark energy took over 8 billion years after the first 379,000 years, that deceleration soon changed to acceleration and that no more galaxies could form.

33. Paul Davies, *The Goldilocks Enigma*, pp47-49. Also http://superstringtheory.com/cosmo21.html, 'What is the structure of the universe? Open, closed or flat?'

34. Barrow, *The Infinite Book*, pxv.

35. Barrow, *The Infinite Book*, p96.

36. Barrow, *The Infinite Book*, p22.

37. Barrow, *The Infinite Book*, p22.

38. Barrow, *The Infinite Book*, p146; Davies, *The Goldilocks Enigma*, p26.

39. For WMAP putting the age of the universe at 13.7 billion years ago, see

http://map.gsfc.nasa.gov/news/ and
http://en.wikipedia.org/wiki/Age_of_the_universe

40. Barrow, *The Infinite Book*, p149.

41. John Gribbin, *The Universe*, p224.

42. Adams, *Origins of Existence*, p66.

43. Adams and Laughlin, *op. cit.*, pxxiii.

44. Adams, *Origins of Existence*, p218.

45. Adams, *Origins of Existence*, p204.

46. Adams, *Origins of Existence*, p63.

47. See http://physicsworld.com/cws/article/print/601

48. Rees, *Our Cosmic Habitat*, pp102-104.

49. See http://home.earthlink.net/~rarydin/

50. For some of the detail in the three paragraphs of (2), see Amanda Gefter, *New Scientist Magazine*, 10 March 2007.

51. See http://en.wikipedia.org/wiki/SLOAN_Great_Wall
     The Sloan Digital Sky Survey, named after Alfred P. Sloan, the former president and chief executive of General Motors whose foundation helped fund the project, mapped the universe from 2000 to 2008 and made its findings public in October 2008. The project used a specially built telescope at the Apache Point observatory in the Sacramento Mountains of New Mexico and used redshift to measure distances between galaxies and also the length and width of large objects. It has determined the exact position of more than 200 million celestial bodies, including one million galaxies and 100,000 quasars.

52. See http://en.wikipedia.org/wiki/Great_Wall

53. See
     http://newsvote.bbc.co.uk/mpapps/pagetools/print/news.bbc.co.uk/1/hi/sci/tech/69621

54. Rees, *Our Cosmic Habitat*, p134.

55. See http://home.earthlink.net/~rarydin/

56. Fred Hoyle, *The Intelligent Universe*, p181.

57. Gribbin, *In the Beginning*, pxiii.

58. John Gribbin and Martin Rees, *The Stuff of the Universe*, p179.

59. Brian Greene, *The Elegant Universe*, pp135-136.

60. See http://home.earthlink.net/~rarydin/bigwave.html

61. *The Edge,* annual question, 2006, see http://www.edge.org
     Quoted in Davies, *The Goldilocks Enigma*, p194.

62. The Leibnizian philosopher Geoffrey Read, in conversation with Nicholas Hagger in December 1993. Leibniz contended that space is relational and that spatial events are primary. Newton held that space and time are absolutes and are primary. Hence Leibniz disputed with the Newtonians. Einstein held that space is relational in being intrinsically bound up with matter and time, but primary in so far as the curvature of space by the distortion of matter is the cause of gravity, and Leibniz/Read therefore cannot accept Einstein's special and general theories of relativity. See Peter Hewitt, *The Coherent Universe*, p39, where Read agrees with Guy Burniston Brown that the physics of the special theory of relativity are "fundamentally wrong". See p204 for Read's disagreement with Einstein's general theory of relativity: "As a pure field theory, general relativity is intrinsically incompatible with Read's theory….General relativity, for example, conflates time and space into a purely mathematical conception….But both time and space in Read's theory are abstractions (from temporal relationship and spatial relationship respectively)."

4. Physics: The Quantum Vacuum and the Expanding Force of Light

1. For early quantum theory, see Duncan and Kennett, *op. cit.,* pp391-392.
2. *Encyclopaedia Britannica*, 1991 ed., micropaedia, 9.841, 'Quantum mechanics'.
3. *Encyclopaedia Britannica*, 1991 ed., micropaedia, 9.840, 'Quantum electrodynamics'.
4. David Castillejo, *The Expanding Force in Newton's Cosmos*, p108.
5. Castillejo, *op. cit.,* p108.
6. Castillejo, *op. cit.,* p109. For ether in the next two paragraphs, see Singh, *op. cit.,* pp93-98.
7. *The Born-Einstein Letters*, trans. by Irene Born, p91.
8. Albrecht Fölsing, *Albert Einstein*, pp587-592.
9. Duncan and Kennett, *op. cit.,* p507. Freeman Dyson questioned whether any experiment in the real universe can detect one graviton. If not, he held that it is meaningless to talk about gravitons as physical entities.
10. Fölsing, *op. cit.,* pp704-705.
11. Greene, *op. cit.,* p15.
12. Bohm, *op. cit.,* pp65-110; and Penrose, *The Emperor's New Mind.,* p362.
13. Penrose, *The Emperor's New Mind,* p362.

14. Bohm, *op. cit.*, pp149, 190-193.
15. Renée Weber, *Dialogues with Scientists and Sages: The Search for Unity*, pp25-27. Also David Bohm in conversation with Nicholas Hagger.
16. Bohm, *op. cit.*, pp189, 193, 211.
17. Bohm, *op. cit.*, p149.
18. Bohm, *op. cit.*, ppix-x.
19. Bohm, *op. cit.*, p190.
20. Bohm, *op. cit.*, p191.
21. Bohm, *op. cit.*, p191.
22. Bohm, *op. cit.*, p208.
23. Bohm, *op. cit.*, pp191-192.
24. Bohm, *op. cit.*, p193.
25. Rees, *Just Six Numbers*, p12.
26. Hewitt, *op. cit.*, p2.
27. Gunzig is Professor of General Relativity at the Free University, Brussels. This and the next paragraph are based on a lecture he gave in April 1991.
28. See http://en.wikipedia.org/wiki/Gravitational_slingshot
29. See http://technology.newscientist.com/article/mg19225771.800
30. Formally "infinite" masks the possibility that zero energy in the vacuum is metaphysically as well as mathematically infinite. See pp60-62.
31. *The Daily Telegraph*, 26 February 2008, p27.
32. *The Sunday Telegraph*, 6 April 2008, p31. For Higgs boson field being like *aither*, see Davies, *The Goldilock Enigma*, pp178-179; *The Times*, 8 April 2008, p18.
33. Barrow, *The Infinite Book*, pp186-188.
34. Hawking, *op. cit.*, p74.
35. Hawking, *op. cit.*, p136.
36. Barrow, *The Infinite Book* pxiv.
37. See Hewitt, *op. cit.*, pp55 and 106, for Geoffrey Read's Leibnizian view that Newton was wrong about the primary nature of time and that Einstein's view of time was closer to Leibniz's than to Newton's.

5. Biocosmology: The Bio-Friendly Universe and Order

1. I have broadly followed, with some refinements, the scheme in Paul Davies, *The Goldilocks Enigma*, pp295-302.
2. Fred Hoyle, 'The universe: past and present reflections', *Annual Review of*

Astronomy and Astrophysics, vol. 20, 1982, p16.

3.   A.R. Wallace, *Man's Place in the Universe: A Study of the Results of Scientific Research in relation to the Unity or Plurality of Worlds*, p256-257.

4.   Robert H. Dicke, 'Gravitation without a Principle of Equivalence', in *Reviews of Modern Physics*, 29, 1957, pp363-376.

5.   Dicke, *Nature*, 192, 1961, p440.

6.   Alan Lightman, *Ancient Light*, p118.

7.   Brandon Carter, 'Large Number Coincidences and the Anthropic Principle in Cosmology', in *IAU Symposium 63: Confrontation of Cosmological Theories with Observational Data*, ed. M.S. Longair, Reidel, Dordrecht, 1974, pp291-298. For Krakow/Copernicus, see http://en.wikipedia.org/wiki/Anthropic_principle

8.   Carter, *op. cit.*, quoted in Barrow and Tipler, *op. cit.*, p1.

9.   Carter, *op. cit.*, quoted in Wikipedia, see http://en.wikipedia.org/wiki/Anthropic_principle

10.   Barrow and Tipler, *op. cit.*, p16.

11.   Barrow and Tipler, *op. cit.*, p21.

12.   Barrow and Tipler, *op. cit.*, p22.

13.   Penrose, *The Emperor's New Mind*, pp560-561.

14.   Barrow and Tipler, *op. cit.*, pp185-204.

15.   For example, John Gribbin and Martin Rees, *The Stuff of the Universe* (1989); John Gribbin and Martin Rees, *Cosmic Coincidences* (1990); Paul Davies and John Gribbin, *The Matter Myth* (1991); Paul Davies, *The Mind of God* (1992); John Gribbin, *In the Beginning* (1993); Martin Rees, *Before the* Beginning (1997); Martin Rees, *Just Six Numbers* (1999); Martin Rees, *Our Cosmic Habitat* (2002); Fred Adams, *Origins of Existence* (2002); Paul Davies, *The Goldilocks Enigma* (2006); and John Gribbin, *The Universe, A Biography* (2006). It could be argued that Paul Davies's earlier books in the early 1980s created the climate in which Barrow and Tipler could write *The Anthropic Cosmological Principle*.

16.   Adams, *op. cit.*, p209.

17.   George Greenstein, *The Symbiotic Universe*, pp257-258; also Adams, *Origins of Existence*, p204.

18.   Rees, *Just Six Numbers*, pp118, 127-128. For gravity being 1,000 billion times weaker than on a neutron star, see Rees, *Our Cosmic Habitat*, p93.

19.   Rees, *Just Six* Numbers, pp3, 150; Hawking, *op. cit.*, pp162-165.

20. Duncan with Kennett, *op. cit.,* p496.

21. Rees, *Just Six Numbers*, pp3, 80-82; Rees, *Our Cosmic Habitat*, p130; Gribbin and Rees, *Cosmic Coincidences*, pp16-18. For a decrease of one part in a hundred thousand million million, see Hawking, *op. cit.,* pp121-122. For increase see http://space.newscientist.com/article/dn11498

22. Lightman, *op. cit.,* pp60-61.

23. Rees, *Just Six Numbers,* pp2-3, 80-82, 91-101; Rees, *Our Cosmic Habitat,* p104.

24. Rees, *Just Six Numbers*, pp107ff; Rees, *Our Cosmic Habitat,* p105.

25. Rees, *Just Six Numbers*, p35.

26. Gribbin and Rees, *Cosmic Coincidences*, pp263-269.

27. Rees, *Just Six Numbers*, pp33-35.

28. Rees, *Just Six Numbers*, pp2, 52-57.

29. Davies, *The Goldilocks Enigma*, pp20-26.

30. Compare Davies, *The Goldilocks Enigma*, pp272-274.

31. Gribbin and Rees, *Cosmic Coincidences*, pp23-26.

32. Gribbin and Rees, *Cosmic Coincidences*, pp88, 241-246. For the temperature of the sun, see http://en.wikipedia.org/wiki/Sun

33. Gribbin and Rees, *Cosmic Coincidences*, pp33-34, 251-254.

34. Rees, *Our Cosmic Habitat*, p81.

35. Rees, *Just Six Numbers*, pp133-135.

36. Rees, *Just Six Numbers*, p134; Adams and Laughlin, *op. cit.,* pp197-199; also see http://www.counterbalance.net/ghc-bb/coinc-body.html; http://indoforums.com/englishklab/viewtopic.php?p=757&sid=2917ff951b76d760301; http://books.google.co.uk/books?id=amXAmOYB3WkC&pg=PA55&lpg=PA55&dq=strong+nuclear+force+5+per+cent+weaker&source=web&ots=d800APnF42&sig=0gZwxqgtoRQR5L-dwpTrto6UuvQ&hl=en

37. Rees, *Just Six Numbers*, pp133-136; Nicholas Hagger's correspondence with CERN expert.

38. Gribbin and Rees, *Cosmic Coincidences*, pp177-178.

39. Gribbin and Rees, *Cosmic Coincidences*, pp247-253.

40. Gribbin, *In the Beginning*, p179.

41. Gribbin and Rees, *Cosmic Coincidences*, p254.

42. Gribbin and Rees, *Cosmic Coincidences*, pp9-10. 43. Lightman, *op. cit.,* p119.

44. Adams, *Origins of Existence*, pp57, 63.

45. Adams, *Origins of Existence*, p66. For 10^{23} grains of sand, see http://en.wikipedia.org/wiki/1000000000000_(number)

46. Gribbin, *In the Beginning*, p255. For 90,000 years, see Jack Phillips, *Suppressed Science*, p32.

47. See http://science.qi.net/

48. See http://www.dailymail.co.uk/pages/live/articles/technology/technology.html?in_article,
    For 228 planets see http://www.physorg.com/news124121930.html

49. Gribbin and Rees, *Cosmic Coincidences*, pp11-14.

50. Barrow and Tipler, *The Anthropic Cosmological Principle*, p385. For the Milky Way containing 200-400 billion stars, see http://en.wikipedia.org/wiki/Milky_Way

51. Greenstein, *op. cit.*, p255.

52. Greenstein, *op. cit.*, pp256, 63-65.

53. Lightman, *op. cit.*, p118.

54. Gribbin and Rees, *Cosmic Coincidences*, p104; Barrow and Tipler, *The Anthropic Cosmological Principle*, pp304-305.

55. Barrow and Tipler, *The Anthropic Cosmological Principle*, p305.

56. Barrow and Tipler, *The Anthropic Cosmological Principle*, p408. As regards light in a Steady-State universe, if the universe were static, light would reach us from every part of the sky and it would be bright at night. As it is, we are in an expanding universe and light has not reached us from some regions of the sky. Because of the non-directional nature of the cosmic microwave background radiation, to an observer on another galaxy our galaxy would look similar to our view of the observer's galaxy.

57. Greenstein, *op. cit.*, pp93-95.

58. Barrow and Tipler, *The Anthropic Cosmological Principle*, p400.

59. Greenstein, *op. cit.*, pp96-97.

60. Barrow and Tipler, *The Anthropic Cosmological Principle*, pp545-556.

61. Rees, *Just Six Numbers*, pp18-19.

62. Barrow and Tipler, *The Anthropic Cosmological Principle*, pp544-545.

63. C.J. Clegg with D.G. Mackean, *Advanced Biology, Principles and Applications*, pp58, 321 and elsewhere; p75 for the water cycle.

64. Adams, *Origins of Existence*, p66.

65. Rees, *Just Six Numbers*, pp94-96; Gribbin and Rees, *Cosmic Coincidences*, p113.

66. Davies, *The Goldilocks Enigma*, pp116-122; Gribbin and Rees, *Cosmic Coincidences*, pp181-182.

67. Davies, *The Goldilocks Enigma*, pp132-137.

68. Rees, *Just Six Numbers*, pp88-89.

69. Gribbin and Rees, *Cosmic Coincidences*, pp87-89.

70. Rees, *Our Cosmic Habitat*, p75.

71. Davies, *The Goldilocks Enigma*, pp169-170.

72. Rees, *Our Cosmic Habitat*, pp107-108, 146. For the observable universe being over 13 billion years and including about 100 billion galaxies see Davies, *The Goldilocks Enigma*, p27.

73. Also see Phillips, *op. cit.*, p35.

74. *Nature*, 438, November 2005.

75. Davies, *The Golidlocks Enigma*, pp27, 37; Björn, *op. cit.*, p238. For 10^{54} see http://abyss.uoregon.edu/~js/21st_century_science/lectures/lec25.html

76. Adams, *Origins of Existence*, pp84-85.

77. Adams, *Origins of Existence*, p55.

78. See http://www.leaderu.com/offices/billcraig/docs/teleo.html

79. Barrow and Tipler, *The Anthropic Cosmological Principle*, pp401-403.

80. *Darwinism, Design and Public Education, op. cit.*, pp241-242.

81. William A. Dembski, *The Design Revolution*, p85: "Computer scientist Seth Lloyd sets 10^{120} as the maximum number of bit-operations that the universe could have formed throughout its entire history (*Physical Review Letters*, June 10, 2002). That number corresponds to a universal probability bound of 1 in 10^{120}. Stuart Kaufman in his most recent book, *Investigations* (2000), comes up with similar numbers." Also Davies, *The Goldilocks Enigma*, pp270-271, which reflect the bit-operations calculation. See Seth Lloyd, 'Computational capacity of the universe', *Physical Review Letters*, vol. 88, 2002, p237901, and his book *The Computational Universe*.

82. *Darwinism, Design and Public* Education, p241 and pp277-278, note 56. For 10^{80} see http://en.wikipedia.org/wiki/Observable_universe

83. As 81.

6. Biology and Geology: The Origin of Life and Evolution

1. For 4.6 billion years ago, see Adams, *Origins of Existence*, p183; Clegg and

Mackean, *op. cit.*, p76; *Encyclopaedia Britannica* (1991), macropaedia, 19.773. For 4.55 billion years ago, see, for example, http://www.talkorigins.org/faqs/faq-age-of-earth.html.
Davies, *The Goldilocks Enigma*, p23, puts the age of the earth at 4.56 billion years.

2. Clegg with Mackean, *op. cit.*, p664.

3. See http://www.accessexcellence.org/bioforum/bf02/awramik/bf02a3.html
    Also http://en.wikipedia.org/wiki/Origin_of_life

4. Adams, *Origins of Existence*, p182. Also as 3.

5. See http://en.wikipedia.org/wiki/Abiogenesis

6. The "Primordial Soup Theory" suggested that life began in a pond or ocean as a result of the combination of chemicals. It was so named after Oparin referred to a "Primeval Soup" of organic molecules. See http://en.wikipedia.org/wiki/Origin_of_life

7. http://en.wikipedia.org/wiki/Origin_of_life

8. As 7.

9. As 7.

10. Clegg with Mackean, *op. cit.*, p664.

11. As 7.

12. Clegg with Mackean, *op. cit.*, pp664-665.

13. For 10^{130} combinations of amino acids see *Darwinism, Design and Public Education, op. cit.* p241. Also see http://leiwenwu.tripod.com/primordials.htm

14. As 7.

15. As 7.

16. Fred Hoyle, Chandra Wickramasinghe and John Watkins, *Viruses from Space*, and Fred Hoyle and Chandra Wickramasinghe, *Evolution from Space, passim.*

17. As 7. See orbituary of Leslie Orgel in *The Times*, 6 December 2007: "The theory that life is RNA-based life was shared by the late Francis Crick."

18. As 7. For a report on the "Lost City" at the bottom of the Atlantic – so-called because of its towering pinnacles and chimneys – see http://www.telegraph.co.uk/earth/main.jhtml?xml=/earth/2008/01/31/scilost131.xml

19. As 7.

20. As 7.

21. Mary Morrison in conversation with Nicholas Hagger.

22. As 21.

23. As 21. The Cancer Council states 10^{15}, see
    http://www.sunsmart.com.au/downloads/schools/project_helpers/what_is_cancer.pdf
    For an alternative, 10^{14}, see
    http://en.wikipedia.org/wiki/1000000000000_(number)

24. As 23.
    http://www.sunsmart.com.au/downloads/schools/project_helpers/what_is_cancer.pdf

25. UN Population Division, see
    http://www.un.org/esa/population/publications/sixbillion/sixbilpart1.pdf-box1.

26. See http://en.wikipedia.org/wiki/World_population

27. See http://leiwenwu.tripod.com/primordials.htm Also,
    http://www.accessexcellence.org/bioforum/bf02/awramik/bf02a.html

28. Gribbin and Rees, *Cosmic Coincidences*, pp33-34: "We are, quite literally, made from the ashes of long-dead stars." Also see
    http://books.nap.edu/openbook.php?record_id=12161&page=9

29. This section draws on *Encyclopaedia Britannica* (1991), macropaedia, 25.883-892. Also on lectures given by, and discussions with, Patrick Abbott, Professor Emeritus of San Diego University, on plate tectonics.

30. *Encyclopaedia Britannica*, 25.889.

31. For Pangaea, Gondwana and Laurasia, see *Encyclopaedia Britannica* (1991), macropaedia, 19.748-876, 'Geochronology', and *Encyclopaedia Britannica*, 25.889.

32. *Encyclopaedia Britannica*, 25.889-890.

33. See http://en.wikipedia.org/wiki/Chicxulub_Crater

34. Patrick Abbott's research and conversations with Nicholas Hagger.

35. Patrick Abbott's research. Also see
    http://en.wikipedia.org/wiki/Earth's_magnetic_field#Magnetic_field_reversals

36. Clegg with Mackean, *op. cit.,* p646.

37. Clegg with Mackean, *op. cit.,* pp648-649.

38. See
    http://books.google.co.uk/books?id=FTtrd9cl8tAC&pg=PA151&lpg=

PA151&dq=3.8+billion+years+ago+fossils&source=web&ots=phr1A6h6la
&sig=-h52e-rCOO3Xko9T2cEeFb56d8E&hl=en
Also see http://www.geocities.com/capecanaveral/lab/2948/onceupon.html

39.   *Encyclopaedia Britannica* (1991), macropaedia, 19.773.

40.   See
      http://campus.udayton.edu/~INSS/ThemeEvol/EvolTimeline.HSM.
      pp#358,33,Slide

41.   See *Encyclopaedia Britannica*, macropaedia, 7. 1066 for eras and periods
      in this and the next paragraph. Also Clegg and Mackean, *op. cit.,* p649.

42.   See http://en.wikipedia.org/wiki/Ice_age

43.   As 42.

44.   For 11,750/11,500-10,000 years ago, and for Greenland and Antarctica, as
      42.

45.   As 42. For the 29 and 10 per cent, Patrick Abbott's research. For the 13
      per cent, Michael Schmid's research.

46.   As 42. For 90,000 years and 12,500 years see Phillips, *op. cit.*, p32.

47.   Patrick Abbott's research.

48.   As 42. See also Phillips, *op. cit.,* pp32-34

49.   As 42. Also *Encyclopaedia Britannica* (1991), macropaedia, 19.783.

50.   As 42. for numbers (1)-(5). In addition, for the Huronian Ice Age being 2.7-
      2.3 billion years ago see http://en.wikipedia.org/wiki/Ice_age.
      For this first Ice Age beginning 2.3 billion years ago, see *Encyclopaedia
      Britannica* (1991), macropaedia, 19.783.
      Also see http://en.wikipedia.org/wiki/Timeline_of_glaciation

51.   See http://www.tqnyc.org/NYC040654/draft_file/SomeFacts.htm. Also
      Patrick Abbott's research and conversations with Nicholas Hagger.

52.   As 42.

53.   Confirmation was given in a paper by J.D. Hays, J. Imbrie and N.J.
      Shackleton, 'Variations in the Earth's Orbit: Pacemaker of the Ice Ages' in
      Science, 194, 1976, pp1121-1132.

54.   For Milankovitch cycles, see http://en.wikipedia.org/wiki/Milankovitch_
      cycles

55.   As 42.

56.   See http://muller.lbl.gov/pages/IceAgeBook/IceAgeTheories.html
      See also http://en.wikipedia.org/wiki/Milankovitch_cycles

57.   As 54.

58. Patrick Abbott's research and conversations with Nicholas Hagger.

59. As 57.

60. As 57.

61. A phrase used by Dr. Willie Soon of the Harvard-Smithsonian Center for Astrophysics at a meeting in Colorado Springs on 27 July 2007.

62. A view expressed by Christopher Booker and Richard North in *Scared to Death*.

63. See http://www.sciencedaily.com/releases/1997/11/971114070632.htm Also see http://books.google.co.uk/books?id=DJxlzuOdK2IC&pg=PA8&lpg=PA8&dq=32000+years+solar+2001&source=web&ots=vZbRD1oI_G&sig=kiqwaFx9-WL6Ea72-P6WnfmpVq0&hl=en

64. A 1980 study by J. Imbrie and J.Z. Imbrie predicts a long-term cooling trend that will last for the next 23,000 years: 'Modelling the Climatic Response to Orbital Variations', in *Science, 2007* (29 February 1980), pp943-953. But since 1980 there has been a better understanding of glacial-interglacial cycles, and this 1980 study may not appreciate that we are in a short-term interglacial of warming within a long-term glacial of cooling.

65. Clegg with Mackean, *op. cit.,* p647.

66. Clegg with Mackean, *op. cit.,* p11.

67. Charles Darwin, *Voyage of the Beagle*, p268-290.

68. Nicholas Hagger's research in the Galapagos Islands.

69. Carl Zimmer, *Evolution*, pp21-22; Jonathan Weiner, *The Beak of the Finch*, p28.

70. Pierre Constant, *The Galapagos Islands and Natural History Guide*, pp143-144; Weiner, *op. cit.,* pp283-286.

71. Constant, *op. cit.,* p138.

72. Constant, *op. cit.,* p80.

73. Constant. *op. cit.,* pp44-48.

74. David Andrew, *Watching Wildlife, Galapagos Islands*, pp118-119.

75. Charles Darwin, *The Voyage of the Beagle: Journal of Researches Into the Natural History and Geology of the Countries Visited During the Voyage of HMS, Beagle Round the World*, p339. Quoted in Andrew, *op. cit.*, p118.

76. Constant, *op. cit.,* p138.

77. Constant, *op. cit.,* pp97-100.

78. Andrew, *op. cit.,* p60.

79. *Encyclopaedia Britannica*, macropaedia, 5.493.

80. Article by Jim Al-Khalili in *The Daily Telegraph*, 29 January 2008.

81. For example, see http://www.crystalinks.com/darwin.html

82. Clegg with Mackean, *op. cit.,* p657.

83. Herbert Spencer, *Principles of Biology*, 1864.

84. As 82.

85. As 82.

86. Mary Morrison in conversation with Nicholas Hagger.

87. As 86.

88. Clegg with Mackean, *op. cit.,* p653.

89. Neil Schubin, *Your Inner Fish: A Journey into the 3.5 Billion-Year-History of the Human Body.*
    Also see http://en.wikipedia.org/wiki/Tiktaalik

90. Clegg with Mackean, *op. cit.,* p669 for apes and for Africa joining with Eurasia 16 or 17 million years ago.

91. See
    http://en.wikipedia.org/wiki/Human_evolution#Comparative_table_of
    _Homo_species

92. Clegg with Mackean, *op. cit.,* p669. Douglas Palmer, *Neanderthal*, pp65-66, 160.

93. Palmer, *op. cit.,* pp157-161, 162. For 400,000/250,000 years ago, see *Encyclopaedia Britannica*, 1991 ed., micropaedia, 6.28.

94. See
    http://en.wikipedia.org/wiki/Human_evolution#Comparative_table_of
    _Homo_species

95. Palmer, *op. cit.,* pp180-181.

96. Palmer, *op. cit.,* pp 165, 182-183. For *Homo antecessor*, see *The Daily Telegraph*, 27 March 2008, p36. For 400,000/250,000 years ago, see *Encyclopaedia Britannica*, 1991 ed., micropaedia, 6.28.
    The fossilised bones of a female *Homo Floresiensis* were discovered on the island of Flores, Indonesia in 2003 and swiftly identified as a new species of *Homo* whose adults were only 1.5 metres tall (and so the species was nicknamed "the hobbit"). Her brain, modelled from her skull, would have been tinier than a pygmy's brain. The species died out 0.012 million years ago (though similar creatures, called Ebu Gogos, were reported to be living on a volcano in the vicinity 300 years ago). There is a speculative view that

800,000 years ago *Homo erectus* advanced to Indonesia, despite strong currents found a way to cross water on a raft, which would have required the use of language, settled in Flores and shrank. Elephants there also shrank to 1.5 metres, but reptiles grew larger. If this theory is true, then *Homo erectus*, not *Homo sapiens*, was the first to cross water and use language 800,000 years ago. However, many scholars do not accept that *Homo Floresiensis* was a new species of *Homo* but rather a pygmy-like variation of *Homo*.

97. Palmer, *op. cit.,* pp157-159.

98. See

http://en.wikipedia.org/wiki/Human_evolution#Comparative_table_of _Homo_species

Also see

http://johnhawks.net/weblog/reviews/genomics/divergence/dawn_ chumans_patterson_2006.html

99. See

http://en.wikipedia.org/wiki/Human_evolution#Comparative_table_of _Homo_species

Also see Frans de Waal, *Our Inner Ape,*p14,

http://johnhawks.net/weblog/reviews/genomics/divergence/dawn_ chumans_patterson_2006.html

100. See

http://en.wikipedia.org/wiki/Human_evolution#Comparative_table_of _Homo_species

Also see de Waal, *op. cit.,* p14.

101. See de Waal, *op. cit.,* pp13-15.

102. Palmer, *op. cit.,* pp160-162.

103. Palmer, *op. cit.,* pp157-159.

104. Palmer, *op. cit.,* p162.

105. Palmer, *op. cit.,* p162.

106. Palmer, *op. cit.,* pp156-159.

107. Palmer, *op. cit.,* p157.

108. Palmer, *op. cit.,* p158. Evidence for interbreeding is within the multi-regional theory.

109. Palmer, *op. cit.,* pp157, 165, 180-181.

110. Palmer, *op. cit.,* p157.

111. See http://en.wikipedia.org/wiki/human_evolution

for 70,000/50,000 years ago. See Bryan Sykes, *The Seven Daughters of Eve*, p278 for 100,000 years ago.

The Genographic project – backed by National Geographic, IBM and the Waitt Family Foundation – has studied DNA samples from a quarter of a million volunteers in different continents to map humankind's diaspora from Africa and migration. It has found that 60,000 years ago African men and women took to the Red Sea in boats and crossed the Mandab Strait to Asia. Some 30,000 years ago a group headed west and emerged in Europe, which had been dominated by Neanderthals, as the Cro-Magnons. In England 70 per cent of men carry the Cro-Magnon genetic signature, the R1b haplogroup. See p16 of *The Observer*, 31 August 2008.

112. See http://familytreedna.com/forum/showthread.php?p=28713

113. See http://www.newscientist.com/article/dn3744-chimps-r-human-gene-study-implies.h

114. As 113.

115. Sykes, *op. cit.*, pp50-51.

116. *Encyclopaedia Britannica* (1991), micropaedia, 4.140-141.

117. See

http://www.newscientist.com/article/dn2317-row-unravels-over-claim-of-oldest-dna.h

118. See

http://news.nationalgeographic.com/news/2007/07/070705-oldest-dna.html

119. See http://news.bbc.co.uk/1/hi/sci/tech/2909803.stm

120. As 112.

Also see file://C:\DOCUME~1\user\LOCALS~1\Temp\j5941bvt.htm

121. See http://en.wikipedia.org/wiki/Toba_catastrophe_theory

122. Sykes, *op. cit.*, ch. 1 and p277.

123. Sykes, *op. cit.*, p278.

124. As 115.

125. Clegg with Mackean, *op. cit.*, p651.

126. Clegg with Mackean, *op. cit.*, pp648-650.

127. Clegg with Mackean, *op. cit.*, pp651-655.

128. Nicholas Hagger's research in the Galapagos Islands.

129. Denyse O'Leary, *By Design or by Chance?*, p86.

130. O'Leary, *op. cit.*, p86; *Darwinism, Design and Public Education*, ed. by John Angus Campbell and Stephen C. Mayer, pp324-326.

131. O'Leary, *op. cit.*, p274, notes 8 and 9.

132. O'Leary, *op. cit.*, p86.

133. Gribbin and Rees, *op. cit.*, p289.

134. O'Leary, *op. cit.*, p93.

135. O'Leary, *op. cit.*, p97.

136. O'Leary, *op. cit.*, pp98-99.

137. See

http://members.iinet.net.au/~sejones/histry2a.html#hstrydrwndshnstys
trtgyws

138. O'Leary, *op. cit.*, pp170-171.

139. The group of intellectuals included the biochemist-historian Charles Thaxton and the professor of engineering Walter Bradley, who published *The Mystery of Life's Origin* (1984); also biologist Dean Kenyon. Michael Denton in *Evolution: A Theory in Crisis* (1985) offered a critique of Darwinism which showed that the problems, from palaeontology to molecular biology, were too severe to offer resolution within the orthodox Darwinian framework. For example, Darwinists trace the five-digit limb through wings, flippers and hands (bats' wings, sea lions' and dolphins' flippers and chimpanzee/human fingers, all of which share five-fingered bone structure) to establish common ancestry, but the limbs do not develop from the same body parts of the embryo: "The forelimbs develop from the trunk segments 2, 3, 4 and 5 in the newt, segments 6, 7, 8 and 9 in the lizard and from segments, 13, 14, 15, 16, 17 and 18 in man." Colin Patterson, curator of the British Museum of Natural History, pointed out that there are no fossils of a transitional form of an animal or species turning into another animal or species. Phillip Johnson in *Darwin on Trial* (1991) pointed out that the force of attrition also created a bacterial cell that produced trees, flowers, ants, birds and humans. Michael Behe, biochemist and author of *Darwin's Black Box* (1996), argued that Darwin could not account for the complex machinery of cells, which could not have arisen by chance. William Dembski wrote *The Design Inference* (1998), *No Free Lunch* (2002) and *The Design Revolution* (2004).

140. *Encyclopaedia Britannica* (1991), micropaedia, 6.312.

141. O'Leary, *op. cit.*, pp172-174.

142. William A. Dembski, *The Design Revolution*, p85.

143. *Darwinism, Design and Public Education, op. cit.*, p242.

144. *Encyclopaedia Britannica* (1991), micropaedia, 4.915.

145. Mario Livio, *The Golden Ratio*, pp214-221.

146. Livio, *op. cit.*, pp4-5.

147. Livio, *op. cit.*, pp203-206.

148. Livio, *op. cit.*, p203. For the 17 different symmetries, see Marcus Du Sautoy, *Finding Moonshine: A Mathematician's Journey through Symmetry*.

149. For nose see http://www.cojoweb.com/phi.html
Livio, *op. cit.*, p10 and ch. 7, e.g. pp162-163, 165 and 173.
Also see http://www.justinhenry.info/docs/phi_research_paper.pdf

150. Livio, *op. cit.*, pp44-45.

151. Livio, *op. cit.*, pp5-6, 72.

152. See http://en.wikipedia.org/wiki/Systems_theory

153. See http://www.bsn-gn.eku.edu/BEGLEY/GSThand1.htm

154. Mark Ward, *Universality*, pp70, 77-78.

155. Ward, *op. cit.*, pp37-38; *Encyclopaedia Britannica* (1991), micropaedia, 2.948.

156. For Chaos Theory, see *Encyclopaedia Britannica* (1991), micropaedia, 3.92.
For universality, see Ward, *op. cit.*, pp101-102.

157. See Arthur S. Eddington, *The Nature of the Physical World: The Gifford Lectures, 1927*, MacMillan, 1929. For a discussion, see
http://www.antievolution.org/people/wre/essays/typing.txt

158. Fred Hoyle, *The Intelligent* Universe, p19; paraphrased in O'Leary, *op. cit.*, p44.

159. O'Leary, *op. cit.*, p44.

160. O'Leary, *op. cit.*, ch. 14.

161. Gribbin, *In the Beginning*, p186.

162. *The Sunday Times*, 23 December 2007, News Review, p9.

163. *The Times*, 18 April 2008.

164. Rupert Sheldrake, *A New Science of Life*, pp133-136 and *The Presence of the Past*, pp140-146.

165. Compare Willis W. Harman, *A Re-examination of the Metaphysical Foundations of Modern Science*, p87, where "integrative" is used as the opposite of "reductionist".

166. Coleridge, *Biographia Literaria*, ch. 13, p356, published 1854, original from the University of Michigan, Digitized 23 November 2005.
http://books.google.co.uk/books?id=5xg5G-4ai3oC&pg=PA356&dq=

esemplastic+power+of+the+imagination

167. For example Geoffrey Read, who attacks the holistic principle in Hewitt, *op. cit.*, pp143-144.

168. Lee Spectner, *Not by Chance!*

169. See http://harvardscience.harvard.edu/foundations/articles/j-craig-venter-named-visiting-scholar

   The Human Genome Project (HGP) began in 1990. Hundreds of scientists in China, France, Germany, Canada, Britain and the United States collaborated to work out the sequences of bases ("letters") of DNA that "spell out" how to make a human being, the human genome. The project ended in 2003.

170. Adams, *Origins of Existence*, p189.

171. Richard Dawkins, *The Blind Watchmaker*, pp21, 317-318.

7. Ecology: Nature's Self-Running, Self-Organising System

1. Clegg with Mackean, *op. cit.*, p45.

2. See http://jackytappet.tripod.com/chain.html. and
   http://www.kheper.net/topics/greatchainofbeing/index.html

3. For a drawing of the hierarchical great chain of being, see Didacus Valades *Rhetorica Christiana* (1579) in
   http://en.wikipedia.org/wiki/Great_Chain_of_Being

4. Clegg with Mackean, *op. cit.*, pp20-21.

5. See http://www.enviroliteracy.org/article/php/58.html

6. See http://www.guardian.co.uk/science/2007/feb/26/biodiversity.taxonomy

7. See
   http://www.nzherald.co.nz/section/2/print.cfm?c_id=2&objected=10466120&pnum=0
   Also
   http://news.webindia123.com/news/printer.asp?story=/news/articles/Science/2007092

8. For the 2 million, see http://hypertextbook.com/facts/2003/FelixNisimov.shtml
   For the 2-30million, see Weiner, *op. cit.*, p134.
   For 10-100 million, see
   http://www.iucn.org/en/news/archive/2007/11/12_pr/bear.htm

9. Charles Wheatley's research.

10. Clegg with Mackean, *op. cit.*, p656.

11. Idea developed in Nicholas Hagger's discussions with Mary Morrison.

12. Nicholas Hagger's research in the Galapagos Islands.

13. Clegg with Mackean, *op. cit.*, p516.

14. Clegg with Mackean, *op. cit.*, pp516-517.

15. Nicholas Hagger's direct observation in the Galapagos Islands.

16. *Proceedings of the National Academy of Sciences*, January 2008, based on research in Rome. Article in *The Daily Telegraph*, 30 January 2008, p3.

17. Nicholas Hagger's research in the Galapagos Islands and Southern Ocean.

18. Nicholas Hagger's research in the Southern Ocean.

19. As 18.

20. Nicholas Hagger's research in the Galapagos Islands.

21. As 20.

22. Clegg with Mackean, *op. cit.*, p518.

23. Clegg with Mackean, *op. cit.*, p520.

24. Clegg with Mackean, *op. cit.*, p518.

25. Clegg with Mackean, *op. cit.*, p519.

26. Nicholas Hagger's research in the Galapagos Islands and the Antarctic. Also, Oscar-winning documentary, *March of the Penguins*, 2005, narrated by Morgan Freeman, 2005.

27. As 26.

28. Nicholas Hagger's observation to Charles Wheatley and subsequent discussions.

29. Clegg with Mackean, *op. cit.*, pp44-45.

30. Clegg with Mackean, *op. cit.*, pp46-47. Also see http://en.wikipedia.org/wiki/Food_chain . Also http://waterontheweb.org/under/lakeecology/11_foodweb.html. For 10^{18} see http://en.wikipedia.org/wiki/1000000000000_(number)

31. Clegg with Mackean, *op. cit.*, p46.

32. Clegg with Mackean, *op. cit.*, pp47 and 46. Also http://www.answers.com/topic/blue-tit-1. Also Robert Brooker, Eric Widmaier, Linda Graham and Peter Stiling, *Biology*, p1263.

33. See http://en.wikipedia.org/wiki/Food_chain; also Brooker, Widmaier, Graham and Stiling, *op. cit.*, p1263.

34. See http://en.wikipedia.org/wiki/Food_chain; http://www.cwmb.sa.gov.au/kwc/programs/Food%20webs%20incl%20

teacher%20notes.pdf

Also Michael Schmid's research (60 million unicellar organisms).

35. See http://www.countrysideinfo.co.uk/talks/centre_intro/foodweb.htm; Clegg with Mackean, *op. cit.,* p47. Also http://www.bbc.co.uk/schools/gcsebiteside/geography/ecosystems/ ecosys temsresourc

36. Clegg with Mackean, *op. cit.,* p47; and http://www.woodlands-junior.kent.sch.uk/Homework/fooodchains.htm

Also

http://weedeco.msu.montana.edu/class/lres443/Lectures/lecture20/ Food Web.JPG

Also

http://www.bbc.co.uk/schools/gcsebitesize/geography/ecoystems/ ecosys temsresourc and Brooker, Widmaier, Graham and Stiling, *op. cit.,* p1263.

37. Lake Michigan food web, see http://www.glerl.noaa.gov/pubs/brochures/foodweb/LMfoodweb.pdf

38. Clegg with Mackean, *op. cit.,* p56. Also http://oceanworld.tamu.edu/resources/oceanography-book/marinefood webs.htm

39. Brooker, Widmaier, Graham and Stiling, *op. cit.,* p1237.

40. Clegg with Mackean, *op. cit.,* pp45, 65.

41. Clegg with Mackean, *op. cit.,* p65.

42. Tricia Moxey's research shared in discussions with Nicholas Hagger.

43. As 42.

44. As 42.

45. See http://en.wikipedia.org/wiki/Bottom_feeder

46. Clegg with Mackean, *op. cit.,* p65.

47. Nicholas Hagger's research in the Southern Ocean.

48. As 47.

49. Clegg with Mackean, *op. cit.,* p66.

50. Clegg with Mackean, *op. cit.,* p67.

51. Robin Woodleigh's research shared with Nicholas Hagger; Nicholas Hagger's own research.

52. Clegg with Mackean, *op. cit.,* p66.

53. As 51.

54. Clegg with Mackean, *op. cit.,* p67.

55. As 51.

56. See Dieter Podlech, *Herbs and Healing Plants of Britain and Europe*.

57. For a more comprehensive list, see http://www.boldweb.com/greenweb/ailplant.htm. For Western herbalism's debt to Hippocrates and early Greek medicine, see http://www.webnb.btinternet.co.uk/course12.htm

8. Physiology: The Ordering of Body, Brain and Consciousness

1. Adams, *Origins of Existence*, p189.

2. Adams, *Origins of Existence*, p189.

3. See Adams, *Origins of Existence*, p202, for 10^{29}. For (7×10^{27}), for 10^{15} (twice) and 10^{12} see http://en.wikipedia.org/wiki/1000000000000_(number)

4. Brooker, Widmaier, Graham and Stiling, *op. cit.*, pp4-5.

5. For homeostasis, see Clegg with Mackean, *op. cit.*, pp502-515; and Brooker, Widmaier, Graham and Stiling, *op. cit.*, pp856-859.

6. Clegg with Mackean, *op. cit.*, pp503-504; Brooker, Widmaier, Graham and Stiling, *op. cit.*, pp858-859.

7. Brooker, Widmaier, Graham and Stiling, *op. cit.*, p858.

8. Jason Cook's research shared with Nicholas Hagger.

9. Clegg with Mackean, *op. cit.*, pp509-515.

10. Clegg with Mackean, *op. cit.*, pp619-620.

11. Adams, *Origins of Existence*, p202.

12. Clegg with Mackean, *op. cit.*, p465.

13. A general, fairly reductionist non-medical book on the brain is Susan Greenfield, *The Human Brain – A Guided Tour*.

14. See http://en.wikipedia.org/wiki/Brain: "Evidence strongly suggests that developed brains derive consciousness from the complex interactions between the numerous systems within the brain. Cognitive processing in mammals occurs in the cerebral cortex but relies on midbrain and limbic functions as well."
    Also see http://en.wikipedia.org/wiki/Human_brain

15. As 14.

16. Clegg with Mackean, *op. cit.*, p467.

17. Clegg with Mackean, *op. cit.*, p465.

18. See http://en.wikipedia.org/wiki/Famine_response

19. Jason Cook's research shared with Nicholas Hagger.
20. See
    http://www.redicecreations.com/specialreport/2006/02feb/quantum.html
21. See http://gtresearchnews.gatechedu/newsrelease/quantum-memory.htm
22. As 20.
23. See
    http://technology.newscientist.com/article.no?id=dn10851&feedId=
    online-news_rss20
24. See
    http://findarticles.com/p/articles/mi_m1200/is_n26_v151/ai_19587608/
    print
25. See
    http://www.trnmag.com/Stories/2002/071002/Photons_heft_more_data
    _7-10-02.html
26. See http://www.sciencedaily.com/releases/2005/12/051213082424.htm.
    Also as 21.
    Also http://www.nerdshit.com/wordpress/?p=1935.
    Also http://oemagazine.com/newscast/2005/120905_newscast01.html.
27. http://arxiv.org/vc/cs/papers/0511/0511070v1.pdf
28. See http://news.softpedia.com/news/How-to-Make-Photons-Talk-to-One-
    Another-and-Possibly-Transmit-Information-59756.shtml
29. See
    http://www.nikon.co.jp/main/eng/feelnikon/discovery/light/chap04/
    sec02.htm
30. See
    http://www.americanscientist.org/template/AssetDetail/assetid/16218? full
    text=true&print=yes
31. See
    http://www.psych.ndsu.nodak.edu/mccourt/Psy460/Anatomy%20and
    %20physiology%20of%20the%20retina/Anatomy%20and%20physiol
    ogy%20of%20the%20retina.html
32. As 30.
33. See http://arxiv.org/ftp/quant-ph/papers/0208/0208053.pdf
34. As 30.
35. As 30.
36. As 30.

37. As 30.

38. See http://en.wikipedia.org/wiki/Photoreceptor_cell

39. As 38. and see
    http://www.ncbi.nlm.nih.gov/books/bv.fcgi?rid=neurosci.section.747

40. See
    http://www.patentstorm.us/patents/7270637-description.html

41. See http://www.biocybernaut.com/about/brainwaves/alpha.htm
    Also W. Grey Walter, *The Living Brain*. Also see diagram of
    electro-magnetic spectrum on p400.

42. Owen J. Flanagan, *Consciousness Reconsidered*, p234; quoted in Malcolm
    Hollick, *The Science of Oneness*, p266.

43. Roger Penrose, *Shadows of the Mind*, p366.

44. Eric R. Kandel, In Search of Memory, p384. For the paper Crick wrote with
    Koch, *What is the function of the claustrum?*, see
    http://publishing.royalsociety.org/media/philtrans_b/crick.pdf

45. Kandel, *op. cit.*, p389.

46. Francis Crick, *The Astonishing Hypothesis*, p268.

47. Kandel, *op. cit.*, pp279-285.

48. William Blake, letter to Thomas Butts, 22 November 1802; in *Poetry and
    Prose of William Blake*, ed. By Geoffrey Keynes, pp859-862.

49. Coleridge, *Collected Letters II*, p709.

50. See http://www.lhc.ac.uk/about-the-lhc/faqs.html
    Some sources put the cost of the LHC at £4.4 billion.

PART THREE
THE METAPHYSICAL VIEW OF THE UNIVERSE
IN THE NEW PHILOSOPHY

1. Quoted in N. Rose, *Mathematical Maxims and Minims*, Raleigh, W.C. 1988;
   and in Barrow, *The Infinite Book*, p115.

9. The Rational Approach to Reality and the System of the Universe

1. See J.J. Williamson, *The Structure of ALL*, lectures delivered at the R.A.F.
   College, Cranwell, Lincolnshire, UK, which defines metaphysics as "the
   science of ALL".

2. For Uncreated Light see Robert Grosseteste, *On Light (De Luce)*, trans. by

Clare Riedl.

3. For Universal Light see Plato, *Republic*, book VII 540a: "The time has now arrived at which they must raise the eye of the soul to the universal light which lightens all things, and behold the absolute good" (trans. by Benjamin Jowett).
   See http://classics.mit.edu/Plato/republic.8.vii.html

4. Augustine, *Confessions*, 7.10, pp146-147.

5. Mechthild of Magdeburg, *The Book of the Flowing Light of the Godhead*. See Hagger, *The Light of Civilization*, pp142-143 for context.

6. Dante, *Paradiso*, canto 33, line 124; p346.

7. Dante, *op. cit.*, canto 30, line 40; p319.

8. Dante, *op. cit.*, p323.

9. Adams, *Origins of Existence*, p37.

10. Adams, *Origins of Existence*, p41.

11. Duncan, *op. cit.*, p405.

12. Duncan, *op. cit.*, p484. For a millionth, see Davies, *The Goldilocks* Enigma, p135. For neutrinos travelling at the speed of light, see Adams, *Origins of Existence*, p44.

13. Adams, *Origins of Existence*, p66.

14. Clement of Alexandria, *Stromateis*, v,104,1. Quoted in Kirk, Raven and Schofield, *op. cit.*, pp197-198.

15. A.A. Long, ed. *op. cit.*, pp47, 79.

16. Kirk, Raven and Schofield, *op. cit.*, pp197-198: "Pure or aitherial, fire has a directive capacity....The pure cosmic fire was probably identified by Heraclitus with *aither*, the brilliant fiery stuff which filled the shining sky and surrounds the world; this aither was widely regarded both as divine and as a place of souls. The idea that the soul may be fire or *aither*...must have helped to determine the choice of fire as the controlling form of matter."

17. Aristotle, *Physics*, 8.6; 260a: "But the unmoved mover..., since it remains permanently simple and unvarying and in the same state, will cause motion that is one and simple."

18. W.O.E. Oesterley and Theodore H. Robinson, *An Introduction to the Books of the Old Testament*, pp24-34; *Encyclopaedia Britannica* (1991), micropaedia, 5.177.

19. Note by Nicholas Hagger: "I announced my Form from Movement theory over breakfast at a Symposium on Reductionism's Primacy in the Natural

Sciences, which was held at Jesus College, Cambridge in September 1992. I had contributed a paper which was circulated among fifteen speakers who included ten Professors of Physics, Biology, Physiology, Neurology, Astronomy and Philosophy in Britain and the USA. Roger Penrose and John Barrow were among these. About ten of the attendees gathered round as I explained my thinking, and a Norwegian mathematician, Henning Bråten, converted what I said to mathematical symbols as I spoke. The mathematics were refined during subsequent discussions, and I was grateful for his expertise."

20. Nicholas Hagger's research in China.
21. Duncan, *op. cit.*, p497; Singh, *op. cit.*, pp430-431.
    See http://www.nersc.gov/news/nerscnews/NERSCNews_2005_02.pdf
22. Lars Olof Björn, *Light and Life*, p17.
23. Björn, *op. cit.*, pp46-47.
24. Björn, *op. cit.*, pp81, 86, 101.
25. Björn, *op. cit.*, p101.
26. Nicholas Hagger's research in the Galapagos Islands.
27. Björn, *op. cit.*, pp81-90, 101.
28. Björn, *op. cit.*, pp81-83, 101-127.
29. Björn, *op. cit.*, ch. 4.
30. Björn, *op. cit.*, ch. 5.
31. Björn, *op. cit.*, ch. 6.
32. Björn, *op. cit.*, ch. 7.
33. See http://www.creationofuniverse.com/html/order_skies_04.html
    Also *Encyclopaedia Britannica*, micropaedia, III.911.
34. Aristotle, *Metaphysics,* pxxviii.
35. Willis Harman in *Global Mind Change*, pp34-35, distinguishes the first three. I have added the fourth.
36. Plato held that all knowledge can be derived from a single set of principles, perhaps even from a single principle. In *Republic* VI-VII, Plato describes the philosopher's knowledge as "synoptic", taking in the whole of reality, and resting on The Good, or knowledge of The Good. See
    http://aristotle.tamu.edu/~rasmith/Courses/Ancient/homonymy-and-science.html
    Aristotle in *Metaphysics* IV, 1/1003a, writes: "There is a science which studies Being as Being....This science is not the same as any of the

so-called particular sciences, for none of the others contemplates Being generally as Being; they divide off some portion of it and study the attribute of this portion, as do for example the mathematical sciences." Also see

http://books.google.co.uk/books?id=8tYcp0vYd5EC&pg=RA1-PA140 &lpg=RA1-PA140&dq=universal+science+plato& source=web&ots=l1ASNzlF8d&sig=VvK5PCVlA_86bDy3jUiRNk WZXeo&hl=en

37. Eleanor Swift, *A Layman's Guide to Neometaphysics*, which is based on the three parts of J.J. Williamson's Cranwell Lectures, particularly on Part Two ('Metaphysical Application') and Part Three ('Metaphysical Analysis').

38. For example, in 1 *Thessalonians* 5.23.

See also

http://books.google.co.uk/books?id=Hk4iTvy4Gx8C&pg=PA353&lpg =PA353&dq=st+paul+spirit+pneuma&source=web&ots=skidBEIBG9&sig =L3XQDhf5hkUyZ5MGzSfDD7zxVtc&hl=en

39. See Mircea Eliade, *Shamanism, Archaic Techniques of Ecstasy*, pp91-93.

40. Z'ev ben Shimon Halevi, *Tree of Life*, p18. Abraham is reputed to have received the Kabbalah from Melchizedek, King of Salem or Jerusalem.

10. The Intuitional Approach to Reality and the System of the Universe

1. Note by Nicholas Hagger: "I have presented the full tradition of the experience of Being as Fire or Light in *The Fire and the Stones*; also in *The Light of Civilization* and in *The Rise and Fall of Civilizations*."

2. See Hagger, *The Light of Civilization*.

3. Whitehead, *op. cit.,* p924. For the philosophy of organism see *The Concept of Nature*. For Reality, see *Process and Reality*. For the totality and lack of bifurcation, see *op. cit.*, p338, and for poetry, p924: "Philosophy is akin to poetry."

4. Shelley's 'Hymn to Intellectual Beauty', 1816, in *Shelley,* sel. by Isabel Quigly, pp71-73.

5. St Augustine, *op. cit.,* 7.10, pp146-147.

6. Hildegard von Bingen, quoted in John Ferguson, *An Illustrated Encyclopaedia of Mysticism and the Mystery Religions*, p77.

7. Dante Alighieri, *The Divine Comedy*, trans. by Allen Mandelbaum, verse turned into prose, pp512, 539-540. The note for "a point" says (on p779):

"God, represented as an infinitesimal point of pure light."

8. For example, in a course taught by Z'ev ben Shimon Halevi.

9. Heidegger, *An Introduction to Metaphysics*, pp81-82, 93-95. Heidegger looks back to Parmenides' poetic intuition of the One. For Heidegger's poetic intuition, see
   http://www.whpq.org/visitor/200212/200212/whpq200212-004-1.htm
   and http://en.wikipedia.org/wiki/Martin_Heidegger

10. Based on Nicholas Hagger's thinking. Many traditions have similar applications, for example New Thought. For the quotation from van Gogh, see Henri Scrépel, *Van Gogh.*

11. Plato, *Phaedrus*, 253d.

12. Coleridge, Notebooks, I: I.561
    http://fds.oup.com/www.oup.co.uk/pdf/0-19-818397-6.pdf

13. For the *chakras* or *cakras*, see *Encyclopaedia Britannica*, micropaedia, II.445. Also K.M. Sen, *Hinduism*, pp61, 69, 70. For a theosophical view of the *chakras*, see C.W. Leadbeater, *The Chakras.*

14. Shelley, 'Adonais', LII.

15. *Encyclopaedia Britannica*, micropaedia, II.176. For photons as bosons see Barbour, *op. cit.*, p191.

16. Renée Weber, *op. cit.*, p48 and also p44 (which reports Bohm referring to G.M. Lewis). For matter as frozen light, see p45. Also David Bohm in conversation on light with Nicholas Hagger. For matter as frozen light, see Weber, *op. cit.*, p45. For matter being formed by collision of photons and therefore by light, see
    http://abyss.uoregon.edu/~js/21st_century_science/lectures/lec25.html

17. Rees, *Just Six Numbers*, p52.

18. Weber, *op. cit.*, p45.

19. Björn, *op. cit.*, p16.

20. See http://en.wikipedia.org/wiki/Dualism_(philosophy_of_mind)

21. Weber, *op. cit.*, pp48-49.

22. John Hunt's research.

23. Eric Chudler, 'Myths about the Brain: 10 per cent and Counting', see
    http://www.brainconnection.com/topics/?main=fa/brain-myth
    quoted in http://en.wikipedia.org.wiki.Human_brain

24. Atomic theory was proposed in the 5th century BC by the Greek philosophers Leucippus and Democritus and was revived by Lucretius in

439

Roman poetry in the 1st century BC. Modern atomic theory began with John Dalton in the 19th century, and his scientific theory was confirmed in the 20th century. See *Encyclopaedia Britannica*, II.346-351.

25. Alfred North Whitehead, *An Introduction to Mathematics*, pp41-42.

PART FOUR
ORDER IN HUMAN AFFAIRS:
APPLICATIONS OF UNIVERSALIST THINKING

11. New Understanding of the Past

1. See Hagger, *The Syndicate*, pp30-32.
2. Arnold Toynbee, *A Study of History*, one-volume edition, p97. Quoted in Hagger, *The Light of Civilization*, pp525-527.
3. Hagger, *The Rise and Fall of Civilizations, passim.*
4. Hagger, *The Fire and the Stones, The Light of Civilization, The Rise and Fall of Civilizations.*
5. See Hagger, *The Secret History of the West, passim;* and *The Syndicate.*
6. Edward Burnett Tylor, *Primitive Culture*, 1871.
7. See Hagger, *The Light of Civilization*, p510.

12. Shaping the Future

1. Clegg with Mackean, *op. cit.,* p89; Jeff Rubin, *Antarctica*, p230.
2. See http://www.gsfc.nasa.gov/topstory/20020926ozonehole.html
3. See http://inventors.about.com/library/inventors/blfreon.htm
4. See http://ozonewatch.gsfc.nasa.gov/facts/ozone.html which states that ozone forms 0.00006 per cent of the atmosphere
5. Rubin, *op. cit.,* p98.
6. Clegg with Mackean, *op. cit.,* p247.
7. Clegg with Mackean, *op. cit.,* p88. Also see http://www.ucar.edu/learn/1_6_1.htm
8. Clegg with Mackean, *op. cit.,* p89.
9. Patrick Abbott's research. It should however be pointed out that in 2002 the ozone hole over the Antarctic was much smaller than in 2000 and 2001, and had split into two holes. The opening was the smallest since 1988. See http://www.gsfc.nasa.gov/topstory/20020926ozonehole.html See

http://www.sciencedaily.com/releases/2006/06/060630095235.htm

10. Patrick Abbott's research.

11. Patrick Abbott's research.

12. Henry N. Pollack, *Uncertain Science...Uncertain World.*

13. See

http://www.ft.com/cms/s/0/c70p7cfc-9ce4-11db-8ec6-0000779e2340.html

14. Pollack, *op. cit.,* pp219-220.

15. Patrick Abbott, quoting James P. Kennett, author of *Marine Geology.*

16. Pollack, *op. cit.* Also see

http://flood.firetree.net/?ll=33.8339,129.7265&z=12&m=7

17. Pollack, *op. cit.,* p224.

18. Pollack, *op. cit.,* pp225-226.

19. Pollack, *op. cit.*

20. See

http://www.un.org/esa/population/publications/sixbillion/sixbilpart1.pdf,
figures of the UN Population Divison.

21. Charles W. Fowler and Larry Hobbs, 'Limits to Natural Variation: Implications for Systemic Management', 2002, see http://www.doaj.org/doaj?fund=abstract&id=122283&openurl=1. Also, 'Is Humanity Sustainable', 2003, see http://links.jstor.org/sici?sici=0962-8452(20031222)270%3A1533%C2579%3AIHS%3E2.0.CO%3B2-9

22. Patrick Abbott's research.

23. See Hagger, *The Syndicate*, pp267-274.

24. See http://en.wikipedia.org/wiki/Little_Ice_Age

25. Patrick Abbott's research.

26. See http://news.bbc.co.uk/1/hi/world/americas/6999078.stm and http://www.timesonline.co.uk/tol/news/uk/article2461996.ece

27. Nicholas Hagger's research in Antarctica. For the Wilkins Ice Shelf, see *The Daily Telegraph*, 26 March 2008, p34.

28. See Hagger, *The Syndicate*, pp43-48, 257-264.

29. See Hagger, *The Syndicate*, p280.

30. Hagger, *The Syndicate*, pp23-24.

31. Hagger, *The Syndicate,* pp2-4.

32. Hagger, *The Syndicate,* pp291-293.

33. Hagger, *The Fire and the Stones*, and *The Light of Civilization.*

34. Apuleius, *The Golden Ass*, p286: "At midnight I saw the sun shining as if

it were noon." (or, "In the middle of the night I have seen the Sun scintillating with a pure light.")

35. Palmer, *op. cit.*, p200.
36. A reference to Dawkins' *The God Delusion.*

EPILOGUE

1. Hawking, *op. cit.*, pp174-175.

POSTSCRIPT

1. Reported on p8 of the London *Times* on 11 September 2008.

APPENDICES

1. Peter J. Mohr and Barry N. Taylor, Codata Recommended Values of the Fundamental Physical Constants: 2002, published in *Review of Modern Physics* 77, 1 (2005). See http://physics.nist.gov/cuu/Constants/Table/awwascii.txt
2. As 1. See http://physics.nist.gov/constants
3. As 1.
4. Based on *Darwinism, Design, and Public Education.*
5. Based on Björn, *Light and Life*; and see http://www.evarudlinger.co.uk/about/html
6. See http://www.akri.org/cognition/images/brstruc.gif
7. See http://en.wikipedia.org/wiki/Pineal_gland
8. See http://www.nsf.gov/news/newsmedia/JarvisNatRevNeuro.pdf
9. See http://www.sistemanervoso.com/images/ventral.jpg
10. See http://webvision.med.utah.edu/sretina.html
11. See http://www.humanthermodynamics.com/eye-diagram.jpg
12. See http://www.dva.gov.au/health/menshealth/images/13_eye_cx.gif and http://static.icr.org/i/articles/imp/imp-388.jpg and http://webvision.med.utah.edu/imageswv/schem.jpeg
13. See http://richarddawkins.net/images/blindwatchmaker_ch1_img.jpg

BIBLIOGRAPHY

Adams, Fred, *Origins of Existence*, The Free Press, 2002.

Adams, Fred and Laughlin, Greg, *The Five Ages of the Universe*, Touchstone, 2000.

Andrew, David, *Watching Wildlife, Galapagos Islands*, Lonely Planet Publications, 2005.

Apuleius, *The Golden Ass*, trans. by Robert Graves, Penguin Book, 1954.

Aristotle, *Metaphysics*, books I-IX, trans. by Hugh Tredennick, Harvard University Press, 1933/2003.

Augustine, St, *Confessions*, trans. by R.S. Pine-Coffin, Penguin, London, 1961.

Barbour, Julian, *The End of Time*, Phoenix, 2000.

Barnes, Jonathan, *Early Greek Philosophy*, Penguin Books, London, 2001.

Barnes, Jonathan, *The Presocratic Philosophers*, Routledge, London, 1979/1996.

Barrow, John, *The Infinite Book*, Jonathan Cape, 2005.

Barrow, John and Tipler, Frank, *The Anthropic Cosmological Principle*, Oxford University Press, 1986.

Bergson, Henri (-Louis), *Introduction to Metaphysics*, 1903, published as *L'Énergie Spirituelle*, Presses Universitaires de France, Paris, 1919.

Björn, Lars Olof, *Light and Life*, Hodder and Stoughton, 1976.

Blake, William, *Poetry and Prose of*, ed. by Geoffrey Keynes, The Nonesuch Library, London, 1956.

Bohm, David, *Wholeness and the Implicate Order*, Routledge & Kegan Paul, 1980.

Booker, Christopher and North, Richard, *Scared to Death*, Continuum International Publishing Group Ltd., 2007

Boswell, James, *The Life of Samuel Johnson*, abridged version, *Everybody's Boswell*, G. Bell and Sons, London, 1930.

Brooker, Robert J., Widmaier, Eric P., Graham, Linda E., Stiling, Peter D., *Biology*, McGraw-Hill, 2008.

Capra, Fritjof, *The Tao of Physics*, Fontana, 1976.

Carr, Brian, *Metaphysics: an Introduction*, Macmillan Education, 1987.

Castillejo, David, *The Expanding Force in Newton's Cosmos*, Ediciones de Arte y Bibliofilia, Madrid, 1981.

Clegg, C.J. with Mackean, D.G., *Advanced Biology, Principles and Applications*, 2nd ed., Hodder Murray, 2000.

Coleridge, Samuel Taylor, *Biographia Literaria*, published 1854, original in the University of Michigan, digitized 23 November 2005.

Coleridge, Samuel Taylor, *Collected Letters II*, ed. Earl Leslie Griggs, Oxford University Press, 2000.

Constant, Pierre, *The Galapagos Islands and Natural History Guide*, Odyssey Books, 2006.

Crick, Francis, *The Astonishing Hypothesis*, Simon and Schuster, 1994.

Dante, *The Divine Comedy,* Book 3: *Paradise*, trans. by Dorothy L. Sayers and Barbara Reynolds, Penguin Classics, 1962.

Dante Alighieri, *The Divine Comedy*, trans. by Allen Mandelbaum, Everyman's Library, 1995.

Darwin, Charles, *Voyage of the Beagle*, Penguin Books, 1839/1989.

Darwin, Charles, *The Voyage of the Beagle: Journal of Researches Into the Natural History and Geology of the Countries Visited During the Voyage of HMS, Beagle Round the World*, 1909, reprinted in Modern Library: New York, 2001.

Darwinism, Design and Public Education, ed. by John Angus Campbell and Stephen C. Mayer, Michigan State University Press, 2003.

Davies, Paul, *The Goldilocks Enigma*, Allen Lane, 2006.

Davies, Paul, *The Mind of God*, Simon and Schuster, 1992.

Davies, Paul and Gribbin, John, *The Matter Myth*, Viking, 1991.

Dawkins, Richard, *The Blind Watchmaker*, Penguin Books, 2006.

Dawkins, Richard, *The God Delusion*, Bantam Press, 2006.

Dembski, William A., *The Design Revolution*, InterVarsity Press, Illinois, 2004.

Descartes, René, *Treatise of Man*, Prometheus Books, New York, 2003.

De Waal, Frans, *Our Inner Ape*, Granta Books, 2006.

Duncan, Tom with Kennett, Heather, *Advanced Physics*, fifth edition, Hodder Murray, 2000.

Du Sautoy, Marcus, *Finding Moonshine*: *A Mathematician's Journey through Symmetry*, Fourth Estate, 2008.

Edelman, Gerald M., *Bright Air, Brilliant Fire*, Basic Books, 1992.

Eliade, Mircea, *Shamanism, Archaic Techniques of Ecstasy*, Princeton University Press, USA, 1964/Routledge & Kegan Paul, 1974.

Ferguson, John, *An Illustrated Encylopaedia of Mysticsm and the Mystery Religions*, Thames and Hudson, London, 1976.

Flanagan, Owen J., *Consciousness Reconsidered*, MIT Press, Cambridge, 1992.

Fölsing, Albrecht, *Albert Einstein*, trans. by Ewald Osers, Viking, 1997.

Gellner, Ernest, *Words and Things*, Victor Gollancz, London, 1959.

Greene, Brian, *The Elegant Universe*, Vintage, 2000.

Greenfield, Susan A., *The Human Brain – A Guided Tour*, Basic Books, New Edition, 1998.

Greenstein, George, *The Symbiotic Universe*, William Morrow, New York, 1988.

Gribbin, John, *In the Beginning*, Viking, 1993.

Gribbin, John, *The Universe, A Biography*, Allen Lane, 2006.

Gribbin, John and Rees, Martin, *Cosmic Coincidences*, Black Swan, 1990.

Gribbin, John and Rees, Martin, *The Stuff of the Universe*, Heinemann, London, 1990.

Grosseteste, Robert, *On Light*, trans. by Clare Riedl, Marquette University Press, 1942.

Guth, Alan, *The Inflationary Universe*, Jonathan Cape, 1997.

Hagger, Nicholas, *The Fire and the Stones*, Element, 1991.

Hagger, Nicholas, *The Last Tourist in Iran*, O Books, 2008.

Hagger, Nicholas, *The Light of Civilization*, O Books, 2006.

Hagger, Nicholas, *The Rise and Fall of Civilizations*, O Books, 2008.

Hagger, Nicholas, *The Secret History of the West*, O Books, 2005.

Hagger, Nicholas, *The Syndicate*, O Books, 2004.

Halevi, Z'ev ben Shimon, *Tree of Life*, Rider, 1972.

Hamlyn, D.W., *Metaphysics*, Cambridge University Press, 1984.

Harman, Willis W., *A Re-examination of the Metaphysical Foundations of Modern Science*, The Institute of Noetic Sciences, California, 1991.

Harman, Willis W., *Global Mind Change*, Knowledge Systems/The Institute of Noetic Sciences, USA, 1988.

Hawking, Stephen W., *A Brief History of Time*, Bantam Press/Transworld, 1988.

Heidegger, Martin, *An Introduction to Metaphysics*, Yale University Press, 1959.

Heracleitus, *On the Universe*, in *Hippocrates*, trans. by W.H.S. Jones, Loeb Classical Library, vol. IV, Harvard University Press, 1931/1992.

Hewitt, Peter, *The Coherent Universe*, Linden House, 2003.

Hollick, Malcolm, *The Science of Oneness*, O Books, 2006.

Hoyle, Fred, *The Intelligent Universe*, Michael Joseph, 1985.

Hoyle, Fred and Wickramasinghe, Chandra, *Evolution from Space*, Grenada, 1983.

Hoyle, Fred, Wickramasinghe, Chandra and Watkins, John, *Viruses from Space*,

University College Cardiff Press, 1986.

Kahn, Charles H., *The Art and Thought of Heraclitus*, Cambridge University Press, 1995.

Kandel, Eric R., *In Search of Memory*, W.W. Norton, 2006.

Kennett, James P., *Marine Geology*, Prentice Hall, 1981.

Kirk, G.S., Raven, J.E. and Schofield, M., *The Presocratic Philosophers*, Second Edition, Cambridge University Press, 195 // 1995.

Leadbeater, C.W., *The Chakras*, The Theosophical Publishing House, Adyar, 1927.

Lightman, Alan, *Ancient Light*, Harvard University Press, 1993.

Livio, Mario, *The Golden Ratio*, Headline Review, 2002.

Lloyd, Seth, *The Computational Universe*, Random House, New York, 2006.

Long, A.A., ed., *The Cambridge Companion to Early Greek Philosophy*, Cambridge University Press, 1999.

Lovejoy, Arthur, *The Great Chain of Being: A Study of the History of an Idea*, Harper & Row, 1960

Lovelock, James, *Gaia: A New Look at Life*, Oxford Paperbacks, new ed., 1982.

Lucas, Mike, *Antarctica*, New Holland Publishers, 1996.

Macann, Christopher, *Being and Becoming*, 4 vols, Online Originals, 1998-2007.

Miller, Jeanine, *The Vision of Cosmic Order in the Vedas*, Routledge and Kegan Paul, London, 1985.

Oesterley, W.O.E. and Robinson, Theodore H., *An Introduction to the Books of the Old Testament*, Macmillan, 1934-1958.

O'Leary, Denyse, *By Design or by Chance*, Augsburg Books, Minneapolis, 2004.

Palmer, Douglas, *Neanderthal*, Channel 4 Books, 2000.

Parmenides of Elea, *Fragments*, trans. by David Gallop, University of Toronto Press, 1984.

Pascal, Blaise, *Pensées*, trans. by W.F. Trotter, Everyman's Library, J.M. Dent, 1931.

Penrose, Roger, *Shadows of the Mind*, Oxford University Press, 1994.

Penrose, Roger, *The Emperor's New Mind*, Vintage, 1989.

Phillips, Jack, *Suppressed Science*, John J Phillips Jr, 2006.

Plato, *The Collected Dialogues*, ed. by Edith Hamilton and Huntington Cairns, Princeton University Press, 1961/1982.

Podlech, Dieter, *Herbs and Healing Plants of Britain and Europe*, Collins, 1996.

Pollack, Henry N., *Uncertain Science...Uncertain World*, Cambridge University

Press, New Ed. edition, 2005.

Rees, Martin, *Before the Beginning*, Simon and Schuster, 1997.

Rees, Martin, *Just Six Numbers: The Deep Forces That Shape The Universe*, Basic Books, 2000.

Rees, Martin, *Our Cosmic Habitat*, Phoenix, 2001.

Rubin, Jeff, *Antarctica*, Lonely Planet, 2005.

Rucker, Rudy, *Infinity and The Mind*, Paladin, London, 1984.

Russell, Bertrand, *Why I Am Not a Christian*, Allen & Unwin, London, 1957.

Schlesinger, George N., *Metaphysics*, Basil Blackwell, Oxford, 1983.

Schubin, Neil, *Your Inner Fish: A Journey into the 3.5 Billion-Year- History of the Human Body*, Allen Lane, 2008.

Scrépel, Henri, *Van Gogh*, Gallimard, Paris, 1972.

Sen, K.M., *Hinduism*, Penguin Books, 1961.

Sheldrake, Rupert, *A New Science of Life*, Granada, 1983.

Sheldrake, Rupert, *The Presence of the Past*, William Collins, 1989.

Shelley, *A Selection* by Isabel Quigly, The Penguin Poets, 1956.

Singh, Simon, *Big Bang*, Fourth Estate, 2004.

Smuts, Jan C., *Holism and Evolution*, published 1926, reissued Gestalt Journal Press, 1996.

Spectner, Lee, *Not by Chance!*, Judaica Press Inc., New York, 1997.

Spencer, Herbert, *Principles of Biology*, 1864, reissued in two volumes by University Press of the Pacific, 2002.

Swift, Eleanor, *A Layman's Guide to Neometaphysics*, The Society of Metaphysicians, Hastings, 1993.

Sykes, Bryan, *The Seven Daughters of Eve*, Bantam Press, 2001.

The Born-Einstein Letters, trans. by Irene Born, MacMillan, London, 1971.

Toynbee, Arnold, *A Study of History*, revised one-volume edition, OUP/Thames and Hudson, London, 1972.

Tylor, Edward Burnett, *Primitive Culture*, 1871.

Wallace, A.R. *Man's Place in the Universe: A Study of the Results of Scientific Research in relation to the Unity or Plurality of Worlds*, 4th ed., George Bell and Sons, London, 1904.

Walter, W. Grey, *The Living Brain*, Penguin Books, 1961.

Ward, Mark, *Universality*, Pan Books, 2001.

Waterfield, Robin, *The First Philosophers: The Presocratics and Sophists*, Oxford University Press, 2000.

Weber, Renée, *Dialogues with Scientists and Sages: The Search for Unitity*, Routledge & Kegan Paul, 1986.

Wegener, Alfred, *Die Entstehung der Kontinente und Ozeane* (*The Origins of Continents and Oceans*), Dover Publications, New York, 1966.

Weiner, Jonathan, *The Beak of the Finch*, Vintage Books, 1995.

White, Alan R., *Methods of Metaphysics*, Croom Helm, 1987.

Whitehead, Alfred North, *An Anthology*, selected by F.S.C. Northrop and Mason W. Gross, Cambridge University Press, 1953.

Whitehead, Alfred North, *An Introduction to Mathematics*, Oxford University Press, 1958.

Williamson, J.J., The *Structure of ALL*, The Society of Metaphysicians, Hastings, 1986.

Xenophanes of Colophon, *Fragments*, trans. by J.H. Lesher, University of Toronto Press, 2001.

Zimmer, Carl, *Evolution*, William Heinemann, 2002.

INDEX

Italicised page numbers indicate tables or illustrations

BOOKS

O is a symbol of the world, of oneness and unity. In different cultures it also means the "eye", symbolizing knowledge and insight. We aim to publish books that are accessible, constructive and that challenge accepted opinion, both that of academia and the "moral majority".

Our books are available in all good English language bookstores worldwide. If you don't see the book on the shelves ask the bookstore to order it for you, quoting the ISBN number and title. Alternatively you can order online (all major online retail sites carry our titles) or contact the distributor in the relevant country, listed on the copyright page.

See our website **www.o-books.net** for a full list of over 400 titles, growing by 100 a year.

And tune in to myspiritradio.com for our book review radio show, hosted by June-Elleni Laine, where you can listen to the authors discussing their books.

MySpiritRadio